ROUTLEDGE LIBRARY EDITIONS:
COMPARATIVE URBANIZATION

I0028288

Volume 1

THE EUROPEAN CITY

THE EUROPEAN CITY
A Western Perspective

D. BURTENSHAW, M. BATEMAN
AND G. J. ASHWORTH

Routledge
Taylor & Francis Group

LONDON AND NEW YORK

First published in 1991 by David Fulton Publishers

This edition first published in 2021
by Routledge
4 Park Square, Milton Park, Abingdon, Oxon OX14 4RN
605 Third Avenue, New York, NY 10017

Routledge is an imprint of the Taylor & Francis Group, an informa business

British Library Cataloguing in Publication Data
A catalogue record for this book is available from the British Library

ISBN: 978-0-367-75717-5 (Set)
ISBN: 978-1-00-317423-3 (Set) (ebk)
ISBN: 978-0-367-77123-2 (Volume 1) (hbk)
ISBN: 978-0-367-77128-7 (Volume 1) (pbk)
ISBN: 978-1-00-316990-1 (Volume 1) (ebk)

Publisher's Note
The publisher has gone to great lengths to ensure the quality of this reprint but points out that some imperfections in the original copies may be apparent.

Disclaimer
The publisher has made every effort to trace copyright holders and would welcome correspondence from those they have been unable to trace.

THE EUROPEAN CITY
A Western Perspective

D. BURTENSHAW, M. BATEMAN

and

G. ASHWORTH

David Fulton Publishers
London

David Fulton Publishers Ltd
2 Barbon Close, London WC1N 3JX

First published in Great Britain by
David Fulton Publishers 1991

Note: The right of David Burtenshaw, Michael Bateman and Gregory
Ashworth to be identified as the authors of this work has been asserted
by them in accordance with the Copyright, Designs and Patents Act
1988.

Copyright © David Burtenshaw, Michael Bateman and Gregory Ashworth

British Library Cataloguing in Publication Data
Burtenshaw, D. (David) *1939-*
  The European city: a Western perspective.-2nd ed.
  1. Western Europe. Cities. Environment planning
  I. Title II. Bateman, M. (Michael) *1944-* III. Ashworth,
  G. J. (Gregory John) *1941-* IV. Burtenshaw, D. (David)
  1939-. City in West Europe
  711.4094

  ISBN 1-85346-030-3

Typeset by Chapterhouse, The Cloisters, Formby, L37 3PX
Printed and bound in Great Britain by
Biddles Ltd, Guildford and King's Lynn

Contents

Preface and Acknowledgements

The nature of this book, indeed the fact that it was written at all, was a result of our joint experiences as teachers, researchers and citizens. Our attempts to explain to students over a number of years in various institutions the nature of change in the contemporary West European city revealed gaps in the literature. When faced with this problem, each of us independently could find no existing book which fitted the needs of our students. There were and are many excellent British, American and Continental works on the problems and policies of individual West European cities and countries, and rather fewer good syntheses of international experiences in particular aspects of urban policy-making. While acknowledging our debt to these writers (we hope adequately in our references), we feel that the comparative view of West European urban planning still merits discussion. The decade since this text was first published has seen several excellent contributions which begin to fill these gaps at a research level (Hall and Hay 1980, Cheshire and Hay 1989, Heinritz and Lichtenburger 1986), at a descriptive level (van der Haegen 1982) and, most importantly in our opinion, at the level of the student (White 1984).

From the outset, we have attempted to provide a comparative view of West European urban development and planning, which has grown out of our previous research into office developments, retailing and urban recreation, particularly applied to Germany, France and the Low Countries. Our fortuitous presence together on the staff of Portsmouth Polytechnic demonstrated to us both the weaknesses of our individual topical approaches and the value of close collaboration on a collective project. We acknowledge that the shape of the second edition owes much to the development of our specialist views. Equally it must be said that we are, all three, intellectual 'children of the 1960s', whose optimism about the efficacy of planning has survived the disappointments of the 1970s and the anti-planning atmosphere of the 1980s. In addition, our writing has been coloured by an increasing awareness, gleaned from contact with our European colleagues, that assumptions and models derived from North American studies are not only simplistic in the European context, but at times quite erroneous. Cities in Europe remain distinct, just as the approaches to their planning remain distinct. We strongly believe that the qualities inherent in this distinctiveness make the city in Europe an attractive and highly amenable place in which to live.

The original delimitation of a geographical area of study and the selection of examples from within it was somewhat arbitrary. We give prominence to the ideas and applications developed in Britain, France, Germany, Italy, the Low Countries and Scandinavia, as forming the core of our West Europe, supplemented by reference to Iberia, Greece, Austria, Switzerland and Ireland. The events of the turn of the decade have served to increase our concern over the exclusion of reference to cities in East Europe, since they shared the heritage of cities in West Europe and made their contribution to the European tradition. Between 1945 and 1990 they became more distinct from cities in West Europe, a fact acknowledged in Hamilton and French's work on *The Socialist City*. However, the changes set in motion in 1989–1990 will take time to have an effect on the cities of East Europe and many of the changes will be to the townscapes of socialism. Over the coming decade the processes outlined in Chapters 3–9 will begin to affect Dresden, Warsaw and Prague as much as they have Paris, Manchester and Barcelona in the past decades. Therefore we have excluded East Europe from this study.

On a more technical note, we have adopted the use of anglicised spellings of place-names where they exist. Although this is at the risk of offending our European colleagues outside Britain, it was our inevitable compromise necessary if we were to avoid the problems of Liege, *Luik* or *Luttich*, or even Milan, *Milano* or *Mailand*!

Our approach to each of the aspects of urban life owes much to the continuing debate amongst geographers on the role of our subject in the making of public policy. We have taken it as axiomatic that the geographical investigations of the city must begin with the attempt to trace and understand the patterns and processes of urban life, and then to move, without a clearly envisaged gap, to consider the impacts of the policies of public and private bodies on these patterns. The optimistic assumptions about both the efficacy of planning, broadly defined to include a much wider range of spatial policies than is handled by most city planners, and the role of an applied geography are obvious to us and we make no apology for them. If this is seen as a naivety, contrary to the spirit of the times, it is in our eyes morally preferable to what has been termed the 'structural trap of informed inactivity'.

In writing the first edition, we acknowledged the assistance given by Portsmouth Polytechnic, the (then) SSRC and the DFG. In revising our text we acknowledge our continuing debt to Portsmouth Polytechnic and the University of Groningen, although in the climate of financial stringency our major financial debt must go to our families, who have tolerated the diversion of household funds into our research travels. Miriam Boyle and Alistair Drummond assisted with the first edition and Yvonne Court provided assistance during the early stages of this edition. Our colleagues in Portsmouth and Groningen have all helped in various ways. The Cartographic Unit in Portsmouth Polytechnic, and especially Rosemary Shearer, drew all the maps. Sundra, Lynn and Angela typed some revisions although in this age one also thanks the unknown boffins at Acorn, IBM, etc!

Our dedication, like the message of the text, is almost the same:

To those who have taught us and the new Europeans we have taught.

References

Cheshire, P. and Hay, D. (1989) *Urban Problems in Western Europe*, Allen and Unwin, London.

Hall, P. and Hay, D. (1980) *Growth Centres in the European Urban System*, Heinemann, London.

Hamilton, I. and French, H. (1980) *The Socialist City*, Wiley, Chichester.

Heinritz, G. and Lichtenburger, E. (eds) (1986) *The Take-Off of Suburbia and the Crisis of the Central City*, Steiner, Stuttgart.

Van der Haegen, H. (1982) *West European Settlement System*, Louvain University, Louvain.

White, P. (1984) *The Western European City*, Longman, London.

List of Figures

List of Tables

Introduction

The basic thesis of this book is that the cities of West Europe have sufficient in common to justify the use of the term 'West European City', and are sufficiently distinct in important respects to render examination of the common elements worthwhile. In this examination of the city in West Europe three broad approaches have been taken, represented by the three parts of this book. In the first part, we concentrate on the philosophies, histories and processes which have made the European city system rich in internal variety yet distinct from that of the rest of western industrialised urban society.

We examine in the first chapter the various guiding themes of urban planning which have stemmed from a wide variety of national cultures and are rooted in varying depths of urban experience. It is against this diverse background that Europe has produced a large number of architects, planners and urban visionaries whose thinking has found physical expression in the cities of Europe. In Chapter 2 we have traced the history of some of these traditions through the works, both literary and architectural, of 'household names' such as Haussmann, Howard, Le Corbusier, Sitte, Geddes and Cerda. In addition Chapter 2 examines the post-1945 shifts in planning ideology that have influenced city morphology. The social composition and patterning of the European city does not fit easily into the models of urban social structure that have been derived from the North American, or even Australian, experience. In Chapter 3 we examine a selection of studies of social morphology, social processes and social change in cities as diverse as Sunderland, Helsinki, Rome and Arnhem. The length of the historical experience, the variety of social welfare policies and differences in the value placed on the various possible lifestyles offered by the city will, again, underpin the explanation of the distinctiveness of Europe.

If we are to sustain the theory of distinctiveness, it is necessary to examine those policies which are tending to maintain the unique flavour of urban life in Europe, or which constitute a peculiarly European response to a more widespread trend. The second, and largest, part of the book examines policies towards various aspects of the city's development. It is not intended to be an all-encompassing inventory of policy-making in cities but an attempt to synthesise the policies from diverse examples in those fields of urban planning and management where West European cities have made significant contributions towards solving the various problems resulting from urban change in the period since 1945.

Although cities have always been service centres there has been a notable shift in the last decade away from employment in the secondary sector so that in most cities between 60 and 70 per cent of the workforce is employed in tertiary and quaternary occupations. The rapid expansion of financial services, the growing scale of company activities and the growth of the national and local civil service to administer the increase in legislation have all contributed to the demand for new locations for high technology, science and business parks.

There have been pressures for locational change of retailing deriving both from changes in the structure of retailing and from social and demographic changes. The new consumption patterns that modern lifestyles have encouraged have occurred at the same time as fundamental change in the economics and management structure of shops. The consequence of such trends in location have been resisted by the established interests in the city-centre high streets, and often also by the conservation interest. Chapter 4 examines various planning responses to such developments in the retailing and office sectors of the urban economy.

Part of the reason for the dispersal shared by the 'suburbanisation' of shops and offices has been the shrinkage of distance effected by the widespread use of the motorcar. Its benefits were more quickly appreciated than its costs, and policies designed to accommodate its use have generally been superseded by policies of restraint on private transport use and support for alternative public transport services. Chapter 5 traces variations in this trend.

Housing policies are the third set of planning policies to be discussed (Chapter 6). Policies have been subject to the shifts in national and local ideology which, combined with the architectural fashions of modernist and post-modernist influences, have produced housing policies that reflect the time and place of their construction. Sarcelles, Sloetermeer and Perlach, or Wapping, Ile St Louis and St Agnes Viertel, are products of these interactions.

An obvious difference between the cities of Europe and those of the new world is that the former contain relict buildings and street patterns which are valued by tourist and citizen alike as giving interest and identity to the urban scene. Many of the changes in the economic functions, housing provision and transport use discussed above were seen to threaten this historic heritage. Chapter 7 traces the growth of a response to this threat mainly in the last fifteen years and argues that conservation planning can no longer been seen as a simple matter of preservation but has implications for many other aspects of urban life, and involves a choice between priorities, lifestyles and values.

It is not now possible to conceive of a West European architectural heritage without considering how the enjoyment of that heritage is made accessible to Europeans. Thus planning for tourism and planning for conservation must be reconciled despite the pressures that visitors themselves can place upon the heritage that they have come to experience. We hope therefore that Chapters 7 and 9 will be viewed together. The rapid growth of the major cities of West

Europe as tourist attractions and the development of extensive tourist districts have posed problems to planners which have only recently been appreciated.

Cities have always been centres of leisure and entertainment for citizens as well as visitors, but, since the war, increased disposable income, leisure time and personal mobility have intensified demands for facilities as diverse as theatres, sports halls, parks and open spaces. Chapter 8 takes the view that recreation planning has ceased to be an incidental embellishment to the town plan, but is occupying an increasing share of urban resources, planners' attention and citizens' expectations. In addition the recreational demands of urban populations provide the most important functional link between the city and its more rural surroundings.

The third part of the book synthesises the sectoral policies described above by means of an examination of the city plan as a whole. Initially in Chapter 10 we return to the *leitmotifs* of history, nationality and ideology in an examination of national urban systems, and the involved question of the relationship of city planning to the regional and national planning policies. This acts as a necessary preliminary to the understanding of the structure plans of individual cities. Chapter 11 examines the outcomes of city plans, devised and implemented in the third quarter of this century, with the benefit of hindsight. It attempts to explain the changing approaches to city planning which reflect the more pragmatic approaches to urban change which have dominated the eighties. Part 3 concludes with a review of what has been West Europe's most distinctive contribution to urban planning, the new town.

It must be emphasised that our attempt to examine the cities of West Europe has not been to create an exhaustive catalogue of urban experience in Europe. Selection of examples from some countries at the expense of others has been necessary; no examination could be totally exhaustive. Our selection of examples, broadened in this edition, should reflect the distinctiveness and variety of the urban system which remains the overriding thesis of the book.

The problems facing the planners of the West European city continue to be formidable and the policies created for their solution are on only rare occasions obviously and immediately successful. The fact that citizens have taken a critical interest in the actions of politicians and planners upon their environment is encouraging. However, the reaction which places all blame in the hands of architects and planners, the servants of the decision makers, rather than the decision makers themselves, is a cause for concern. Such *laissez-faire* attitudes might equally destroy the European city as a congenial environment for living whose distinctive qualities are worth considerable effort to preserve. This is the *raison d'être* for revising the book and maintaining our own philosophy into the new decade.

CHAPTER 1

The Role of History, Nation and Ideology

The use of the term 'the West European City' strongly implies that the cities of the sub-continent have characteristics in common which give them a distinctiveness that sets them apart from cities developed elsewhere. It is a paradox that, in a large part, this distinctiveness is a result of the variety of historical experiences which have contributed to the physical fabric in which citizens live, work and play. In many cases it is the history of the nation or, occasionally, the nations in which a particular town has been situated that have shaped the town's development from age to age. Each stage in national development has, in its turn, brought about corresponding changes in the form and functions of towns and cities. In this way the city in West Europe becomes a palimpsest representing the physical expression in the built environment of successive ideologies at each period of development. What the events of 1989–1990 might cause can only remain conjecture at this time. However, the changes have not occurred simultaneously throughout Western Europe and therefore part of this urban distinctiveness has resulted from the differential effects of policies and social movements.

The role of history

The historical experience has been sharp and dramatic for many cities. In some cases, for instance, it involved the loss of an imperial role with consequent effects on the imperial city. In 1919 as a result of the Treaties of St Germain and Trianon, Vienna was transformed from the capital of an empire of 52 million persons to the capital of a dismembered country with approximately one-tenth of its former population. Past capitals abound in Europe: Braunschweig, Munchen, Turku. More recently we can point to West Berlin as an exclave of the Federal Republic, divested of its capital status by the events of the 'cold war', and divided and distinct from East Berlin, Hauptstadt Berlin, the former capital of the German Democratic Republic. The last vestiges of Berlin's pre-war role remain in the Tiergarten and in the boundaries of 1920 Berlin even if the past was less than glorious. By 1990 the map of Berlin appears to be in the process of redrafting and one can only speculate on the impact of a reunified

1

city on the form and function of twenty-first century Berlin. For other cities the dramatic change took the form of the truncation of a trading hinterland, as in Londonderry in 1920, or Hamburg and Lubeck in 1945, or Trieste at the same time. In other cases, the dramatic changes were more frequent as the nation states of Europe fought over the cultural shatter belt between the French and German peoples. Saarbrucken's national allegiance oscillated for a past century. Only since 1960 has the city or its near neighbours in the Saarland been an integral part of West Germany. From 1945 until 1960 French laws, customs, education and urban development fashioned the Saarland towns and built a university modelled on the French pattern of higher education. This was not the first period of French domination since, as cities in the League of Nations Territory of the Saarland, the cities were subject to French administration from 1920 until 1935. Before that Saarlouis and Saarbrucken had been under French control for a short time in the Napoleonic Wars. The history of other cities in Alsace-Lorraine, such as Metz and Strasbourg, has also been affected by the oscillations of the international boundary over the past century. Boundary changes have left Vyborg with a similar imprint of changed national allegiance. The more one penetrates history from the recent past the greater are the number of cities affected by the changing political map of Europe.

In contrast, other cities have had greatness thrust upon them by the course of history. The relatively sudden promotion of Dublin from the role of a provincial capital to that of the capital of a new republic in 1920 is a case in point. Bonn enjoyed a similar fate a quarter of a century later when it became the temporary capital of the western zones of Germany. While the former was the logical choice, the latter was a mere university town or 'A Small Town in Germany' according to Le Carré, of no great urban status compared with its rivals such as Kassel, Frankfurt or Stuttgart or its near neighbour Cologne. Yet today Bonn-Bad Godesberg is a maturing mixture of government offices, nestling in a suburban sprawl surrounding the old town of Bonn and its elegant spa neighbour, the home of the diplomatic corps. But the radical changes of 1989–90 might leave a new relic of history, an abandoned twentieth-century capital, the first of its kind for a half a century.

Violent change as the consequence of war and revolution is a strong theme in the development of the city in Western Europe. Frequently the fear of war or revolution has had a lasting effect on the form of our cities. The changes in military technology in the medieval period resulted in equally rapid alterations to the defence of cities. The imprint of these defences on the present urban fabric is unmistakable. As the cities grew in population and as the technology of war improved so new defences were planned and built while the old walls frequently vanished to be sites for parks and boulevards. Even as late as the nineteenth century Palmerston's fear of a French attack on Portsmouth dictated fortifications which affected subsequent city form and city development. The forts and defences remain prominent features in the townscape, serving varied roles as recreational open spaces, museums, sites for industrial estates and more recently superstores, whilst a select few retain their military function. Similarly the French mistrust of the Prussians produced similar military works around Verdun.

The imprints of past defence are indeed commonplace over much of urban Europe, often in towns which were never directly involved in military action. Without question, however, this has destroyed the fabric of many cities twice in this century and has been the catalyst from which new approaches to urban growth or changes have emerged. The whole concept of the Lijnbaan pedestrianised shopping centre of Rotterdam was formulated during the darker days of the Second World War. Abercrombie was conceiving his plan for Greater London in a period when large areas of that city were being razed to the ground by enemy incendiary bombs. Frequently forty per cent and more of the buildings in German cities were destroyed by allied bombing in the latter stages of that war. Other cities throughout Western Europe bore the scars of war to a greater or lesser extent. Le Havre, Coventry and Antwerp were among some of those that were severely devastated. Spanish cities had suffered a similar fate in the Civil War so that the opportunities to build anew existed in part of Iberia.

On the other hand, the neutrality maintained by Sweden and Switzerland meant that cities and towns in those two countries remained unscathed during the twentieth century. It was left to the violence of the bulldozer and the developer to alter the fabric of Stockholm or Zurich. Perhaps neutrality has been a prominent factor in the success of building conservation in these two countries.

It is out of the ashes of 1945 that many ideas for urban development sprung as a part of the new free Europe that was to grow from the ruins of war. The development of so many fundamental attitudes to urban life, urban development and future urban change are embodied in that period of the new Europe. The ideas that were embodied in Gravier's seminal work Paris et le Desert *Français*, which was published in 1947, were conceived in Vichy France. In Britain the founding of the welfare state, the new town movement and the self-congratulatory Festival of Britain epitomised the optimism born of years of sacrifice.

When one is examining the imprint of history, it is not solely the landmarks of national history that have fashioned the modern city. At the more local scale the redefinition of urban boundaries in areas of strong allegiance to particular towns can have lasting impact. The creation of modern Hamburg in 1938 is a case in point. The modern city is the product of a major boundary change which absorbed the Prussian city of Altona. To this day Altona has a separate identity; the trains to Hamburg invariably terminate at Altona and the shopping centre is still an important regional centre within greater Hamburg. In Italy many of the regional planning boundaries, which in turn affect city planning, were derived from the pre-existing states of the Italian peninsula rather than from areas specifically designed for effective planning.

History can also be viewed as an important aspect of our urban cultural heritage from the point of view of the buildings that were the product of each age. The variety of architectural responses to social conditions in the past are not the subject of this text although they are an integral part of the variety of forms that comprise the city. As Kormoss (1976) stated:

> It is one of the consequences of the historic, geographic, linguistic and philosophic diversity of Europe which turn it into a very rich region with a very important heritage

It is somewhat surprising to note authors who have criticised approaches to the reconstruction of previous building forms in the years after 1945 as being 'anti-progressive and anti-innovation' (Holzner, 1970). Such a view represents the extreme application of cultural blinkers to the understanding of Western European urbanism, with the resultant misunderstanding of the role of the built environment in our distinctive European heritage.

The development of cities has followed a far from common path through time. Industrialisation, for instance, came earlier to Britain and Belgium with consequent effects on town growth and form that are not found to the same degree in France, Germany or Spain. Late industrialisation, the advent of railways prior to industrialisation and the coming of electric power so quickly after the 'take-off' to industrialisation have made German cities very distinct from their British counterparts. Industries that migrated to the coal-field cities and towns in Britain remained *in situ* in German towns. The railways penetrated closer to the medieval core of German cities and linked the pre-industrial towns and states rather than focusing on the needs of industrial exploitation. German cities developed tram systems very early in their period of industrial growth thus reducing the need to concentrate the workforce close to the point of employment. Thus the industrial colonies, while being both a distinctive characteristic and a product of their times, were more 'open' than their British or Belgian counterparts. Industrialisation in Mediterranean Europe was even later and, in the case of Portugal, a product of the mid-twentieth century. Here and in southern Italy, the problems of rapid urbanisation have only just begun to recede. For these states urbanisation was often achieved through emigration to other countries (Gaspar, 1984).

The role of nation

A recurring theme in this book is the interplay between the city as an expression of the nation state, and the city as a product of international forces. The former condition tends towards the production of distinctly national city systems, whilst the latter leads towards increasing uniformity across political boundaries. Hall contends this to be a common process with a simple time-lag between different systems, with all countries moving inexorably towards a single model, a model replicated by Cheshire & Hay (1989) (Hall, 1977). In the case of housing policies (chapter 6) the national view is so dominant in our opinion that the policies cannot be compared unless the national variations are noted. Similarly the consideration of city planning policies seems incomplete without some reference to the relationships between national, regional and urban planning (chapter 10). On the other hand the international aspects of producer services location are important factors in any consideration of economic change (chapter 4) or tourism (chapter 8). Obviously the international role of cities is a major consideration in any consideration of the planning strategies for both the national capitals and other cities that traditionally aspire to international status, such as Barcelona, Frankfurt, Zurich, Geneva and Strasbourg.

The impact of the national political system is unmistakable in its effect on the national urban system. Centralised nation states have created strong, dominant primate cities such as Athens, Lisbon, London and Paris. In Portugal, Lisbon and Oporto to a lesser extent dominate the state with 40 per cent of the population living in them. Perhaps the most dominant city is Athens which, in 1950 contained 15 per cent of Greek population and by 1980, 57 per cent. Architectural and planning innovations tended to spread from the core city to towns on the national periphery. The primate city is the centre of perceived wisdom and through the activities of the inhabitants and organisations housed there, not least of which is central government, proceeds to reinforce the perceived centralised wisdom. In France the ideas of Haussmann, Le Corbusier and the more recent new town planners were initially Parisian ideas which later spread to the provincial towns. The architectural ideas of Georgian England originated in London and were copied, modified and improved on throughout the Kingdom. Ideas emerged from Copenhagen to be instituted in the towns of the Jutland peninsula. On the other hand, countries with a strong tradition of provincial separatism produced a less clearly defined urban hierarchy. In Italy the relict capitals of Turin, Venice and Naples are serious rivals to the status of Rome, while Milan has acquired many of the trappings of an economic, if not a political, national capital city. Similarly the separate national identities of Castile, Catalonia and the Basque country have produced Madrid, Barcelona and Bilbao respectively, competing with the older products of a Moslem culture, Seville and Grenada, which dominated the Spanish urban system until the recent past. Such symbolism of international status is a powerful stimulus to urban reform. As long ago as 1915 Geddes was advocating civic exhibitions as a stimulus, but more powerful than civic self-agrandissement are the benefits which accrue directly and indirectly from being on the world's television sets. The Olympic games brought an improved transport infrastructure to Munich and are stimulating a surge of development in Barcelona for 1992. Birmingham, Manchester and Sheffield have all looked unsuccessfully, to the Olympic movement to boost their flagging reputations, as have Amsterdam and Paris in competition with Barcelona. In Rome the 1990 World Cup has brought potential investment in the shape of a metro line extension, a rail link to Fiumicino airport and fifteen underground car parks. In Belgium the severity of the cultural schism has led to both restrictions on the growth of Brussels and the identification of many Belgians with either Antwerp or Liege rather than the legal capital.

If Athens, London and Paris are the products of a history of centralised government, the cities of Austria, West Germany and Switzerland reflect the role of federalism. Five cities dominate the urban hierarchy of Switzerland which only became a federal republic in 1848. Berne, the national capital, is the fourth largest city and administrative hub. Zurich, Geneva and Basle are larger and Lausanne only slightly smaller. Apart from these there are also Lugano, the cultural capital of the Italian-speaking Ticino, which is the third largest financial centre, an offshore financial centre for Milan, the industrial centres of Biel, Winterthur and St Gallen, and important cantonal capitals.

The German federal tradition and the partition and occupation of the Reich encouraged the similar development of a multi-centred urban system. In this case, the tradition has a long history stretching back to the small princedoms which characterised the central European region prior to German unification, each with its own capital such as Bamberg, Brunswick, Fulda and Oldenburg. The old capital of the Reich, Berlin, was divided and both politically and physically isolated from the Federal Republic. Both Hamburg and Munich could lay claim to being the largest city whilst Kiel, Hannover, Dusseldorf, Mainz, Wiesbaden, Bremen, Stuttgart and Saarbrucken are all *Land* capitals, many much larger than Bonn, the federal capital. In addition, Frankfurt can, with justice, claim to be the financial capital and major centre of international communications. In contrast, Austria does possess a dominant capital city but, like its neighbour, its Federal status has enabled the Land capitals such as Bregenz, in particular, and Linz, Salzburg and Klagenfurt to develop a broader range of employment opportunities than similar sized cities in France.

The 'world city', fulfilling international functions and shaped by forces beyond the nation state, is not a recent phenomenon. Government of the world, or a substantial portion of it, has been exercised from Athens, Rome, Madrid, Lisbon, Paris, Brussels and London. International trade and its financing has been concentrated in Venice, Antwerp, Amsterdam and London in successive centuries from the fifteenth to the nineteenth. Financial prowess in the world today is claimed by London, Paris, Frankfurt, Zurich and Luxembourg City. Similarly, more recently, the movement in Europe towards establishing inter-national governmental institutions and the contemporaneous growth of the multi-national corporations has increased the international role of selected cities. The antecedents for some centres are now rooted deep in history; Vienna was a congress centre in 1814–1815, Berlin in 1870 and Paris in 1918. The International Court of Justice at The Hague, the League of Nations in Geneva, the Council of Europe in Strasbourg, the various organs of the European Community in Brussels, Strasbourg and Luxembourg City are only the largest of a multiplicity of similar international organisations to be found in Western Europe's world cities. To these must be added the efforts of cities such as Vienna and Helsinki to establish themselves as neutral international conference centres to rival the traditional position held by Geneva, the home of the ill-fated League of Nations and more recently the United Nations. The prominence given to the SALT negotiations held in Helsinki and Vienna is but one example of the new role for the 'neutral world cities'. Vienna is also the headquarters for OPEC. Political neutrality has its benefits because Swiss cities now house the head-quarters of 150 international organisations, such as the World Council of Churches and World Wildlife Fund in Geneva.

Neutrality and stability have attracted investments into Switzerland for years. The emergence of Lugan illustrates this in the 1980s more so than the larger financial centres of Zurich and Geneva. The 1980s saw seven foreign banks open offices in the city, most managing Italian investments. There were 50 banks and 75 finance companies in a city of 30,000 people in 1988. Major Italian companies have holding companies in the city and Benetton organises its foreign operations here rather than in Italy. The freedom of capital movement in the European Community after 1992 might threaten Lugano's financial activities and so destroy an economic base founded upon the national reputation.

Besides the international organs of government the world cities have set about attracting the multi-national office complexes, the international banks and financial activities that are part of the world city image. The dreams of de Gaulle to create a centre to rival Manhattan are enshrined in the whole concept of the new office mode of La Défense. The airport which bears his name at Roissy and the modernised telecommunications system based on Paris are part and parcel of the same dream. All the main financial centres have seen the development of new office buildings each frequently more lofty than its pre-decessor housing the major international banking groups. The national Westminster tower in London and the Deutsche Bank, Dresdner Bank and Commerz Bank towers in Frankfurt illustrate the demand for prestige city-centre buildings in the world of banking, a trend that has been backed by the city authorities. The most spectacular growth in this sector has been reserved for Luxembourg City where the number of banks represented in the city rose from 15 in 1960 to 114 in 1966.

No visitor to Dusseldorf can fail to notice the presence of Japanese interests in the city, for the Japanese chose that city as a centre for their West European activities. A short walk north from the Bank of England in London will introduce the pedestrian to a whole new set of banks. The international role of the major cities as tourist centres for the wealth of the Middle East countries and Japan is unmistakable on any visit to Paris or London. Hotels can be patronised, thanks to package tours, and clientele from particular countries. Even tourist 'souvenir honeypots' have attracted retail activities and banking concerned almost exclusively with the needs of the tourists. These are the trappings of world city status.

It was the new international status that produced a powerful backlash in many cities against the type, form and social consequences of development. The take over of the West End district of Frankfurt by tertiary businesses encouraged by the city government because it raised the revenues of the city was violently opposed by protesters. Bloody clashes did have their effects because the city relented at the same time as the gains from property development disappeared and the process was halted. Similar protests have taken place in other cities throughout Europe making cities more conscious of the heritage that has to be sacrificed for international status. Other powerful protest lobbies have gained strength from these successes and therefore anti-motorway and anti-rail line lobbies and anti-nuclear power movements have opposed city and national development policies. Symbolic of this growing strength is the success gained by the ecologist 'Green Party' in the 1979 elections in Bremen, when they achieved in excess of the five per cent of the votes required to guarantee some proportional representation in the Land government. Ecology against urbanism is the classic dichotomy now seen by researchers in France (Mathieu, 1977).

The role of ideology

Alongside the experience of history and the role of national and international identity the third theme that emerges in cities throughout the continent is the impact of ideology, whether conservative, ecological, feudal or socialist. Past ideologies have created cities that are memorials to the divine monarch (Versailles), to the imperial mission (Vienna), and to utilitarianism and the pursuit of profit (Bradford). It has been suggested that the morphology of the city is not only a product of the civilisation that it houses but also a factor in the creation of that civilisation (Sacco, 1976). At a more prosaic level it is clear that in cities such as Stockholm, Gothenburg and Helsinki attitudes towards conservation, social housing provision and public transport reflect the contemporary dominant social-democratic ideology of the Scandinavian countries. In contrast, the development of many West German cities in the immediate post-war period occurred within the framework of a social-market economy and a certain rejection of planning resulting from the experience of twelve years of National Socialism. The result was weak central direction and wide scope for local initiative. In liberal-conservative (Christian Democrat) cities such as Cologne, little attempt was made to control urban development for many years while in other cities such as Essen and Bochum the reaction of local government was quite different (Blacksell, 1968). Nevertheless the rebuilding of Cologne was supervised by Schwarz who had supervised rebuilding in occupied Alsace-Lorraine during the war. Architecturally he was a modernist and therefore it is no surprise that modernist buildings such as the new Opera House emerged from the ruins. In almost all countries it was hoped that rebuilding would promote national values, what Diefendorf calls modernism with a conservative aesthetic.

It is not solely in the modern period that relationships can be drawn between the dominant ideologies and value systems and the nature of urban development. Sjoberg (1960) saw cities in pre-industrial Europe as the product of their societies whether they be the community of merchants at a market point noted by Jones (1966) and Pirenne (1956), or an agricultural-based primary civilisation discussed by Fourastie (1963), or the quarters of a medieval city created by the guilds and political rivalry (Vance, 1978). No matter what forces created these towns, wealth was concentrated at the centre and declined towards the periphery (Langton, 1978). The pre-capitalist modes of production produced a distinctive urban morphology.

By the eighteenth century many of the pre-industrial towns had become the homes of the absolute ruler, while in some cases new residential towns had been created for the princes, such as Karlsruhe, and the productive functions were dispersed to the rural areas for the convenience of the Land (Lichtenberger, 1972). The Zaehringer dynasty built Freiburg, Fribourg and Berne. During the industrial period many of the same towns developed industrial suburbs and altered in character as the new *laissez-faire* ideology dominated the ideas of city builders. Thus the Baroque heart of Nancy was surrounded by industrial housing dating particularly from the years between 1871 and 1914. At the same period new industrial towns sprung up throughout Western Europe to exploit the new resources for industry and the economies of scale that technology had made available, given a concentration of the labour force. Vance's depiction of the 'generalised' suburbs of the Lancashire mill towns, the *villes minières* and the Ruhr, as the products of this liberal industrial phase of city growth, seems to be most apt. All of these towns and cities today are the cores of the late twentieth-century welfare state cities with their peripheral social housing, diffuse peri-urban fringe and their inner area redevelopment and rehabilitation programmes approaching fruition as the spotlight shifts full circle to the peripheral social housing areas. Thus many cities have acquired the buildings and street plan from periods of contrasting ideologies and possess a richness derived from this variety. The pre-industrial burger town could develop into the capital of a principality and a Land capital like Munich, or remain a museum like Arundel in Southern England. Industrial towns have also developed to achieve regional importance like Birmingham or Essen, or a locally important status such as Merthyr Tydfil or Volklingen. Alternatively they can become museums to an ideology such as Millom in Cumbria or Thiers in Central France. Only the new towns and satellite towns are the product of the mid-twentieth century welfare state. Some of these have grown, accumulating the artefacts of changing ideologies, while others look set to remain as museums of mid-twentieth century urban ideals. The contrast between Milton Keynes, Bracknell and Marne La Vallée on the one hand and Aycliffe, Wulfen and Val de Reuil is a contrast between the ideology of market led growth and floundering economic decline. What is certain is that the imprint of ideologies on cities is far from even.

Other particular features of West European urban morphology can be viewed as the product of particular ideological conditions. Lichtenberger (1976) notes that the apartment is a product of the Italian renaissance cities that spread initially to France and Austria. Today it is a form of housing common throughout Europe, praised in some quarters and castigated in others, although the nature of the apartment has been modified to accommodate the contrasting housing needs of the stratified central Parisian society and the social housing needs of the lower socio-economic groups in the inner suburbs of London or the outermost fringes of Cologne. The crowded housing conditions of the nineteenth century industrial city are viewed in a similar vein by Vance (1977). For him the back to backs of Leeds, the Glasgow tenements and the single room apartments of the European mainland were all responses to the needs for a labour force located adjacent to the mill. This strengthened class segregation between the mean terraces of the workers and the more substantial villas or town houses of the bourgeoisie.

It is notable that since 1945 the dominant ideology of government has been more stable in some countries than others. It is possible to recognise countries such as Britain, which has suffered from relatively violent oscillations in its dominant political ideology compared with greater stability of the coalition governments of the continental countries. While there have been changes in the dominant political ideology in Germany, Spain and Sweden, they have been a single change rather than a succession of oscillating and damaging swings. Consequently, British policies on housing tenure, new towns, inner-city regeneration and regional development have often become political symbols to be changed when power changes hands. Some would go as far as to suggest that this lack of coherent goals for society in general, and British cities in particular, has slowed down the urban change and development process compared with that of France or Scandinavia where the continuity of environmental policies is more apparent, despite changes in the details of policies. Even at the local level, sudden changes in local political control can lead to sharp reversals of policy in public transport as is illustrated in Chapter 5.

It must be acknowledged, however, that there is an influential school of thought which claims that the oscillations between more or less reformist parties, or between reformist and conservative parties, does little to alter the fundamentally central role of the capitalist mode of production and urban development. National and local politicians, whether they are labelled social democrat, christian democrat or liberal, will still depend on the institutions of the capitalist money market or the 'Property Machine' for the financing of all developments, whether public or private (Ambrose and Colenutt, 1975). Thus commentators such as Castells, Harvey and Lefebre adopt a very critical stance on urban change even when such change is ostensibly beneficial, such as urban renewal in central Paris. Castells saw the role of the express metro (RER) as providing commuters/workers from the poorer suburbs of the east of Paris for the multinational offices clustered in western Paris (Castells, 1978).

On the other hand, we would maintain that the various shades of ideological commitment contribute towards those very differences in national, regional and local policy which make the character of the city in Western Europe so distinct. Rome was ruled from the mid-1970s to the mid-1980s by the political left, who concentrated their attention on legalising the illegal building in the outer suburbs and so neglected the historic centre. They also established summer festivals to re-use remains, ports and villas. Since 1985 the political right has held power and attention switched to making the city move in time for the 1990 World Cup. The policy focuses upon controlling private vehicle access to the centre and improving the rudimentary metro; unfortunately there are too many permits for cars and investment in the metro had not begun in 1989. If Castells' view is simplistic, that of Berry, who sees Western Europe as a group of identical redistributive welfare states, does no justice to the widely differing social and political experiences of countries as diverse as Ireland and Sweden or Portugal and Austria (Berry, 1973). The simple dichotomies of the Marxists and Americanised view of the world are valuable teaching devices and theoretical constructs from which to begin analysis. However, the variety of built environments, policy goals and responses, whether for separate sectors of the city, such as housing, or for the city as a whole in the form of structure plans, are proof of the distinctiveness of the city in Western Europe. It is distinct because it is the product of a variety of histories in a large number of countries each with its own evolving ideological position.

It can be argued, or believed as an article of faith, that a European Act will bring about a set of more uniform urban policies within the European Community. The changes which ushered in the nineties in East Europe may act as a break on this process although, in the long run, the uniform policies which we saw stretching from Gibraltar to Tromso might arguably also extend from Brest to Bucharest. Whilst numerous references in the following pages to the publications of international organisations bear witness to the steps being made in this direction, there is also another tide flowing in Europe towards the smaller rather than the larger political unit. The city is frequently at the heart of such regionalist movements, as is Bilbao for the Basques, and Bozen for the Tyroleans. Whether regionalist movements can succeed in producing cities that reflect the strength of local ideologies and cultures is a problem which might depend on the dominance of the core over the whole territory.

Three interrelated themes of historical experience, national identity and political ideology together account for both the variety and the continuing distinctiveness of the West European city. The approach is unashamedly 'very historicist, very European and very French; the city is the product of history, the reflection of society, the action of man (people) upon space' (Castells, 1976).

References

Ambrose, P. and Colenutt, B. (1975) *The Property Machine*, Penguin, Harmondsworth.

Berry, B. J. L. (1973) *The Human Consequences of Urbanisation: divergent paths in the urban experience of the twentieth century*, St Marins, New York.

Blacksell, M. (1968) Recent changes in the morphology of West German townscapes, in *Urbanisation and its Problems* (eds M. Beckinsale and J. Houston) pp. 199–217, Blackwell, Oxford.

Castells, M. (1976) Is there an urban sociology?, in *Urban Sociology: critical essays* (ed. J. Pickvance), pp. 35–39, Tavistock Publications, London.

Castells, M. (1978) *City, Class and Power*, Macmillan, Basingstoke.
Cheshire, P. and Hay, A. (1989) *Urban Problems in Western Europe*, Allen and Unwin, London.
Fourastie, J. (1963) *Le Grand Espoir du XXe Siècle*, Gallimond, Paris.
Gaspar, J. (1984) Urbanisation: growth, problems and policies, in *Southern Europe Transformed*, Harper and Row, London.
Gravier, J. F. (1947) *Paris et le Désert Français*, Flammarion, Paris.
Hall, P. (1977) *Europe 2000*, Duckworth, London.
Harvey, D. (1973) *Social Justice and the City*, Edward Arnold, London.
Holzner, L. (1970) The role of history and tradition in the urban geography of West Germany, *Ann Assoc Amer Geog*, 60(2), 315–39.
Jones, E. (1966) *Towns and Cities*, Oxford University Press, London.
Kormoss, I. (1976) Urban extensions in Belgium, in *The Environment of Human Settlements, Vol 1* (eds P. Laconte *et al.*), pp. 177–86, Pergamon, Oxford.
Lichtenberger, E. (1972) Die Europäische Stadt-Wesen Modelle Probleme, *Raumforsch. Raumord*, 16, pp. 3–25.
Lichtenberger, E. (1976) The changing nature of European urbanization, in *Urbanization and Counter-Urbanization* (ed. B. J. L. Berry), pp. 81–107, Sage, Beverly Hills.
Mathieu, H . (1977) L'écologie contre l'urbanisme?, *Urbanisme*, 46(160), pp. 45–49.
Pirenne, H. (1925) *Medieval Cities* (translated by F. Halsey), Princeton University Press, Princeton.
Sacco, G. (1976) Morphology and culture of European cities, in *Europe 2000, Project 3, Volume 1* (ed. M. van Hulton), pp. 162–87, Nijhoff, The Hague.
Sjoberg, G. (1960) *The Pre-Industrial City*, Free Press, New York.
Vance, J. (1977) *This Scene of Man*, Harper and Row, London.
Vance, J. (1978) Institutional forces shape the city, in *Social Areas in Cities* (eds D. T. Herbert and R. J. Johnston), pp. 97–126, Wiley, London.

CHAPTER 2

The European Tradition of Urban Planning

Over the centuries the cities of Europe have inspired men from Plato to Castells to describe their deficiencies and to eulogise about the means of providing the best form of urban life for their future citizens. This tradition of concern for the built environment, unmatched elsewhere in the world, has produced a whole series of approaches to urban planning. These approaches are as diverse as the backgrounds of their authors but it is possible to trace several lineages within Western Europe that still have an important impact on policies and plans.

Such masterly reflections on urban architecture, history and planning as the works of Curl, Gutkind, Peter Hall, Thomas Hall, Lavedan, Mumford and Vance do not agree on all the relationships through time and it is not our intention to replicate their work (Curl, 1969; Gutkind, 1969–72; Hall, P. 1989; Hall, T. 1986; Lavedan, 1959; Mumford, 1961; and Vance, 1977). However glorious the urban past might have been, we are not attempting to describe the heritage of the built environment but rather to highlight those movements, theories and theorists that have had an influence on the urban planner in the final quarter of the twentieth century. Most European nations have been blessed with urban theorists or urbanists and their ideas have been diffused from Austria, France, Great Britain, Spain or Italy with increasing rapidity especially in the past hundred years. In the past century, in particular, a handful of such theorists have become famous beyond Western Europe and their ideas modified and translated into bricks and mortar by subsequent architects. Unlike the earlier city planners the greatest urban theorists of the nineteenth and twentieth centuries have few of their own designs as living memorials to their influential theories.

It is possible to trace five separate urban planning traditions within Western Europe, namely authoritarian, organic, romantic, technocratic utopian, and utopian. These traditions, while obviously distinct at various times, do blur into one another especially in the period since 1945. Others have discussed the deficiencies but have not actually planned the future city and these theorists can also be given a rightful place in the tradition of urban thinkers as socialists. It is the underlying thesis of some commentators that many movements are the product of their time and that emphasising the individual who fashions a movement is placing the wrong emphasis on the individual. The traditions of European urban theorists and their lineages that are traced in this chapter are but one way of viewing what is a complex history. However, this particular perspective does aid the comprehension of some of the policies that can be seen in cities today.

The authoritarian tradition

Perhaps the longest tradition is that of the authoritarian planner, although this is the most maligned today. Rather than reason on the layout of the city this tradition has tended to follow the views of one man, or an élite that seemed essential given the conditions of the time. Little scope was permitted for debate, or if there was debate it was to no avail. The earliest traditions were that of the grid-iron layout credited to Hippodamus by Aristotle and the radial pattern described by Vitruvius in the first century AD (Kirk, 1953). From the dawn of the Renaissance radial forms were utilised in conjunction with the new technologies of fortifications by Alberti (1485) – who said 'once the site is chosen everything *must* conform' – Scamozzi, Martinus, and Avelino ('il Filarete'), who, in conceiving of the city as a perfect whole, describe the imaginary city of Sforzinda. The grid-iron appeared also as an authoritarian creation in the 'bastide' towns such as Villeneuve-sur-Lot in south-west France (Figure 2.1) and Edward I's implantations in Wales, such as Flint. Grid-iron layout was also used by Adolf in designing Jonkoping for Gustav II; other grids within walls were laid out in Gotenburg, Landskrona, Karlskrona and Fredrikshaven, all like the bastides in a disputed frontier zone (Hall, T., 1986). Hookes also prepared a grid design for London in 1666.

The high point of geometric authoritarianism came in the sixteenth century. In Italy Scamozzi designed a grid-iron town within a polygon of defensive walls which was eventually built as the Venetian Frontier town of Palma Nova. Vascari developed this design further to incorporate both grid-iron and radials within the defensive polygon. The ideas spread into France in the form of increasingly stellar shaped defences that were employed by Vauban as the perimeter to the plans of Longwy, Neuf Brisach and Montlouis and elements of plans for Strasbourg, Lille, Brest and La Rochelle. Military needs overshadowéd all other considerations at least until the advances of military technology made each in its turn obsolete.

The tradition of the grand design which had diffused northwards was embodied in Wren's plan for London after the Great Fire of 1666 which was to replace wooden London, but little was to be realised in comparison with the grand designs that had emerged in Place Royal (Place des Vosges) in Paris or the baroque grand design of Place Stanislas in Nancy (Lavedan, 1959). Louis XIV and Le Notre had begun the great east-west axis of Paris from the Louvre to Etoile, the Champs Elysées. At Versailles Louis XIV, with the assistance of the architect Le Vau, transformed a village that his father had popularised into an urban monument to the greatness of France. The monuments to the greatness of the regime or the ruler appeared in Karlsruhe, Berlin, Copenhagen

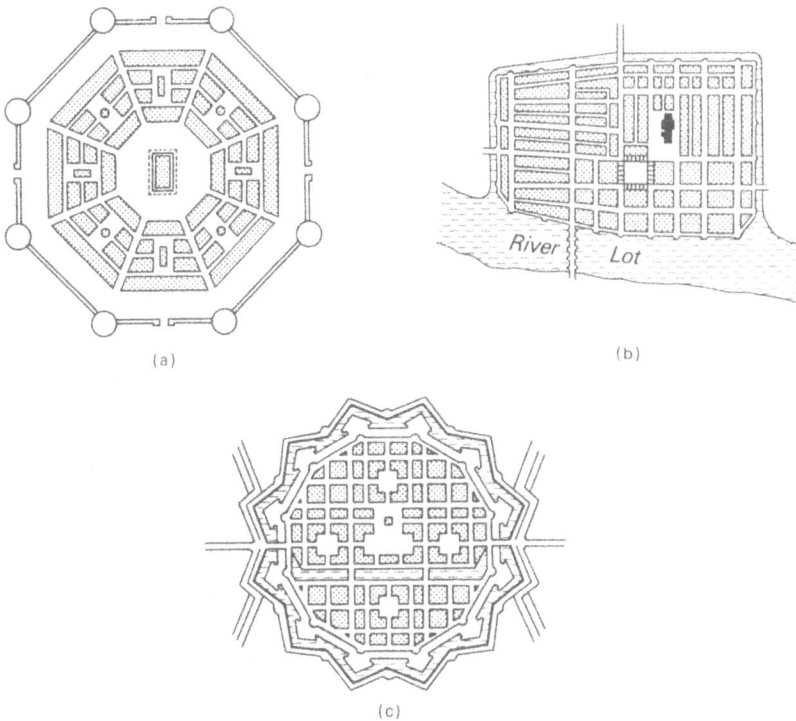

Figure 2.1 Early town plans in the authoritarian tradition: (a) Vitruvius; (b) bastide – Villeneuve-sur-Lot; (c) Scamozzi's ideal town, 1615

(Amalienborg) and Lisbon, where Santos's plan for the development of the city after an earthquake of 1755 gave the city a grid-iron plan with an imposing square as its focus (Williams, 1984).

By the nineteenth century perhaps the best example of this strong tradition comes from France in the work of Haussmann for the Third Empire. Haussmann's imprint on central Paris was never completed but the scale of his enterprise is still amazing. It has also been suggested that Napoleon III actually ordered much of the work himself and used the ideas of Napoleon I, which were culled in turn from the Artists Plan of 1780. Nevertheless, the works attributed to Haussmann were motivated by a mixture of military considerations (control of the 'mob'), the need to provide employment for the mob and the vanity which provided monumental embellishments. Indeed barracks were located at key access points to the boulevards. The new boulevards which cut across the streets of medieval Paris can be seen clearly in the area of Arc de Triomphe/Etoile. Other radials such as the Boulevards Magenta and St Michel and ring routes such as the Boulevard St Germain were conceived as part of his plan (Figure 2.2). Haussmann was able to use power of compulsory acquisition in order to create the wide boulevards and twenty-one urban parks in a city that

Figure 2.2 Haussmann's imprint on Paris (reproduced by permission from Les Travaux d'Haussmann, *Les Grandes villes du Monde, Paris*, no. 3, 483, 23rd April 1986)

had none in 1850. The Bois de Boulogne, Bois de Vincennes beside Parc de Monceau and Buttes Chaumont were transformed under his direction and new landscaped squares such as the Square des Arts et Metiers, Square du Temple and the Square des Invalides were created. Water supplies were improved and mature trees were transplanted into the boulevards, instantly reducing their harshness (Sutcliffe, 1970 and 1981).

Haussmann's influence spread to other French cities. In Lyons, Vaisse built Rue Imperiale and other new streets while other more modest schemes were initiated in Bordeaux, Lille, Le Havre, Marseilles, Montpellier, Rouen and Toulouse. His influence spread to Brussels where Anspach masterminded a new two-kilometre 32m-wide boulevard, transformed the Notre Dames des Nieges quarter and acquired the Bois de la Chambre for the city. The Haussmann tradition was reflected in Viviani's 1882–3 plans for Rome, now obliterated, whilst in Stockholm Lindhagen's 1866 plan for the area between the bridges was plainly influenced by the same source. The Spanish architect Cerda and engineer Castro developed Haussmann-like plans for the expansion of Barcelona and Madrid respectively in 1859. In the latter case security against revolution was a prime consideration in the location of barracks at key points in the plan. At the same time similar autocratically inspired plans were being developed in Berlin by Hobrecht and in Vienna under the auspices of Franz Joseph I. There an urban development commission began to prepare plans for the city's 'Ringstrassenzone' which was to embody considerations of security for the Emperor. In all cases the boulevards, avenues and esplanades were between 22–80 metres wide, so creating a distinct element in the townscape which later generations could exploit commercially. The perspective view focusing upon a landmark, such as St Marie at the end of Rue Royale in Brussels or the Kaiser-Wilhelm-Gedachtniskirche at the end of Tauentzienstrasse in Berlin, were a further legacy of this style or urban design which has been exploited by the tourist industry. The large squares such as Place des Nations were replicated by Hobrecht (Pariser Platz), Viviani's Piazza della Republica and Vaisse's Place Bellecour.

More comprehensive than Haussmann's plans were those of Cerda whose *Teoria General de la Urbanizacion* (1867) developed his principles of urban science that he had already illustrated in his 1859 plan for Barcelona, which remained the basic structure plan for the city for almost a century.

In Germany, Hobrecht's plan for Berlin (1862) was more far-sighted and its wide streets surrounding large blocks as the fundamental units of the layout remained the basis of planning until 1919 (Sutcliffe, 1981). It was a cheap, efficient method of housing the masses which was copied in Dusseldorf by Stubben, and in Aachen and Cologne. As with other more authoritarian plans, Hobrecht's plans led to the formation of societies concerned for building and environmental hygiene. Such a response from middle and working-class urban social movements was a similar reaction a century later (Castells, 1978).

By the inter-war period geometric authoritarianism had been superseded in most of Western Europe by other traditions that we will examine later. But during the Fascist era the tradition reappeared in some of the schemes of the time. In general the authoritarian tradition was reinforced by the monumentalism of the building schemes proposed by Fascist architects. The construction of Via della Conciliazione, Rome and the demolition of Spina dei Borghi, a much more appropriate route to the Piazza San Pietro, is one example of the ill-conceived schemes designed to replace entire districts with showpiece monumentalism. Broad avenues and outsize civic centres replaced the traditional old neighbourhoods in Bologna, Bergamo and Turin where the Via Roma area was demolished so that vast new commercial buildings could be constructed (Calabi, 1984). The Pontine Marshes programme resulted in the building of five small new towns, Littoria, Sabaudia, Pontinia, Aprilia and Pormezia, all of which comprised residential apartments in large blocks surrounding the commercial centre with its public buildings and market, with recreational land on the outskirts (Marian, 1976).

In Berlin similar monumentalist plans were laid by Speer and known as 'Die grosse Strasse', a seven-kilometre north-south avenue leading from the Tiergarten to Schoneberg lined by the 'grandoise façadism' and 'debased neo-classicism' of the new monuments to the Fascist state (Curl, 1969). Only a small section was realised before destruction and the second east-west axis with its shops and offices was never realised (Dahne and Dahl, 1975). Likewise the new towns of Nazi Germany, Wolfsburg and Salzgitter Lebenstad, contained formal layouts to house the workforce for the Volkswagen works and the Hermann Goering steel works respectively. The spacious layout with ample garden areas was in keeping with the 'blood and soil' ideology of the regime. Speer also inspired the Portuguese president Salazar to a more monumentalist architecture in classical style after his visit in 1941.

Madrid's plan which followed the end of the long civil war in 1939 was very much in the authoritarian tradition. Long processional avenues for military and religious functions along which elements of the falangist city were arranged were a feature of the 1941 *Plan General de Urbanización de Madrid*. The working class were to be firmly segregated in this plan into suburban belts of misery which otherwise adapted ideas from earlier plans (Wynne and Smith, 1978; Wynne, 1984).

Since 1945 the strictly authoritarian tradition has had very little influence within Western Europe although it is present in many of the plans of the post-war East European states. Only in Spain were plans produced with the same authoritarian ideals including schemes to decentralise people to new towns around six cities. These ran into problems of coordination between physical and economic planning that were characteristic of most Spanish planning. On the other hand some French planners have expressed the view that on a limited scale the authoritarian tradition was very much present in Gaullist France. The creation despite opposition of a rival business node to Manhattan (La Défense) or the building of the Charles de Gaulle airport at Roissy were pressed through. The same planners have strongly suggested that President Pompidou's desire to create a new arts centre in the heart of Paris (Pompidou Centre, Beaubourg), at the expense of Nanterre where it was originally proposed to locate the centre, is another example of the authoritarian tradition in French urban planning. The ability of successive French presidents to impose their planning and architectural ideas on the capital has continued almost unabated. Giscard d'Estaing had a major influence upon the redevelopment of Les Halles (Bateman and Burtenshaw, 1983). Although Mitterand was unable to realise a commemorative exhibition to mark the second centenary of the French Revolution using a site at La Villette, his patronage has extended to the rehabilitation of the former abbatoir district (Burtenshaw and Moon, 1985), the redevelopment of the Bercy area and the construction of the new Opera at the Bastille, the Louvre pyramid, the Arab Institute and most significantly Tête Défense, the final major building in La Défense. The authoritarian tradition in Western Europe can be seen as the expression in the built environment of the political ideologies of the right ranging from the autocratic heads of states to Fascists and to the liberal conservative governments.

The utilitarian tradition

The utilitarian tradition was a product of *laissez-faire* economics and the rise of nineteenth century capitalism. It resulted in urban developments which were predominantly functional and subservient to the demands of an industrial economy born of free enterprise. It is more difficult to identify individuals connected with this urban tradition because its very essence, derived from nineteenth-century social attitudes, was the lack of concern that created the 'tunnel backs', 'villes minières', 'zijnwijken' and 'kolonien' of the nineteenth-century city. Such developments came later in Lisbon as a result of rural-urban migration in the form of high density infilling in gardens or patios, terraces facing alleyways and large suburban residential units for working classes, called *villas*. In Oporto the *ihlas* or slums were very utilitarian and crude infillings (Williams, 1984). In many cities the worst examples of *laissez-faire* building abutted onto the architecturally better products of the age. Thus in London the poor conditions of Somers Town were separated from the better housing at Fig's Mead by a changed street layout. Others built gates and barriers into their schemes to keep out unwanted incursions from the slum areas (Choay, 1969).

The contribution of the utilitarian tradition in the nineteenth century was twofold. First, it gave birth to legislation to control the worst environmental evils that had been created by the prevalent social attitudes. In Great Britain the 1875 Health Act was the first significant step in compulsory legislation, setting up the Boards of Health whose activities raised suburban housing standards. Belgium introduced laws to enable property for redevelopment to be compulsorily purchased, so controlling the unhealthy housing environments. Similar laws were enacted in Italy between 1865 and 1885, and in France after the turn of the century. Secondly, the movement gave birth to its antithesis, the tradition of utopian thought that reached its zenith by the beginning of the twentieth century.

In the twentieth century the utilitarian tradition continued to have an impact albeit in changed form. The *lotissements* of Paris and the unregulated spread of suburban housing, often in ribbons along arterial routes, in the London suburbs, were examples of the new forms of *laissez-faire* urban growth. The unplanned sprawl that characterised the Bonn urban area between 1945 and 1970 was another example of a continuing tradition which is more frequently observed in the commentaries and reports of the plans of physical planners and architects who pay allegiance to other planning traditions.

Utilitarianism has re-emerged particularly in Britain during the nineteen-eighties. Concerted attempts were made to circumvent the planning system which was perceived as an unnecessary bureaucracy by the new right. In particular, the removal of planning controls from Enterprise Zones and Urban Development Corporations created the opportunity for developers to propose 'a monument to wrong thinking' (HRH The Prince of Wales, 1989). The abolition of the Greater London Council has given West Europe its first capital city not subject to unified planning control now that planning is a district responsibility. The United Kingdom is not alone in abolishing city region planning because the same thing has happened in Hannover, Braunschweig and the Ruhrgebiet in West Germany where planning has been returned to the Regierungsbezirk and Stadkreise. Constraints on the action of planners, such as the instruction to see all development in a positive light, the potential penalties for losing a planning appeal and ministerial statements concerning the release of land in the Green Belt, are all manifestations of the new right utilitarianism.

The utopian tradition

Utopian thinkers have been contributing to urban planning ever since towns were first built. This particular tradition has been exceedingly strong in Western Europe thanks to the influence of one man, Ebenezer Howard. Howard, however, acknowledges the fact that the ideas which he developed in *Tomorrow: a Peaceful Path to Real Reform* (1898) were based on ideas gleaned from a long tradition of utopian thinking. Normally Plato's ideals for city government and the Socratic moral values from which they are derived are regarded as one of the earliest statements in the utopian tradition. In 1516 Thomas More's *Utopia* discussed the ideal city of material abundance linked to work and participation and including a novel scheme of rotating city and country living. The second home boom of the twentieth century had been forecast. In addition, More gave some attention to the recreation function of the urban periphery (Meyerson, 1961). A century later Tommaso Campanella's *Cittá del Sole* (1623) and Francis Bacon's *The New Atlantis* (1622) also searched for the elusive ideal city.

As early as the late eighteenth century, idealism in urban planning had returned as a response to the social and environmental conditions produced in the early years of the industrial revolution. Owen's scheme published in 1816 concentrated on the physical structure of the community as a compromise solution within the confines of the industrial system that he abhorred. His model creation which failed at Orbiston, Lanark, like many others that followed, was small. Fourier's *phalanstères* was a similar creation for France which was utilised in Godin's work at Guise, so giving birth to the technocratic utopian tradition which we will discuss later.

Owen's ideas led to a series of new settlements designed to alleviate the horrors of life in the industrial city. Salt's Saltaire (1851), a grid of housing, almshouses, public baths, washhouses, schools and an alcohol-free institute, surrounded the Congregational Church in this community giving self-respect in a healthy environment. Salt even fitted a smoke-burning device to the Tuscan campanile-style chimney of his Italianate factory, so reducing pollution; a century later pollution controls on chimneys are still not universal in Britain, emphasising the far-thinking nature of Salt's comprehensive plan (Bradley, 1986). Ackroyd's Ackroyden, Richardson's Bessbrook, near Newry, Lever's Port Sunlight, Cadbury's Bournville and Rowntree's New Earswick are others

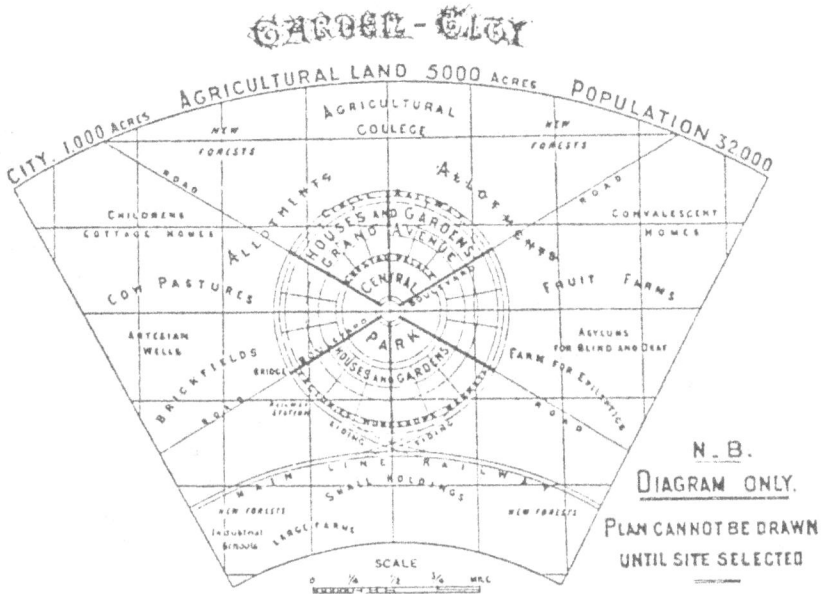

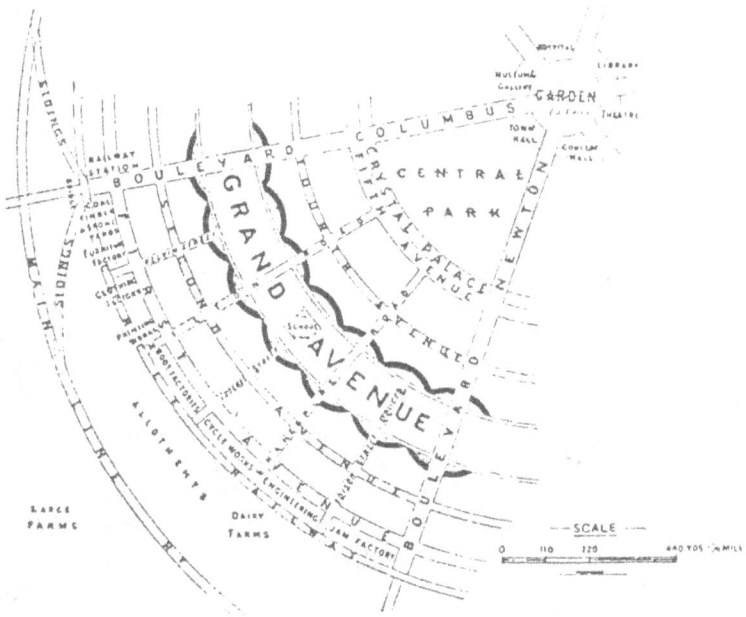

Figure 2.3 Howard's garden city proposals, 1898 (reprinted by permission of Faber and Faber from *Garden Cities of Tomorrow* by Ebenezer Howard)

in Britain. In Prussia the workers' quarter developed by Hannoverschen Maschinenbau A.G. in Linden, Hannover, in 1870 also attempted to follow new ideals for urban living. In contrast J. S. Buckingham's Victoria was a very austere urban design that never reached fruition. Others such as Cabet, Bellamy, Morris, Ruskin, Disraeli, Engels, H. G. Wells and Zola contributed directly, and indirectly through novels, to the growing body of literature that acknowledged the problems of urban life in nineteenth-century Europe.

Howard acknowledged Ruskin's influence on his own thoughts as well as the impact of Buckingham's unsuccessful Victoria which, like the Garden City which it preceded, was to be set in 'a large agricultural estate' (Howard, 1946). To Howard the Garden City was the ideal combination of the benefits of rural and urban living. Howard then describes his Garden City in great detail including the most publicised diagrams in the history of planning (Figure 2.3). The Garden City, which would have grown as a colony of 32,000 in the countryside, was also the vehicle through which Howard was able to forecast both the extension of municipal enterprise in all fields and the need to have teamwork to create fluid plans. The final chapter illustrates the way in which London's problems would be alleviated by his urban programme to achieve a so-called 'polynuclear city'. In fact two years before Howard, Fritsch (1896) had specified the characteristics of the Garden City in *Die Stadt der Zukunft*.

Letchworth was the city that transformed much of Howard's dream into reality although it lacked any municipal enterprise, being developed by private developers. Here, Parker and Unwin soon grasped the essence of Howard's ideals and built what Osborne had described as 'a faithful fulfilment of Howard's essential ideas' (Howard, 1946). Welwyn Garden City was begun in 1919 on Howard's own initiative and remains as a museum to his genius and a permanent memorial to his undoubtedly fundamental contribution to West European urban thinking. In Germany the first Garden City was built at Hellerau (near Dresden) in 1908, one year after Howard's work was published.

The ideas of Howard were gradually diffused in Britain partly by Raymond Unwin in his book *Town Planning in Practice* (1909) where he admitted his admiration for the German planners, and by Geddes who, in his work, acknowledges a debt to Unwin to whom he was related. Unwin also advocated the application of Garden City principles to London (Unwin, 1912). Governmental reports such as those of Marley 1932, Barlow 1940 and Reith 1946 saw the need for new settlements to provide for the growing population and new industries although the *raison d'être* of the New Town had diverged increasingly from the basic Howardian concept. Abercrombie's Greater London Plan 1943, which owed much to ideas developed by Unwin as technical advisor to the Greater London Regional Planning Committee between 1929 and 1933 (Miller, 1989), also contained similar proposals for new towns. Garden Cities, Gartenstadt, and Cité Jardins began to be built throughout Western Europe following the translation of *Garden Cities of Tomorrow* into French in 1917. Letchworth, Welwyn and later Hampstead Garden Suburb were the models on which the architects and planners based their own garden cities adjusted to each nation's need. Beniot-Levy and Forestier had begun to campaign for Garden Cities and more open space around French towns by 1906. Schmitthener planned a garden suburb at Stauken near Berlin in 1913, the same year that the International Garden Cities and Town Planning Association was founded. In 1925 Floreal, a garden city, was built in Boitsfurt near Brussels. Bernoulli was responsible for developing a Swiss garden suburb which was built in the Hirzbrunnen area of Basle between 1924 and 1930 (Gubler, 1975). In Germany Metzendorf acknowledged his debt to Howard when he designed Margarethenhohe (Essen) for the Krupp family, while in Frankfurt-Romerstadt May was building a Gartenstadt and in Munich-Perlach a similar garden suburb was developed. Howard's ideas were used in Finland by Saarinen. In Mediterranean Europe a garden suburb, Milanino, was built eight kilometres outside Milan as early as 1910 and Schiavi founded a tenants' society which adopted many Garden City ideas (Calabi, 1984). Other Italian garden cities such as Tiepolo and Campo del Fiori were developed after the First World War (Whittick, 1974). As early as 1912 Montoliu had founded a garden city movement in Spain which eventually influenced plans developed for Madrid in 1924 and the 1929 *Plan de Extension* designed by Zuazo, Mercadal and Jansen, one of Sitte's disciples. The towns that come closest to Howard's ideals are in Merlin's opinion those in Scandinavia (Merlin, 1969).

The Garden City ideal was also hijacked by the authoritarians in the pursuit of town planning and architecture for the objectives of race hygiene of the Third Reich. In Great Britain Francis Galton's fictitious state of Kantsaywhere and the concept of eugenic utopia depended upon improvements in environmental conditions similar to those being promulgated by the Garden Cities Association (Voigt, 1989).

Whether the new town can be seen as a further link in the utopian tradition is a matter of debate. Certainly, the New Town is not the same as the Garden City in terms of its ethos, financing or size. The British new towns were more controlled though not strictly municipal ventures. Decentralisation from London to the eight towns around the capital was very much in the Howardian tradition that Abercrombie, Barlow and Reith had followed. The plans that emerged after the 1946 Act did represent the idealistic attempts by the new planners of post-1945 socialist Britain to create urban utopias. Perhaps the most publicised of all the plans were those of Gibberd for Harlow (1947), which set a pattern for urban planning that wed transatlantic ideas on movement to social planning concepts from Europe in the context of the site conditions of Harlow (Figure 2.4).

The ready adoption of the new town formula for urban growth in the third quarter of the twentieth century is expanded in chapter 12. Modifications of the formula to suit the needs and ideas of almost all the nations were designed and built. Many cases, as we shall see, were taken from other traditions and fused to provide the new town utopias suited to that culture.

Howard's utopian vision is a landmark in West European urban thinking. There are few urban thinkers whose ideas have been copied and modified as

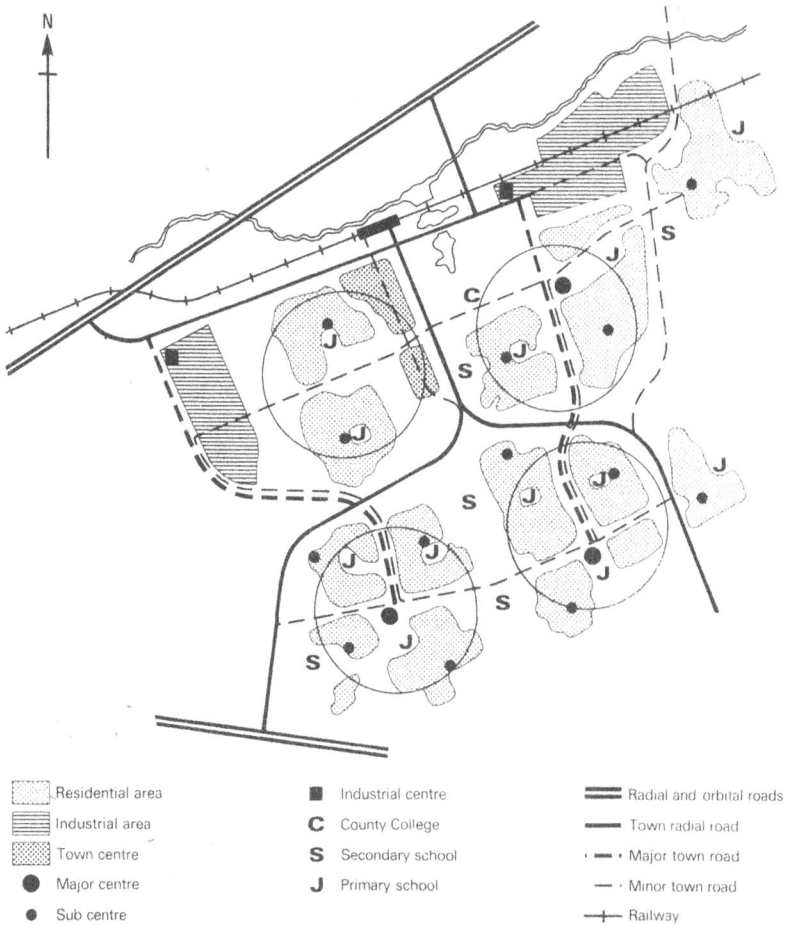

N

Residential area	■ Industrial centre	Radial and orbital roads
Industrial area	C County College	Town radial road
Town centre	S Secondary school	Major town road
● Major centre	J Primary school	Minor town road
● Sub centre		Railway

Figure 2.4 Gibberd's plan for Harlow new town, 1947 (reproduced by permission of Lund Humpries Publishers Ltd.)

extensively. The themes of controlling metropolitan growth, decentralisation, self-contained urban community and green belts are all part of what he called his 'unique combination of proposals'. His work represents the high point in a distinctly European contribution to urban planning whose origins are part of the European urban tradition and whose effects are world-wide. The utopian tradition had its offshoots, of which the technocratic utopian movement was the strongest.

The romantic tradition

A tradition whose zenith was contemporaneous with that of Howard was the romantic tradition of urban planning. It was a direct result of the feeling of horror on the one hand at the Industrial Revolution's impact on urban form and on the other of the authoritarians' impact on the beautiful capitals of Europe. Its main authority, whose work was published a decade before Howard's, was Camillo Sitte. He wrote his classic *Der Städtebau* in 1889 partly as a result of the emperor Franz Joseph's 'Haussmannisation' of the Ringstrasse in Vienna, an act that Sitte condemned. By analysing the spatial organisation and aesthetic values of cities and city planners in the past he began to identify the fundamental role of space in the city. He attempted to analyse the connection between elements in townspaces and saw the need for enclosed spaces which are characterised by irregularity, imagination and symmetry. In so doing he challenged abstract city planning and by reference to examples culled from his extensive travels, such as the work of Alberti (1485) or Louis XIV, he tried to introduce aesthetic values into urban planning. Streets, squares and plazas are analysed before he prescribed his solutions for modern city plans so that 'the monumental centre of a large city could be redesigned artistically in accordance with the teachings of history and the model of beautiful old towns' (Sitte, 1965). Choay (1969) sees the work of Sitte as 'culturalist' because it is based on nostalgia and inspired by a vision of a past community and its culture. His legacy in Thomas Hall's (1986) opinion is that the square or plaza is a major European contribution to urban design whether it has a church as a focus, (Zionskirchplatz, Berlin), or a monument (Trafalgar Square column), or a formal garden (Place de la Nation, Paris).

Sitte's emphasis on aesthetic values was readily adopted in Germany and Austria as an antidote to autocratic planning. His ideas were utilised directly in plans for Olmutz (1894), Salzburg and Vienna (Figure 2.5) and indirectly in plans for Altona, Munich (1892) and Flensburg by Henrici, by Pulzer in extensions to Darmstadt, Wiesbaden and Mainz, and Nystrom and Sonck's 1902 plan for Helsingfors. Sitte's work was translated into French in 1902 but his sole influence in that linguistic region was the preservation of the total townscape of the Grande Place, Brussels by Buls (Choay, 1969), who had himself published a brochure entitled 'Esthétique des villes' in 1893 (Hall, 1986). Le Corbusier mocked the man and his thoughts perhaps accounted for his poor reception in France.

In Britain Sitte's ideas were first used by Unwin in designing Letchworth after the translation of his *Städtebau* into French because the English edition did not appear until 1965. Unwin brought together both the aesthetic values of the Romantic tradition and those of the utopian Garden City. Geddes also recognised Sitte's contributions to urban planning. Perhaps the most deliberate use of Sitte's work has always been in Scandinavia, by Hallman in Stockholm. In Finland, for instance, the German architect Engel incorporated Sitte's ideas in his plans while the early twentieth century plans for many Swedish cities, such as new developments around Goteburg, placed great emphasis on detail that owes much to Sitte's ideas. Van Eesteren's plan for Amsterdam also possessed all the detailed studies that are the hallmark of Sitte's followers. In

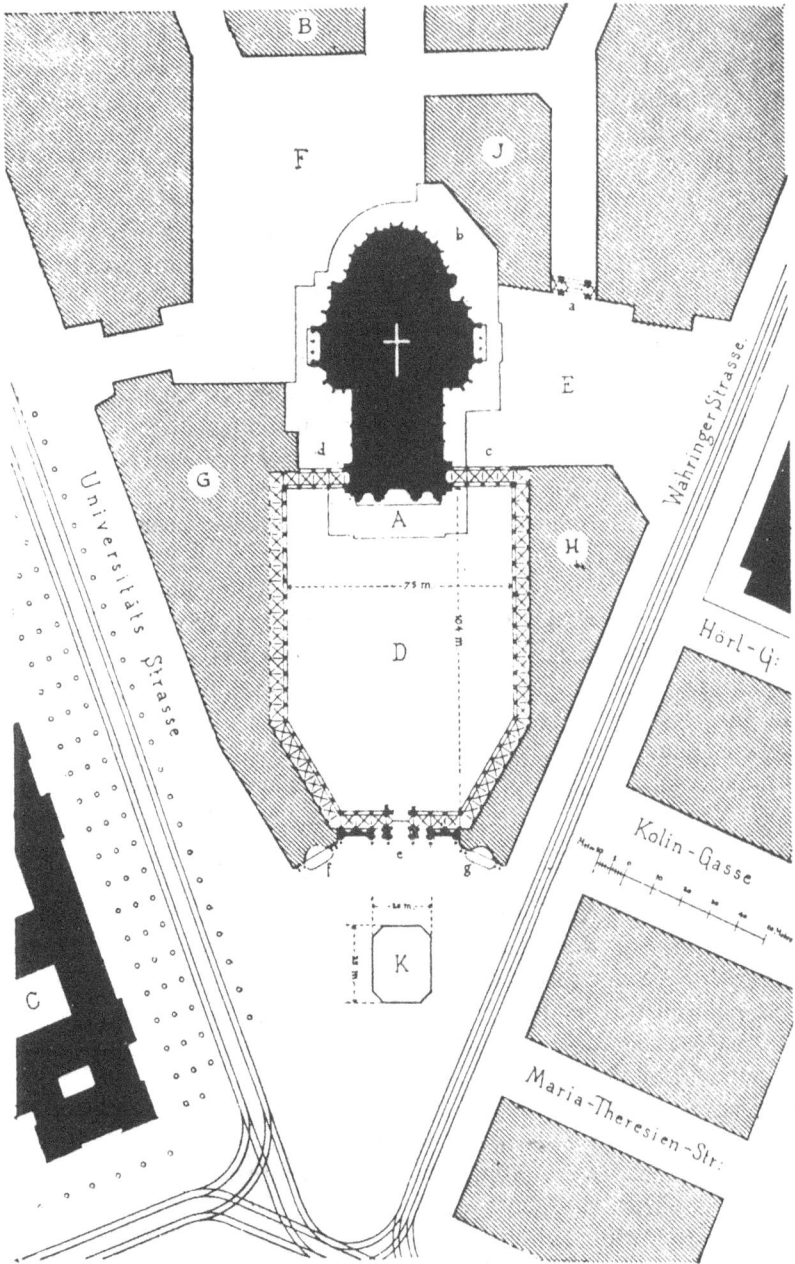

Figure 2.5 Sitte's plan for part of the Western *Ringstrasse*, Vienna. (Letters refer to suggested buildings)

Spain Jansen was able to bring Sitte's ideas to bear on the 1929 Madrid plan. Henrici developed his ideas in cooperation with Stubben for the ring zone around Cologne, a development problem which mirrored that which Sitte had discussed in Vienna. The outcome straddled this and the grid-iron authoritarian traditions.

Sitte's emphasis on town planning as a creative art had dragged urban planning out of the narrow confines of authoritarianism and uncoordinated utilitarian policies. It represented a parallel reaction to that of Howard especially in central Europe and, like Howard's work, it has continued to influence West European urban planning theory almost a century later. Once again building and planning could be an art form again.

Like Sitte, HRH The Prince of Wales was horrified by the nature of urban development, not of Haussmannisation but the functionalist modern architecture approved by planners in late twentieth-century Britain (HRH The Prince of Wales, 1989). Having followed his grandmother's dictum 'look and observe', which might have come from the lips of Geddes's first principle of planning, the Prince castigates much post-1945 planning and architecture. Some of his principles such as enclosure, art, decoration and community embody a romantic view of European planning. His likes are those post-modernist structures which are considered by many to be pastiches of the past although he seems to prefer past creations such as Sitte's Vienna, and Sienna. At the level of a plan, he approves of Leon Krier's plan for Atlantis, an echo of ancient Greece and Turkey. It is unfortunate that the attempt by Krier to turn the principles of late twentieth-century romanticism into practice in 350 acres in Dorchester on Duchy of Cornwall land have been scaled down as a result of public protest and the planning system. The consequence is that the scale of the proposal is not large enough to develop the principles.

The technocratic utopian tradition

The origins of this tradition are inextricably bound with the utopian plans of the eighteenth and nineteenth centuries in which the concept that every urban activity is open to a technical analysis is developed. In the same way the machines that form the heart of a town's industrial production were considered to be the proper core to a town. Technological power has formed the basis of this tradition fron Francis Bacon onwards. Thus the earliest scheme, Ledoux's Chaux-les-Salines (1773), placed the factory at the centre of the industrial town. A second more influential origin to the technocratic tradition was Fourier's *phalanstère* (1829), a vast chateau-like structure to house his utopian community. Fourier's work attracted a similar interest in France to that accorded to Owen in Britain and his influence was acknowledged by subsequent writers including Marx. In 1791 Godin attempted to turn Fourier's ideas into reality at Guise but the rigidity of the system was not designed for the rapid social and technological changes of the nineteenth century despite the fame

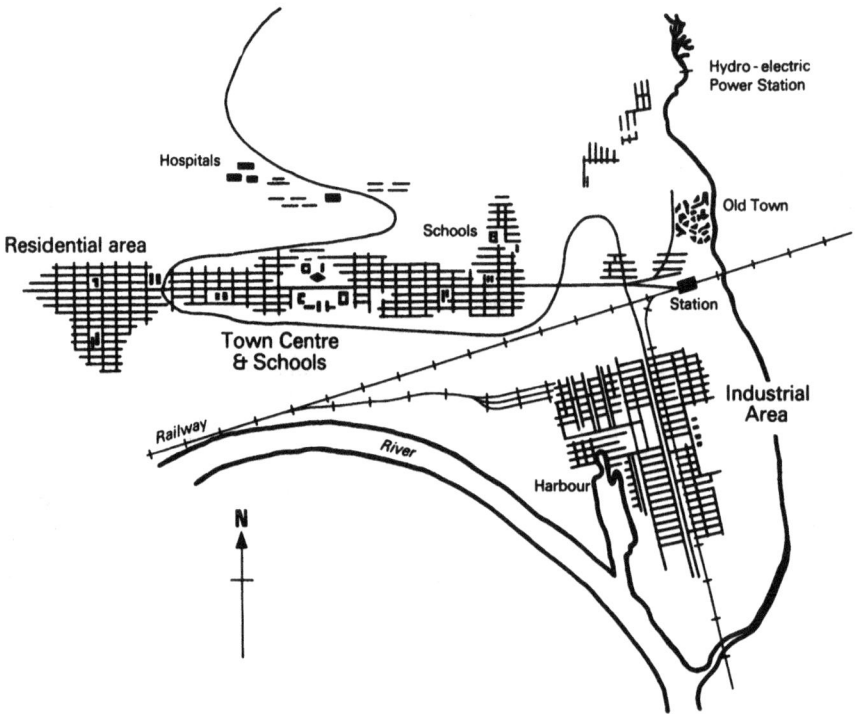

Figure 2.6 Garnier's plan of the *Cité Industrielle*

accorded to the Guise Familistère. Goransson's steel works village at Sandviken is another parallel, less gradiose scheme which dates from 1860.

By the turn of the century another design for the *cité industrielle*, a utopian vision of the industrial age seen through French eyes, had been propounded by Garnier. In his city the land-use zoning was strict and separate industrial, residential, hospital and central service areas were separated by green zones (Figure 2.6). On the plans the built forms were monotonously repetitious and lacking the grandiose scale of the *phalanstère*. It was a pragmatic design that made the maximum use of known building and transport technologies. One plan in existence that acknowledges Garnier's direct influence is Perret's redesigning of Le Havre after 1945.

It was Le Corbusier who translated the vision of Fourier, Garnier and Godin into modern terms and who represents the zenith of the technocratic tradition although others would call it functionalist (Choay, 1969). Le Corbusier, Swiss by birth, came to live permanently in Paris in 1916 and by 1922 he was publicising his *Ville Contemporaine*. This plan and his later *Ville Radieuse* rank alongside Howard's plans as the most frequently published schemes in the history of urban planning (Figures 2.7 and 2.8). Le Corbusier embraced the

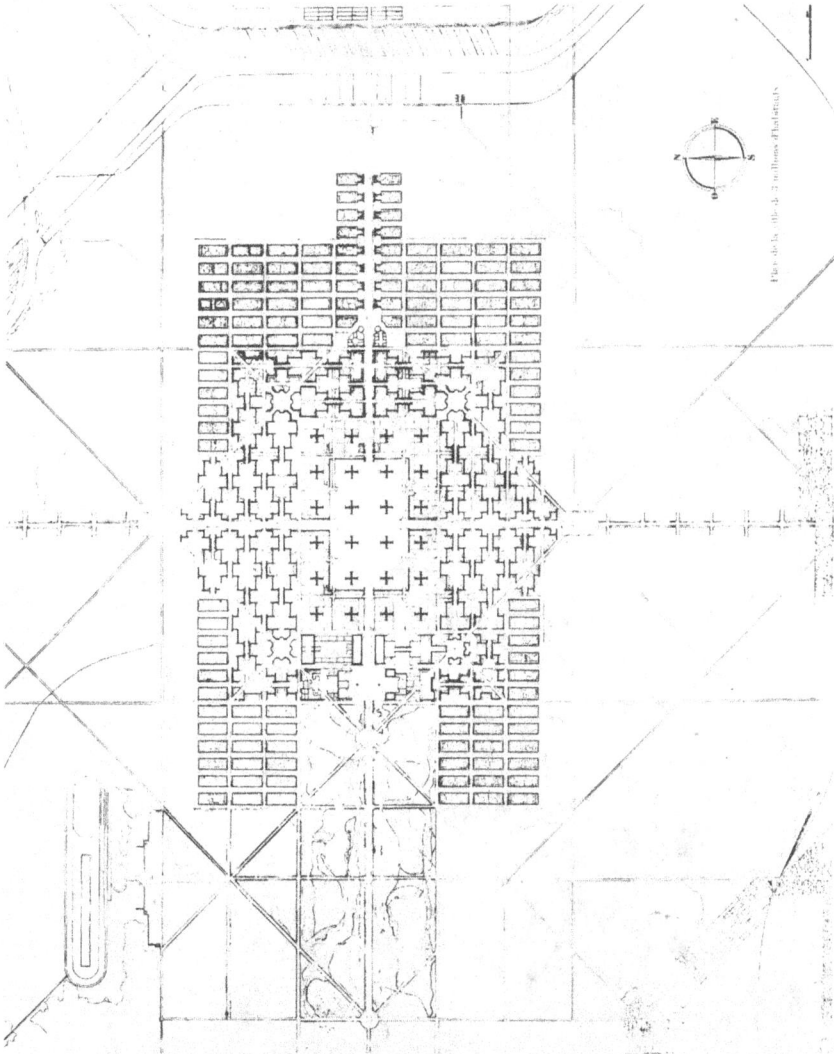

Figure 2.7 Le Corbusier's *La Ville Contemporaine*

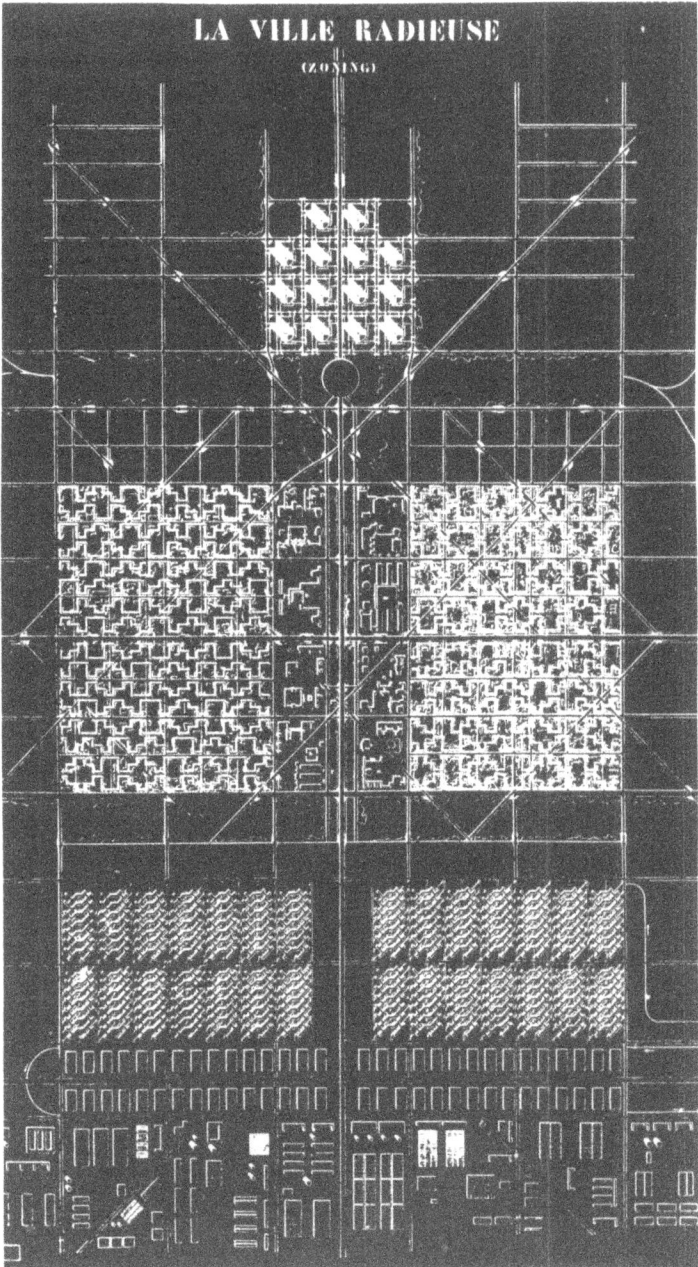

Figure 2.8 Le Corbusier's *La Ville Radieuse*

technocratic ideology of the large-scale corporations led by the new technical élite, satisfying the demands of the population for improved living conditions. Thus the élite were to reside near the city centre in high-rise apartments surrounding the administrative, cultural and entertainment centre, while the rest of the population lived in satellite towns on the outskirts. The grid-iron city, which is characterised by green space with 85 per cent of its surface area devoted to parks, was to be a city of leisure as well as production, the physical embodiment of a social utopia that paid homage to the power of the giant machines that mesmerised Le Corbusier.

In an attempt to give his ideas a focus, Le Corbusier turned to the heart of Paris rather than to the suburban areas which had given birth to the plans of Howard and Sitte. In his *Plan Voisin* he put forward a radical design for the right bank facing the Ile de la Cité (Fishman, 1977; Le Corbusier, 1978). Eighteen skyscrapers would fill the 200 hectares with the headquarters of international corporations but neither they nor anyone else were interested in this vision that owed as much to the authoritarian antecedents of Paris planning as it did to the need to control the technology of the automobile and steel-framed building structures.

Disillusionment with the reception of the *Plan Voisin* pushed Le Corbusier away from capitalism to the syndicalist movement where he began to design a new city for a syndicalist society. *La Ville Radieuse* (1935) was a product of a more coordinated and directed society. The city centre is now the residential district which replaces administration with new 'Unités', housing 2,700 residents, each apartment now being allocated according to need rather than position in the status hierarchy. The services of each 'Unité' like those in the *phalanstère* were to be in common. Otherwise the emphasis on green space and on accommodating transport remained. 'L'Unité d'Habitation' in Marseilles stands as the sole memorial to the total ideal which many see as having not only influenced a generation of French housing built since 1945 (Vance, 1977) but also the work of architects and planners throughout the world through CIAM (Centre Internationale de l'Architecture Moderne). For instance, the 1945 Plan AR for Milan was prepared by a group affiliated to CIAM.

While not producing an ideal city, the Bauhaus architects Mies van der Rohe and Gropius were also concerned with society and the use of the new designs in concrete, steel and glass to serve society. They were committed in particular to socialism and their ideas had a definite influence on Le Corbusier's design for 'Unités'. The Bauhaus school designed long north-south residential blocks separated by plentiful open space. These were built between 1927 and 1931 at Dammerstock near Karlsruhe and at Siemenstadt Berlin to illustrate their theories of density and light. Having left Germany in the thirties Gropius did plan the Berlin suburb that bore his name, Gropiusstadt, which embodies many of his ideals that influenced similar urban developments in the sixties (see Chapter 6). Gropius's influence also came to bear on May's urban community built in the garden city tradition at Romerstadt. Sant Elia's La Nuova Cittá (1914) also utilised high-rise buildings in association with concepts of split level living. At the same time Sartoris (1931) introduced to Italy plans for a belt city incorporating Le Corbusier's ideas. The acceptance of modernist plans based on the principles of CIAM continued well into the mid-twentieth century with some countries such as Portugal only embracing the ideas after 1955. In Spain Le Corbusier's ideas were promulgated by GATEPAC (GATCPAC in Catalonia) in the early thirties.

Sartoris's ideas were not entirely new because the concept of a belt city was a technical solution to the problems of movement that had been discussed in Spain by Soria y Mata in 1882. *Cuidad Lineal* could be a link between existing cities, a new urban zone of any length or a ring around an existing city; part of the latter did materialise in a suburb of Madrid. Communication was seen by both Soria y Mata and Cerda as a formative element of the city, a technocratic solution that Le Corbusier and Sartoris used. However, despite pre-dating Howard and Sitte the ideas of Soria y Mata were not disseminated beyond Spain until much later. Castillo attempted to link the linear city to the Garden City idea and in 1919 planned a linear city in Belgium which never progressed beyond the drawing board (Wynne, 1984).

The tradition of technocratic plans based on various transport technologies has remained as a guiding principle of many plans. Henard's proposals for transforming Paris, published between 1903 and 1906, were based on the spread of the new underground lines within the city (Sutcliffe, 1981). Schumacher's plan for Hamburg (1921) based its fingers of growth on communication lines (Albers, 1977). The MARS (Modern Architects Research Society) plan for London (1943) made use of an elaborate transport plan to divide post-blitz London into sixteen urban tentacles spreading north and south from a central axis (Figure 2.9). This was not surprising because many of the MARS group were architects and future planners who had embraced the Corbusian tradition. The MARS plan was not unlike the Ruhr plan proposals although on a vaster scale. More recently Ling's plan for Runcorn is an extension of the same tradition married to other principles of new town development. The plans for Greater Copenhagen, Greater Stockholm and Evry (Paris) are other more recent examples where great emphasis has been placed on the impact of transport technologies. The emphasis given to technocratic planning is not that unusual given the fact that the new profession of planning that arose after 1945 had to recruit from pre-existing disciplines that designed our urban areas. Architects and civil engineers thus dominated the post-war burgeoning planning offices and their technocratic vision dominated city developments.

In the final decades of this century a new force is beginning to emerge, that of high technology, no longer the technology of the civil engineer but that of the electronic engineer. Micro electronics processes and generates information so that the form of production is process-oriented rather than product-oriented, which in its turn is demanding new space in and around cities and new building forms to accommodate the technologies (Castells, 1985). In addition, because the new activities are knowledge intensive and attract the best brains, it is inevitable that the scarce labour force is located in the most amenable working

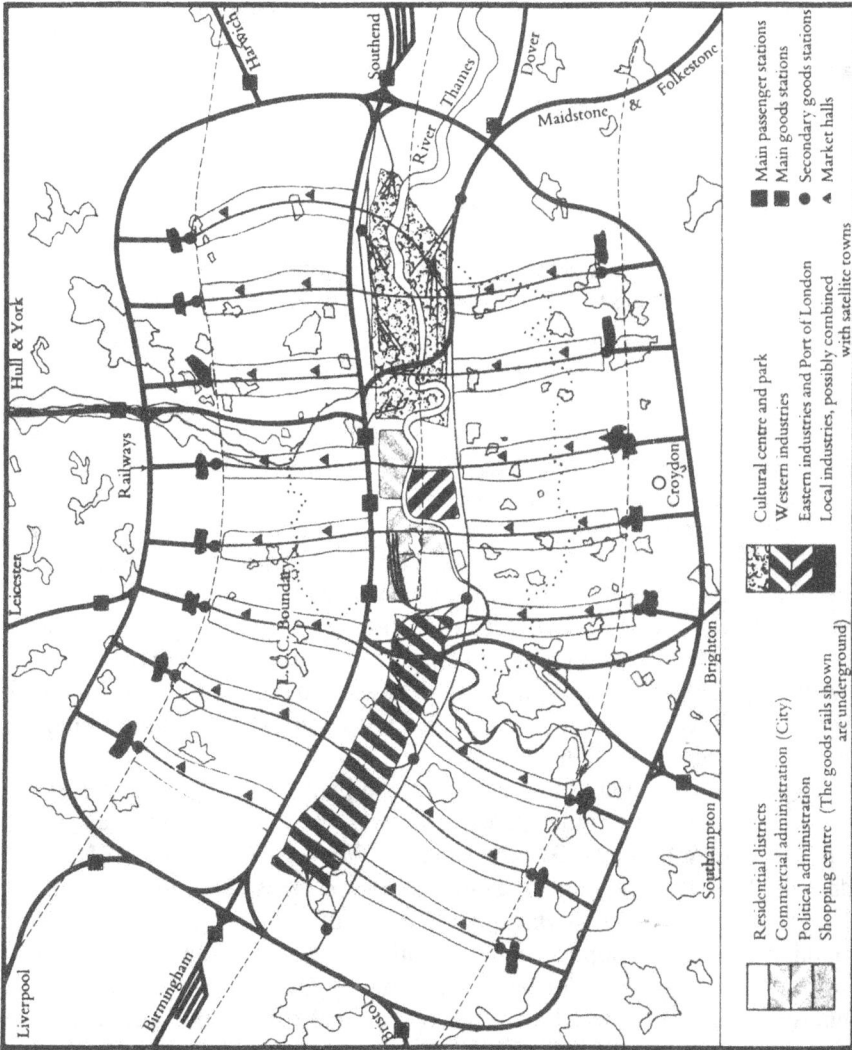

Figure 2.9   The MARS plan for London, 1942 (reproduced by permission of the Architectural Press)

and living environments. Therefore the development of new technology tends to polarise further both social groups and the cities themselves into those with the conditions and those without. French planning has developed the theme of *technopoles* but by far the most successful are those in southern France despite government attempts to promote those in the north such as Caen. The Cambridge Science Park and the M4 chain of R&D locations, the Chilworth Science Park, Southampton and the economic prosperity of Winchester are all related to the impact of planning for the electronic age. When high technology comes into the city either new locations designed for the cabling are planned, for example Canary Wharf, London and Tolbiac, Paris, or old buildings with a greater volume are converted, for example Billingsgate and La Villette.

The tradition of organic planning

Organic planning has arisen from a better understanding of our urban past which came from the works of the utopians and romantics. The main emphasis was on the achievement of an intelligible order in cities. Therefore building uses may change harmoniously as they have in the Piazza San Marco in Venice (Mumford, 1974). Parker was aware of organic planning when he designed unused spaces in Wythenshawe (1927). However, it was left to Geddes to be the tradition's most articulate spokesman. No doubt his biological training had a considerable part to play in his emphasis on the constant reappraisal of needs. He also stressed participation and the need to make the city a living organism in all senses (Herbert, 1980; Meller, 1981). Geddes also considered the city within its region in *Cities in Evolution* (1915) although it has to be acknowledged that Leonardo da Vinci discussed overspill within a city region context four centuries earlier and Mackinder had used the concept in *Britain and the British Seas*. Although Geddes wrote much on the basis of experiences in Germany, he had very little direct influence on mainland Europe; his works of note were in the Empire, in India (Sutcliffe, 1981).

City region plans rather than city plans are plentiful today. Among the earliest to present a regional perspective was Saarinen's 1910 plan for Helsingfors. Abercrombie's *Greater London Plan* (1945) and Peter Hall's *London 2000*, besides the official south-east region plans (*South-East Study*, 1964; *Strategic Plan for the South-East*, 1970), are all attempts at organising the metropolitan growth of London while permitting organic change through the idea of study areas (*Strategy for the South-East*, 1967). Hillebrecht's model of Regionalstadt (1962) which has been partly realised in the concept of Grossraum Hannover, and for a few years Grossraum Braunschweig, and Boustedt's concepts at Stadtregionen (1970) are German derivatives of this tradition. The 'Randstad' concept (1966) and the Greater Stockholm plans are all of the same genre, merging the best of the other traditions in a distinctly Dutch or Swedish scheme that permits continuing modification. Continuously monitored structure plans, such as the *South Hampshire Structure Plan* (1974) and the *Greater Manchester Plan* (1968), and the post-Skeffington participation exercised in British planning, are developments in the organic tradition. In France the concept of the 'germ' being utilised by the planners of Val de Reuil new town and the openness of the plans for the third sector of Marne-La-Vallée were part of the same tradition that produced flexibility for the future acknowledging the inevitability of social, economic, political and technological developments. In this sense the organic tradition could be seen as the attempt to merge the best of the earlier traditions. It also stems from the recognition of the inevitability and unpredictability of change among the modern planners.

Decelerating population growth in the late twentieth century has reduced the need to reserve green field sites for future growth even if the pressures from developers are for such developments. The growing ecological awareness combined with a reaction to technocratic planning has forced planners and politicians into a reconsideration of redundant space in the city. The past decade has seen a growing trend to re-use and rehabilitate sites and buildings rather than develop comprehensively. The regeneration of former dockland areas, for example Watersted, Rotterdam, by the port cities is the largest pan-European initiative (Hoyle, Pinder and Husain, 1988) although the re-use of other buildings is progressing, notably markets (Covent Garden) and a steel works administrative complex (Rheinhausen). At a larger scale, railroads (Tolbiac, Paris), military (Neuherberg, Munich) and institutional land (Shenley, Hertsmere) are other cases where the use of a site is evolving or, perhaps in keeping with the medical analogy, re-emerging with a virulent/dynamic new use.

The socialist tradition

There is one further tradition that deserves inclusion in this overview and that is the socialist tradition which has arisen from the works of Marx, Engels and their interpreters. It is a distinct tradition in one major sense and this is that neither Marx nor Engels, nor the other socialist philosophers, actually produced plans for cities. Engels' work on the housing of Manchester's poor and Marx's *Critique of the Gotha Programme* describe the causes of city problems and prescribe the solutions in terms of proletarian revolution, their utopian vision. Their vision grew out of utopian thought but they distinguish the 'utopian socialism' of Owen, Fourier and Garnier as 'castles in the air' and dialectically opposed to their 'scientific socialism' (Manuel 1966).

The Marxists' criticisms of all city developments in Western Europe are presented in the form of an analysis of spatial structure and its relationships to the influence of the dominant social structures. The major proponents of this view of the city today look very much to the lead provided by the works of Castells, perhaps the foremost urban political-economist in Western Europe. In *Monopolville* he described the social processes at work in Dunkirk in the nineteen sixties and interprets their impact on the spatial form of the city

(Castells and Goddard, 1974). Therefore 'Master plans whatever their scale have an underlying social and political logic which varies for each plan in exact correspondence with the situation of political hegemony within the institutional apparatus on which the planning agency in question depends' (Castells 1977). To Castells the city, like the state, is a tool of the dominant classes and urban space is a material element whose relationship to the dominant class is an important field of study. To Castells the term 'spatial structure' is a description of the particular way in which the basic elements of social structure are spatially articulated (Pickvance, 1976). However, neither Castells (Castells, 1972) nor any other of the growing number of planners and social scientists embracing Marxism has moved forward to any view of the city that might result from the elimination of the worst dichotomies that are discussed. Socialist cities do exist within the Italian political framework and socialist architect-planners did work within the constraints of other traditions, like the Milan AR group, many of whom were communists. Bologna has had a Communist municipal government for the past four decades and might be regarded as the closest urban expression of a Marxist ideology given the constraints of Italian government policies. Nevertheless there is still no evidence of Euro-communism or socialism producing a visionary to match Howard, Le Corbusier, Sitte or Geddes. As Pahl has stated 'In theory "socialism" and "modern big cities" are incompatible. In practice of course socialists must make do with the aspirations of territorial justice which often lead to further inequalities' (Pahl, 1977).

Similar criticisms of the lack of visionaries can be attributed to the Weberian socialist tradition. The Weberian tradition separates economic and other aspects of society and enables the distinction to be made between the political and economic theories of the city which, in turn, enables the multifaceted phenomena that comprise a city to be disaggregated. Studies that have focused on concepts such as urban managerialism have attempted to look at one major group in urban society and the way in which they structure space (Pahl, 1969).

Figure 2.10 is an attempt to relate the five major planning traditions and integrate the present-day socialist tradition into the same structure. It is by no means definitive but it does serve as an apt summary of the traditions of urban planning. The West European urban tradition has very deep roots within the nations. The traditions which emerge have frequently been reactions to dominant ideologies in various states disseminated by translation and built examples, albeit in imperfect form. These traditions of urban planning and urban thought have been of immense importance in moulding the form of the West European city that we see in the late twentieth century. Indeed, will the last decades of the present century give birth to such a range of idealists as did the same period one century before? 'Progress is based on dreams which mobilise the mind, cause discussions, start movements and lead to realisations' (Doxiadis, 1968).

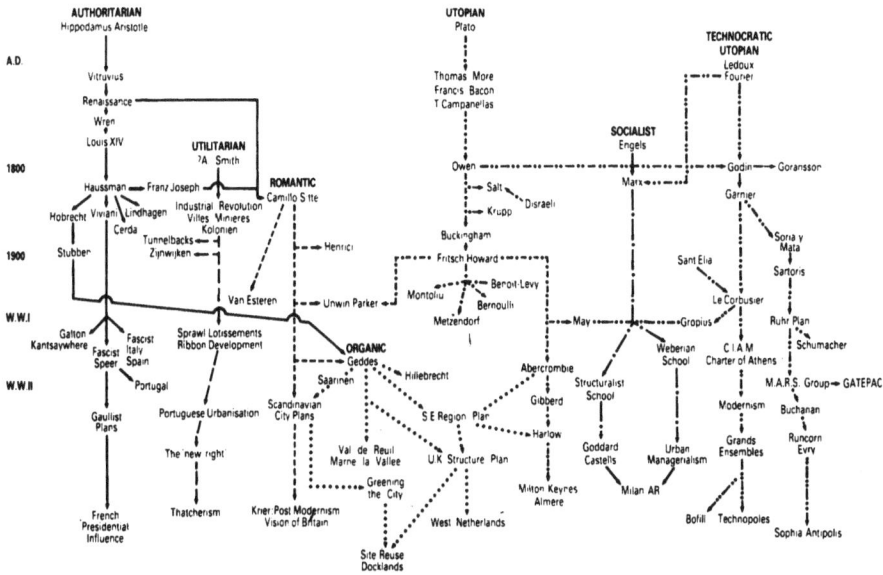

Figure 2.10 The European traditions of urban planning

Post war changes in urban development

Since 1945 at least four periods of urban development can be distinguished in European cities. Each period of development existed for a different length of time in each country although it is possible to identify key periods of change as one dominant philosophy of planning and development replaced another. The four periods have been: the period of rebuilding and recovery; the period of technocratic planning; the flight from modernism; and most recently, the rise of technopolis.

The age of recovery and rebuilding

After the destruction of wartime it was no surprise that the main preoccupation after the war was the rebuilding of the cities. Destruction had to be cleared away and a fresh start made by all. The greatest task faced the citizens of those cities which had borne the brunt of air attacks or the advance of the allied forces into the Third Reich. Berlin and Cologne, both of which were 60 per cent destroyed in their cores, are but two examples. Elsewhere the devastation in Le Havre, Rotterdam, Coventry and Plymouth, while not on the same scale, needed to be cleared and redeveloped. Redevelopment was a matter of pride and a means of re-establishing the role of the state often by rebuilding the past in the form of replica buildings as in the case of Wesel. Much rebuilding was not of a high quality because the need, especially in Germany in the cold winters of 1945 and 1946, was for shelter.

In the liberated and victorious states the mood was one of a new dawn and in the new day a fresh start was going to be made to provide a better built environment for all which would involve the planning of future developments. There was a need to plan because people were only too aware of the need to provide housing, jobs and the social infrastructure demanded by the population which had sacrificed much in the defeat of Fascism. In addition the predictions of a baby boom and the subsequent demand for housing for the new families convinced people of the need to regulate the growth of cities. It was in this atmosphere that the new profession of planners came to exist to organise the development of cities for an era of peace. Understandably the first planners had to come from disciplines most closely allied to planning – civil engineering and architecture.

The age of technocratic planning

Given the academic and professional roots of the new planners, it was not surprising that they turned to the leading ideas of their professions as they began the task of developing the cities. The architectural influence of Le Corbusier's CIAM permeated the MARS groups proposals for London, as we have seen. New housing designs were to take on the economists' interpretation of the Corbusian ideal. The high rise, high density developments began to spring up throughout Europe; Park Hill, Sheffield; Sarcelles, Paris; Chorweiler, Cologne and San Blas, Madrid are all the products of the modernist movement inspired by the Swiss architect. Government aided and abetted many of the solutions to the accommodation crisis by subsidising the 'no-fines' concrete social housing of the English council estates. Comprehensive redevelopment was the byword in those cities whose older housing stock was regarded as below standard. The old street lines were abandoned to provide the new roads fringed by the walkways in the sky.

For their part the civil engineers saw the car and road transport as the technology which would liberate the citizen. New roads, flyovers, underpasses and pedestrian segregation would all ease movement and access to the centre could be provided on urban motorways. Such was the gospel of Buchanan in the UK in the sixties, which was replicated in many German cities as the plans for the *satellitestadten* were developed. Areas in the path of the proposed routes soon suffered from planning blight and became rundown in advance of construction.

Where planning controls were lax or non-existent the problems of spontaneous settlement were to be found. The planners, especially in the Mediterranean countries, were faced by rapid urbanisation which they could not easily contain because of the lack of effective controls. Here the movement into the cities was to continue for much longer and, as a consequence, the technocratic response continued well into the 1980s.

The turning point came about 1968–1970 and was marked by the student uprisings in Paris, West Berlin and Frankfurt. Although the unrest itself was not directly linked to urban conditions in the modernist developments the riots and strikes were part of a general questioning of the nature of urban development. The politicians and planners had failed to deliver the technocratic utopia.

The flight from modernism

From approximately 1970 there was a marked shift in planning policy which may or may not have been influenced by the fact that the first trained planners, who may have originated from other disciplines such as geography and social science, had reached the zenith of their careers and could influence policy. Among the electorates there was both a demand for a greater say in the planning process and an increased awareness that the city as they knew it was being radically altered without any obvious environmental benefit.

The reaction was threefold. First there was the flight from the city first seen as the movement of people to the commuter villages around the cities, a fact noted on Boustedt's model (Figure 3.3a). Censuses later revealed this flight as counter-urbanisation (Fielding, 1989) as people and jobs decentralised away from the congested, dirty and environmentally displeasing city to the rural, healthy countryside. This movement from cities has increased and has affected smaller cities so that by the 1980 census cities and towns of only 125,000 were losing people to the smaller settlements of the peri-urban fringes. As Hall and Hay (1980) have shown the downturn in urban population has moved inexorably southeastwards across western Europe and now affects all but the southernmost areas of the continent where the forces of urbanisation continue to attract people to the cities.

The second reaction was one that focused upon the pace of change in the city. Conservation and rehabilitation replace redevelopment. The city now found itself protected as never before by a range of conservation legislation. Policies also aimed to improve the housing stock rather than demolish it. Almost everywhere gentrification became a highly visible component of a fundamental transformation within the city core, in turn mirroring wider changes in the reshaping of the advanced capitalist countries (Smith and Williams, 1986). The private sector was providing the investment stimulus that state intervention had encouraged. A start had been made to bring the city back into fashion as a place to live.

The third strand to the reaction was the changing political climate. The rise of the green parties, particularly in West Germany, and the growth of ecological awareness also ensured that the major political forces at least began to acknowledge the concerns for the environment which were being expressed very forcefully in some ballot boxes. In some instances the areas which had gentrified were those who then elected the green party member to the city council and then saw further measures enacted to protect their new enclaves. An examination of the electoral records of Cologne has shown that the gentrified areas became the heartlands of green voting very quickly after rehabilitation.

The rise of technopolis

Most recently a new direction to urban development can be detected. The rise of electronic engineering and its application to a variety of secondary and tertiary employment has opened up a whole range of potential scenarios for urban change. The globalisation of the service economy which the computer has enabled has heralded an unprecedented rush to ensure that cities are at the forefront in the new technological era. New technologies demand buildings which can cope with the cabling and ducting. The groundscraper of post-modernism replaces the skyscraper of modernism because the large trading floor has replaced the department as the unit of organisation. The new units of production involving robotics require even larger single storeyed factories; FIAT therefore established new plants beyond Turin. The science park or *technopole* houses the burgeoning research and development organisations that seek educational links. Military research spawns further allied technological industries beyond the city within the new growth regions. Sophia Antipolis, outside Nice, the Cambridge Science Park, and the technological park outside Barcelona are all manifestations of the new electronic utopia. Antenna farms and teleports, the technological equivalents of ports and harbours, are appearing in London Docklands and Hamburg. Cellphones, the French Minitel system in every household, and the new technologies of retailing are all changing the spatial relationships between work, residence and leisure (Friederichs, 1985).

At the same time the city of leisure has developed partly as a result of the release of time produced by the new technologies and partly as a result of the new social order generated by medical advances. Retailing is to become a leisure pursuit if the developers of the Gateshead Metrocentre are to be believed. Leisure was the theme of the large retail complex which was rejected by the Ruhr planners in Mulheim. Leisure in the form of the Olympic Games has been the trigger for the rehabilitation of large areas of Barcelona and is to transform the fourth sector of Marne-la-Vallée when Eurodisneyland opens. The new transport technology represented by the TGV has enabled Eurodisneyland to be at the heart of the new high-speed Europe of the nineties. At a more modest scale Sheffield prepares for the World Student Games in 1991 unsupported by the private commercialism that backs the two developments noted above.

As in previous periods of post-war development the politicians have not been slow to utilise these groundswells in urban development. The state has become proactive in attracting the private investor. The emphasis in the new right ideology is that of leverage, the use of minimal public resources to prime the private sector pump (Shurmer-Smith and Burtenshaw, 1990). Profitability decides what will happen to a building or an area rather than social good. The French sociétés d'économie mixte are used constantly to execute developments whether it be at La Villette in conservative Paris or in socialist Rennes.

What is certain from this brief review of the phases of urban development since 1945 is the fact that the citizens of Europe still crave for utopia and each generation uses the political philosophy of the time combined with the technological and social advances to approach that goal. Nevertheless European urban thinkers are more aware of the imprint of the past today than in any period since 1945 and at the same time they are more conscious of the need to preserve our built heritage for our grandchildren.

References

Abercrombie, P. (1945) *Greater London Plan*, 1944, HMSO, London.

Albers, G. (1977) Stadtbauliche Konzepte im 20 Jahrhundert – ihre Wirkung in Theorie und Praxis *Berichte zur Raumforschung und Raumplanung* Pt 1 pp. 14–26.

Bateman, M. and Burtenshaw, B. Commercial Pressures in Central Paris, in Champion, A. G. and Davies, R. (eds) *The Future for the City Centre*, Academic Press, London.

Boustedt, O. (1970) Stadtregionen, in *Handwortenbuch der Raumforschung und Raumordnung*, pp. 3207–37, Akademie fur Raumforschung und Landesplanung, Hannover.

Bradley, I. (1986) Salt of Saltaire, *New Society* 28 Feb., pp. 360–361.

Burtenshaw, D. and Moon, G. (1985) La Villette: Problems of Land Use Planning and Change in Paris, *Geography*, (70(4)) pp. 356–359.

Calabi, D. (1984) Italy, in Wynne M. (ed.) *Planning and Urban Growth in Southern Europe*, pp. 37–69, Mansell, London.

Castells, M. (1972) *La Question Urbaine*, Maspero, Paris.

Castells, M. (1977) Towards a Political Urban Sociology, in Harloe, M. (ed.), *Captive Cities*, London.

Castells, M. (1978) *City, Class and Power*, Macmillan, London.

Castells, M. (1978) Urban social movements and the struggle for democracy: the Citizens' movement in Madrid, *International Journal of Urban & Regional Research*, (2), pp. 133–146.

Castells, M. (1987) *High Technology Space and Society* (2nd Edition) Sage, Beverly Hills.

Castells, M. and Goddard, F. (1974) *Monopolville*, Mouton, Paris.

Choay, F. (1969) *The Modern City Planning in the Nineteenth Century*, Studio Vista, London.

Curl, J. S. (1969) *European Cities and Society*, Leonard Hill Books, London.

Dahne, R. D. and Dahl, H. (1975) *Die Berliner Strasse*, Senator für Bau und Wohnungswesen, Berlin.

Doxiadis C. (1986) *Between Dystopia and Utopia*, London, Faber.

Fielding, A. J. (1989) Counterurbanisation: threat or blessing, in Pinder, D. A. (ed.) *Western Europe: Conflict and Change*, Belhaven, London.

Fishman, R. (1977) *Urban Utopias in the Twentieth Century*, Basic, New York.

Friedrichs, H. (1985) *Die Stadte in den 80er Jahren*, Westdeutscher Verlag.

Fritsch, T. (1896) *Die Stadt der Zukunft*, Leipzig.

Greater Manchester Council (1978) *Greater Manchester Structure Plan, Draft Written Statement*, Greater Manchester.

Gubler, J. (1975) Hans Bernoulli et le 'Modèle Helvetique' de cité-jardin, *Werk/Oeuvre* (12), pp. 1049-1051.

Gutkind, E. A. (1969-1972) *International History of City Development*, Collier-Macmillan, London.

Hall, P. (1969) *London 2000*, Faber and Faber, London.

Hall, P. (1989) *Cities of the Future*, Blackwell, Oxford.

Hall, T. (1986) *Planung europäischer Hauptstädte*, Almqvist and Wiksell, Stockholm.

Hall, P. and Hay, D. (1980) *Growth Centres in the European Urban System*, Heinemann, London.

Hants County Council (1974) *South Hampshire Structure Plan* Hants CC, Winchester.

Herbert, M. (1980) Patrick Geddes reconsidered, *Town & Country Planning* (49), pp. 15-17.

Hillebrecht, R. (1962) *Die Stadtregion – Grossstadt und Stadtbau*, Schwartz, Göttingen.

HMSO (1964) *The South-East Study*, HMSO, London.

HMSO (1967) *Strategy for the South-East*, HMSO, London.

HMSO (1970) *Strategic Plan for the South-East*, HMSO, London.

Houghton-Evans, W. (1975) *Planning Cities: Legacy and Portent*, Lawrence and Wishart, London.

Howard, E. (1946) *Garden Cities of Tomorrow*, with an introduction by F. J. Osborne, Faber and Faber, London.

Hoyle, B. S., Pinder, D. A. and Husain, M. S. (eds) (1988) *Revitalising the Waterfront: International Dimensions of Dockland Redevelopment*, Heinemann, London.

HRH The Prince of Wales (1989) *A Vision of Britain*, Doubleday, London.

Jurgens, O. (1926) *Spanische Stadte*, Hamburg.

Kirk, W. (1953) The geographical significance of Vitruvius' 'De Architectura', *Scottish Geographical Magazine* (69), pp. 1-10.

Lavedan, P. (1959) *Histoire de l'Urbanisme, renaissance et temps modernes*, Vincent, Freal, Paris.

Le Corbusier (1978) *The City of Tomorrow*, (translated by F. Etchells), The Architectural Press, London.

Manuel, F. E. (1966) *Utopias and Utopian Thought*, Houghton Miflin, Cambridge, Mass.

Mariani, R. (1976) *Fascismo e città nuove*, Feltrini, Milan.

Meller, A. (1981) Patrick Geddes, in Cherry, G. (ed.) *Pioneers in British Planning*, Architectural Press, London.

Merlin, P. (1969) *Les villes nouvelles*, Presses Universitaires de France, Paris.

Meyerson, M. (1961) Utopian Tradition and the Planning of Cities, in Blowers, A. T. *et al.*, (1974) (eds) *The Future of Cities*, Hutchinson, London.

Miller, M. (1989) The elusive background: Raymond Unwin and the Greater London Regional Plan, *Planning Perspectives* (4), pp. 45-78.

Mumford, L. (1961, new ed. 1974) *The City in History*, Penguin, London.

Pahl, R. E. (1969) Urban Social Theory and Research, *Environment and Planning* (1), pp. 143-153.

Pahl, R. E. (1977) Managers, Technical Experts and the State: Forms of Mediation, Manipulation and Dominance in Urban and Regional Development, in Harloe, M. (ed.) *Captive Cities*, Wiley, London.

Pickvance, C. G. (1976) *Urban Sociology: Critical Essays*, Tavistock, London.

Shurmer-Smith, L. and Burtenshaw, D. (1989) Urban decay and rejuvenation, in Pinder, D. (ed.) *Western Europe: Challenge and Change*, Belhaven, London.

Sitte, C. (1965) *City Planning According to Artistic Principles*, (translated by G. R. Collins, and C. C. Collins), Phaidon, London.

Smith, N. and Williams, P. (1986) *Gentrification and the City*, Unwin Hyman, Boston.

Sutcliffe, A. (1970) *The Autumn of Central Paris: the Defeat of Town Planning 1850-1970*, Edward Arnold, London.

Sutcliffe, A. (1981) *Towards the Planned City*, Blackwell, Oxford.

Unwin, R. (1912) *Nothing Gained by Overcrowding*, Garden Cities & Town Planning Association, London.

Vance, J. (1977) *This Scene of Man*, Harper and Row, New York.

Voigt, W. (1989) The Garden City as eugenic utopia, *Planning Perspectives* (4), pp. 295-312.

Whittick, A. (1974) *Encyclopaedia of Urban Planning*, McGraw-Hill, New York.

Williams, A. M. (1984) Portugal, in Wynne, M. (ed.) *op. cit.*, pp. 71-110.

Wynne, M. (1984) *Planning and Urban Growth in Southern Europe*, Mansell, London.

Wynne, M. and Smith, R. (1978) Spain: urban decentralisation, *Built Environment*, 4(1), pp. 49-55.

CHAPTER 3

Socio-Spatial Structures of Cities

A consequence of a distinctive history, national identity and ideology is a distinct social and spatial structure of the city in West Europe. Besides, the study of urban society has many of its origins in West Europe in the work of Tonnies, Weber, Booth, Engels and Marx. Studies of the social patterning of cities have evolved as rapidly as the paradigms in the social sciences, so that the descriptions of phenomena based on simple mapping techniques and simple causal relationships have given way to methodologies which, hopefully, clarify the complexities of spatial variations between and within areas of the city both at a given point in time and through time. Underlying the recent studies of social geography has been the fundamental search for a model of the social and spatial structure of the city in the western world. The result of this quest in West Europe has been a reaffirmation that the city possesses certain distinct social and spatial relationships particularly in the light of comparisons with the cities of North America. All too often such studies have been more concerned with perfecting techniques rather than making appropriate generalisations about the social structure of the city. Nevertheless, the outcome has been a move among research workers towards a greater understanding of the subjective nature of the urban environment by its citizens rather than the emphasis on supposed objectivity of the earlier studies. A more contemporary paradigm in the evolution of social area studies has been developed by those who see all social and socio-spatial divisions of the city as the product of the processes of social formation and the interplay of wealth, status and power. Despite the deep underlying differences between the methodologies of basic descriptive analysis, positivist approaches, behavioural studies and structuralism, they each draw attention not only to the common elements of the West European and other cities but also to the obvious differences.

Spatial models of city structure

The simplest models of spatial structure evident in recent years are those based on the physical form of the city along an imaginary cross-section of the city. The technique has been used by Elkins (1975) and Ashworth (1978) to provide an easily understood picture of the German and Dutch cities respectively

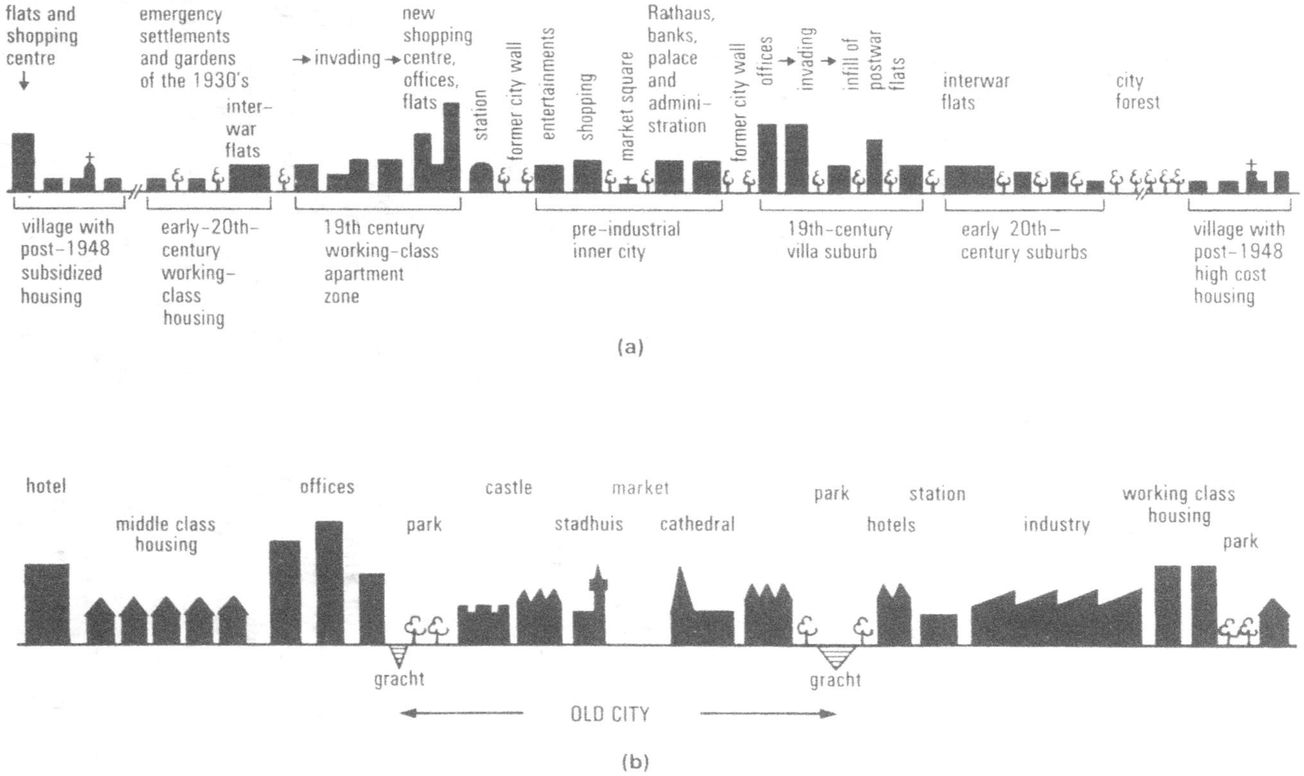

Figure 3.1 Descriptive models of West European cities: (a) Elkins' German city (reproduced from *The Geographical Magazine*, London); (b) Ashworth's Dutch town

(Figure 3.1). Both models emphasised the impact of history on the land use in the relict form of city walls or canals, as well as in some of the unique design features present in the townscape of both cultures.

Others have taken as their point of departure the work of Burgess and Hoyt, and attempted to relate their own formulations, which were the product of more rigorous analysis, to the earlier models. In Great Britain, Mann (1965) produced a model of the northern English city based on a study of three cities. The resulting model resembled a cross between its predecessors containing elements of both the concentric zonation and sectoral pattern (Figure 3.2a). Richardson, Vipond and Furbey (1975), in a study of housing in Edinburgh, were able to show that the sectoral organisation of housing areas is stronger than any concentric zonation, partly as a result of the distinct physical geography of the city but, more notably, as a result of the peculiarly Scottish legal tradition of feuing. An earlier study by Gordon (1971) of the high-status residential areas in the same city over the century from 1855 to 1962 had already noted a factor in the city's socio-spatial structure that is present in many other cities, namely the presence of a relatively permanent high-status area, the Georgian 'new town', close to the city-centre.

Robson (1975), in summarising social, housing and physical spaces in a British city, also drew attention to the mixed concentric and sectoral form (Figure 3.2b). The model is based on the more detailed quantitative studies that had been undertaken in British cities. These studies had shown how the age of housing was concentrically patterned whereas the socio-economic groups were sectorally organised. In many ways, Robson's model is more developed and tested than the other descriptive models from mainland Europe.

In German-speaking West Europe, similar models were developed at approximately the same time by Boustedt (1967 and 1975). Boustedt (1975) did attempt to examine the city within its urban region (*Stadtregion*), defining no more than a set of zones around a core city and the relationship between these zones and nearby satellite and commuter towns (Figure 3.3a). In contrast, Lichtenberger's model of the ecological structure of a West European city was based unreservedly on Vienna (Figure 3.3b). Her model, like that of Mann, recognises the mix of sectors and concentric zones, partly determined, in the case of Vienna, by the course of the Danube and partly by major communications links to the west and south linking the city with the rest of Austria. More importantly, Lichtenberger did attempt to describe the major differences between European and North American cities. First, she stressed the function of the city core as a high-status residential area although there are, in addition, peripheral high-status areas. The old city within the walls maintained its residential function in conjunction with the development of commerce and retailing. Second, the inner city areas are very mixed zones of economic activities and residences with a more concentrated old industrial zone separating the inner city from the suburbs. Finally, the suburban area still contains pockets of intensively used agricultural land apart from allotment gardens and weekend houses (Lichtenberger, 1972).

Nellner based his model (Figure 3.3c) of an urban agglomeration on empirical studies, Karlsruhe and Bonn–Bad Godesberg (Nellner, 1976). Nellner's zones are based on the residential and working population of each zone, the age of buildings, the proportion of rented property, the size of residences and the degree of change in form and functions. While it is possible to question the

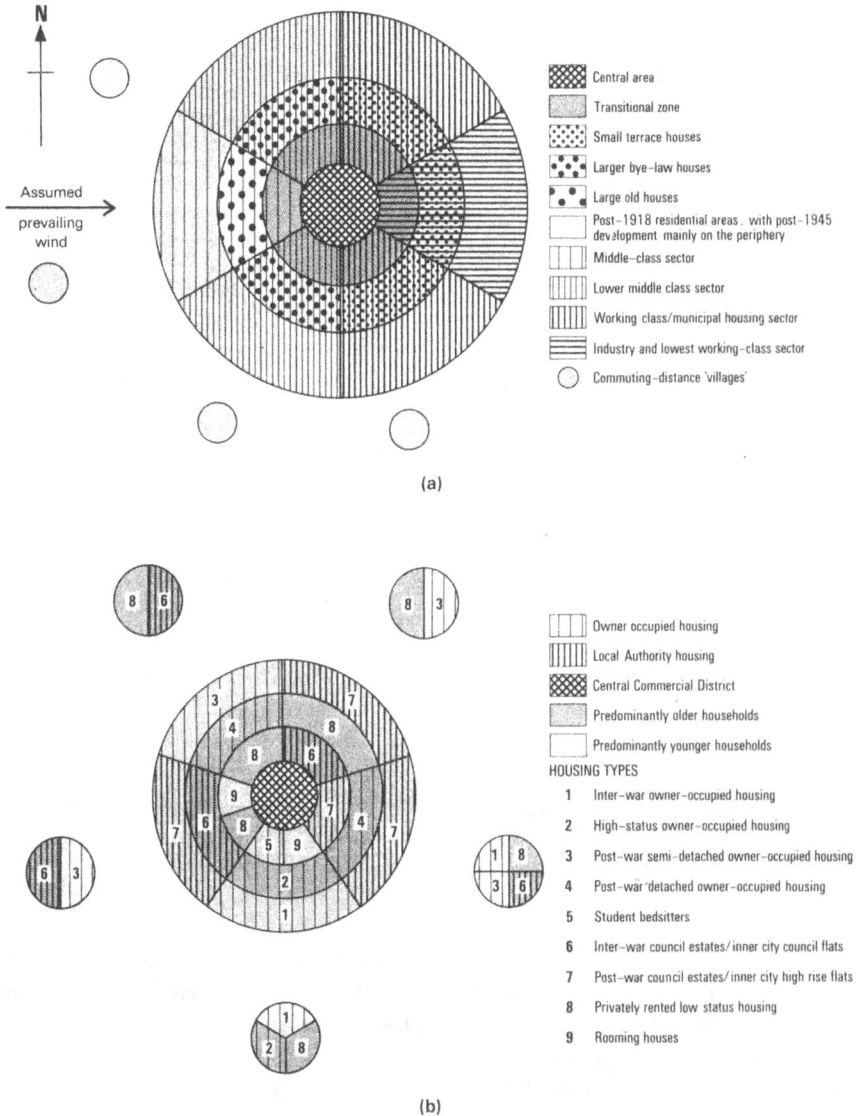

Assumed
prevailing
wind

Central area
Transitional zone
Small terrace houses
Larger bye-law houses
Large old houses
Post-1918 residential areas, with post-1945 development mainly on the periphery
Middle–class sector
Lower middle class sector
Working class/municipal housing sector
Industry and lowest working-class sector
Commuting–distance 'villages'

(a)

Owner occupied housing
Local Authority housing
Central Commercial District
Predominantly older households
Predominantly younger households
HOUSING TYPES
1 Inter-war owner-occupied housing
2 High-status owner-occupied housing
3 Post-war semi-detached owner-occupied housing
4 Post-war detached owner-occupied housing
5 Student bedsitters
6 Inter-war council estates/inner city council flats
7 Post-war council estates/inner city high rise flats
8 Privately rented low status housing
9 Rooming houses

(b)

Figure 3.2 Models of the English city: (a) Mann's northern English city; (b) Robson's British city (reproduced by permission of Oxford University Press)

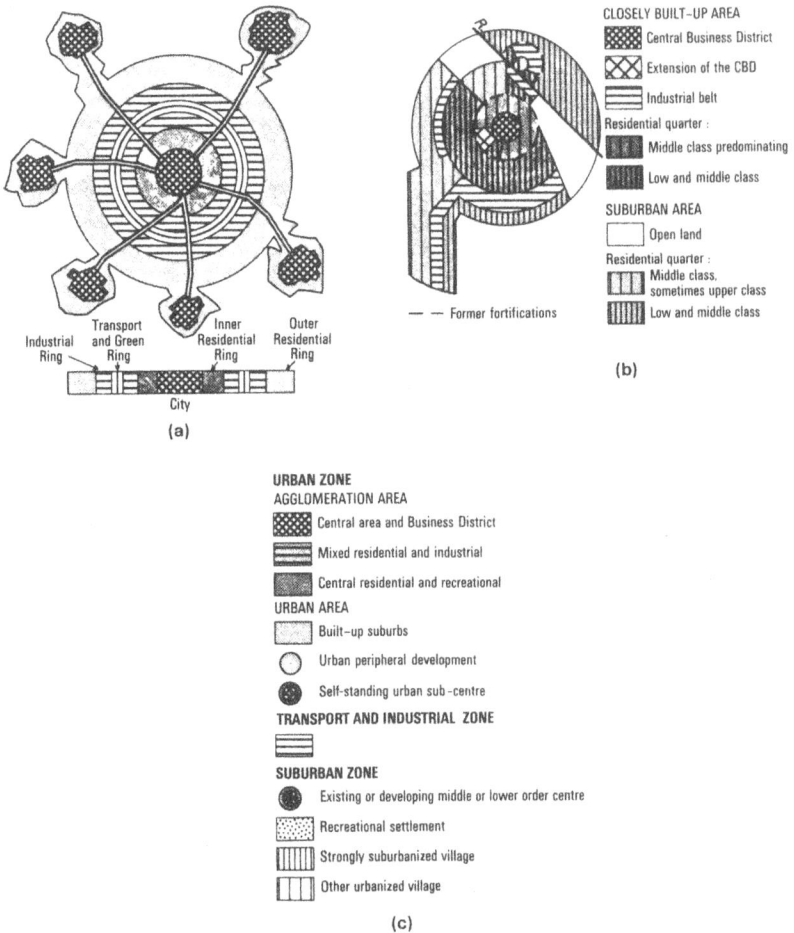

Figure 3.3 (a) Boustedt's city-region, 1970; (b) Lichtenberger's European city, 1970 (reproduced by permission of Pergamon Press); (c) Nellner's urban agglomeration, 1976

exceptional nature of the two empirical case studies, Nellner has utilised examples from other cities such as Frankfurt and Kassel to show its applicability to both large and small agglomerations.

Buursink (1977) has produced a similar model of the structure of the Dutch town which, like the other models discussed so far, discusses the structure in the context of urban growth phases and the distinctively Dutch aspects of urban structure (Figure 3.4). Particular emphasis was placed in his work on the hierarchy of service provision in the post-war extensions but little or no commentary was made on the social compositions of the extensions.

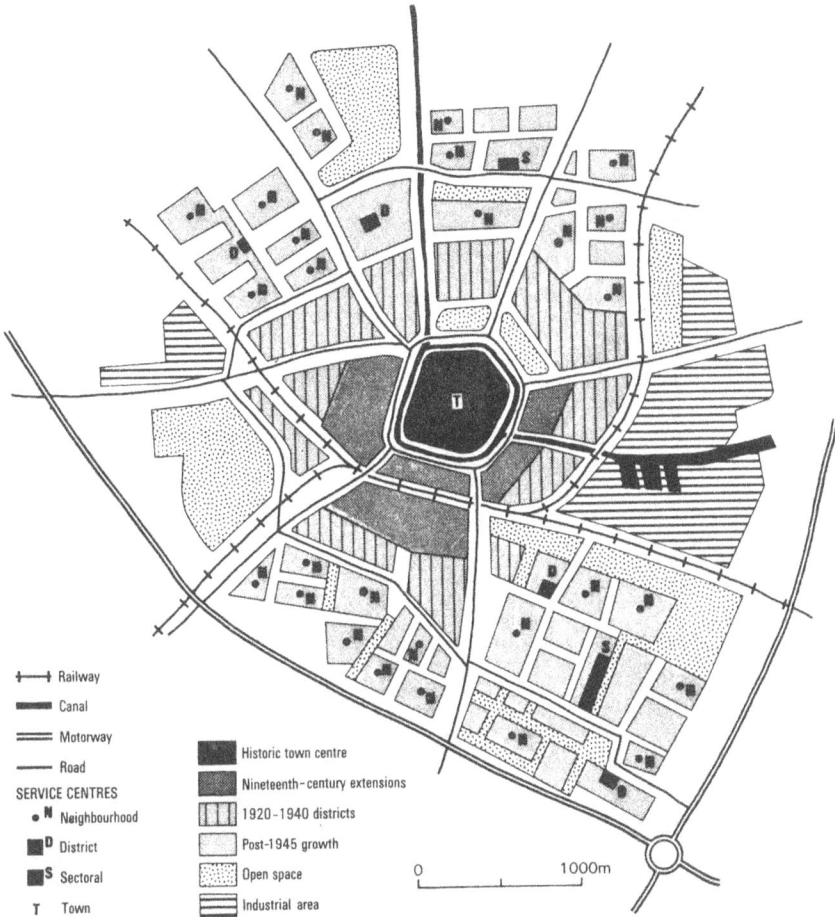

Figure 3.4 Buursink's Dutch town, 1977 (reproduced by permission of Rijksuniversiteit Groningen)

White (1984) attempted to merge together the various national models into one composite model which approximates to the underlying common elements in each city. Although White described it as a model of the West European city, its focus was essentially upon the housing or what he termed 'the residential kaleidoscope'. In order to develop his model we have added to the kaleidoscope some of the economic functions that also characterise the city in the late twentieth century (Figure 3.5). The historic core, surrounded by the relics of the city walls, is not only the central area containing retailing, commerce and the central recreation district, it is also the focus of a middle class residential district whose status is rising due to the forces of both gentrification and reurbanisation

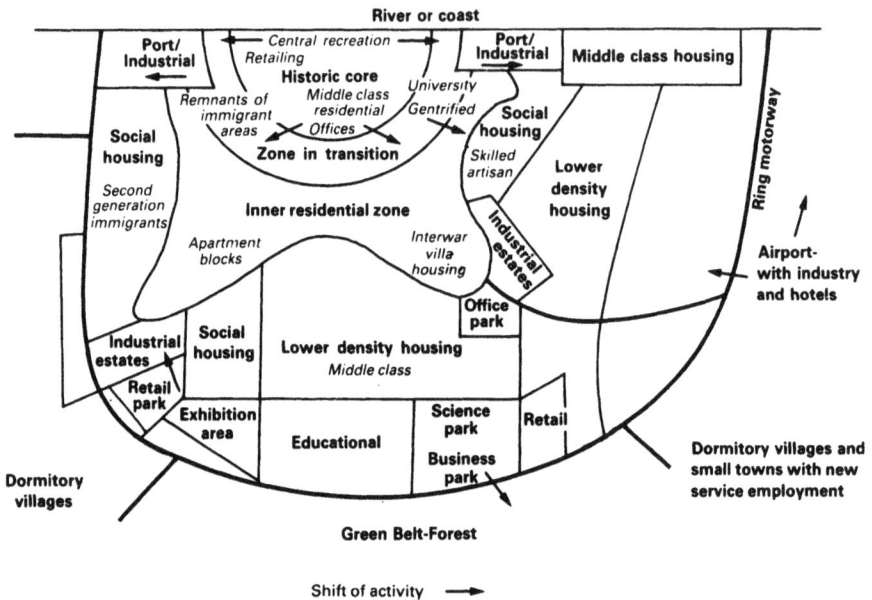

River or coast

| Port/ Industrial | Central recreation — Retailing | Port/ Industrial | Middle class housing |

Historic core
Remnants of immigrant areas | Middle class residential | University | Gentrified
Offices
Social housing | Zone in transition | Social housing | Skilled artisan | Lower density housing
Second generation immigrants

Apartment blocks | Inner residential zone | Interwar villa housing

Industrial estates
Office park

Industrial estates | Social housing | Lower density housing | Middle class

Retail park

Exhibition area | Educational | Science park | Retail
Business park

Dormitory villages

Airport- with industry and hotels

Dormitory villages and small towns with new service employment

Green Belt-Forest

Shift of activity ➔

Ring motorway

Figure 3.5   Model of the structure of the West European City 1990

(Breheny, 1987). Industrial zones, associated with the city's port function and lining part of the river or coastal strip, are frequently in decline and being replaced by waterfront marina, recreation areas and middle class reurbaniser housing (Hoyle *et al.*, 1988). Surrounding this historic core, the late nineteenth century zone in transition is increasingly characterised by gentrified housing areas which are squeezing the characteristic immigrant housing out except for the remnants of immigrant commercial activity. Such a process has over the last decade transformed the Agnesviertel in Cologne's inner north. Beyond this zone is an inner residential zone dating from the inter-war period comprising apartments and villa residences. The next zone, whose spatial extent around the periphery is greater, contains lower density, middle class housing in one sector, often close to new educational and employment nodes in the outer fringes of the city. The industrial estates of the third quarter and associated social housing are also found in the zone with some of the social housing areas now being the new location for the second generation labour migrants. The limit to the contiguous built-up area is frequently the ring motorway beyond which green belt and forest belt have been maintained except in those cities where an airport has provided a focus for industrial and office development. As Boustedt had demonstrated twenty-five years previously, the outermost zone contains dormitory villages and expanding small towns, housing the counterurbanising population of the late twentieth century together with new service employment either in research and development establishments or small business parks, or even working from the home.

Social area analysis and multivariate techniques

With the introduction of the more sophisticated methods of analysis from the English-speaking world, the analysis of socio-spatial structure of cities has relied very much on the work that has emanated from North America. What is ironic is that most of the English literature in the field is based almost entirely on non-European studies and the only West European studies are either by Americans or by Britons working on Britain or the New World. However, those studies which have followed the North American positivist methodology as closely as the data constraints permit have drawn attention to important variations from the North American models besides enabling the researcher to modify existing theory.

One of the earliest studies of social areas by McElrath (1962) attempted to apply Shevky and Bell's principles of social area analysis to Rome. A second study was that by Herbert (1967) of Newcastle-under-Lyme. The deficiencies of the technique and the shortcomings of McElrath's and Herbert's work were discussed in detail by Robson (1969) and are not the subject of investigation here. It is noteworthy that similar French studies are absent mainly due to the lack of suitable French data sets. At the time, there were no data for Rome that would have enabled McElrath to look at Shevky and Bell's third dimension, segregation, and therefore only two of Shevky and Bell's dimensions or combinations of variables, the dimensions of socio-economic rank and urbanisation, were compared with contemporary North American studies. Segregation data were not available, presumably because Italian society did not regard ethnic segregation as an important part of society and worthy of census investigation. This in itself suggests that, from the outset, social area studies based on the North American model might have problems in adjusting to European cultural norms. Other notable differences from the North American model were observed in the social areas of Rome. In the American studies, social rank was shown to rise with distance from the city-centre whereas, in Rome, social rank was highest in the central district, a fact that was explained as typical of older European cities where prestige centres on a central plaza. The urbanisation dimension was noted as a sectoral phenomenon in Rome, unlike the American studies, but no explanation for this difference was offered other than the scale of Italian and North American society.

A second pioneering work was Sweetser's (1965) study of the factorial ecology of Helsinki. Taking as his starting point earlier factorial ecological studies of Boston, Sweetser hypothesised that the three fundamental variables were (a) socio-economic status, (b) progeniture or young familism, and (c) urbanism. He also hypothesised that a fourth factor relating to home ownership would be present and that ethnicity would fail to appear as a separate dimension despite the presence of a Swedish-speaking minority group comprising 14.5 per cent of the city's population. The resulting study did confirm that socio-economic status, progeniture and female careerism were three factors that explained 58.9 per cent of the total variance. The third factor, female careerism, was a specific aspect of urban life which identified working women in high-status white-collar occupations. The analysis had already identified a developing feature of West European urbanisation that becomes the subject of further analysis among Scandinavian workers in particular. Although Sweetser recognised female careerism as his third factor, it was in effect the fifth ranked, explaining only 7.6 per cent of variance; his hypothesised fourth factor related to home ownership was the third largest, explaining 14.6 per cent of variance (Table 3.1). Despite the inclusion of five of the 42 variables specific to the Swedish minority the ethnic factor did not appear; in contrast to the American studies, the minority were found to be more numerous in high-status areas. Although not a factorial ecology, De Lannoy's (1978) study of residential differentiation in Brussels also notes the tendency for the European Economic Community nationals (excluding Italians) and Americans to concentrate in the outer south and south-eastern suburbs such as Ukkel and St Pieters-Woluwe. These are the high social status areas of the city. Sweetser concluded that 'the lesson is plain: variables which are of paramount importance in one urban socio-cultural setting may be of no consequence in another'. Subsequent studies by West European workers have reinforced this opinion although few English-speaking authorities have been prepared to note this distinction.

The third pioneering study was that of Gittus (1964) who introduced principal components analysis as a method of grouping small area data in contrast to its earlier use by Moser and Scott (1961) for their study of British towns. Her studies of Hampshire and the Merseyside and South-East Lancashire conurbations established the validity of the technique which was developed utilising more appropriate areal units by later workers.

Some of the West European studies that have used multivariate techniques are listed chronologically in Table 3.1. It is not our intention to review the methodology or to draw attention to the problems of data or size of areal units that are common to all such studies (Johnston, 1978). Nevertheless, there are several common elements in the results of the studies that merit comment.

It is very noticeable that, in all the studies that we have cited, the first two factors are generally social rank, explaining between 24 per cent and 30 per cent of total variance, and urbanisation, explaining approximately 20 to 29 per cent of total variance. In one study of Amsterdam a second social rank factor is ranked fourth (van Engelsdorp-Gastelaars and Beek, 1972). Thus far, the West European studies, with the exception of an interesting study by Friedrichs (1977) of the city of Hamburg, replicate the North American studies. It is at the level of the third and subsequent factors that the major distinction from other studies can be found. Home ownership, housing, migration and potential mobility, agglomeration of age, and building density are descriptions of the third and fourth factors in these studies. Only in Friedrichs's study does segregation appear as a separate factor; it is his fifth factor, explaining 10 per cent and 8 per cent of variance in his analysis of the city and the city region respectively. Friedrichs, like Sweetser, Hamm (1975) working in Bern, and Mischke (1976) working in Pforzheim, demoted segregation from the role of an important factor in North American cities to a minor factor in West European cities.

It is possible to suggest several reasons for this intercontinental variation in findings. The first is the cultural factor identified by McElrath and Sweetser and stressed by Robson and most other European studies. A second factor is that the ethnic minorities have not been as large or as easily identifiable in the network of areal units in West European cities because the minorities arrived in significant numbers later than the censuses. While there is some substance to this explanation for the 1960s data, research into the distribution and immigration of foreign workers would suggest that these minorities were present in large enough numbers by 1970 and might have been identifiable in factorial analyses. It is more probable that other factors such as Matti's agglomeration, Schaffer's migration and mobility and Friedrichs's mobility are, in fact, surrogates for the concentration of ethnic minorities because the areas with high scores on those factors are concentrated in the city districts with the largest numbers of foreign-born. On the other hand, some would say that the absorption of immigrant groups, whether Huguenots, Jews, Turks, Poles, Algerians or Indians, is a part of the European urban tradition and that there is evidence that ethnic minorities can and do disperse both naturally and as a result of government policies. A final explanation that can be offered is that the results are a direct reflection of the data put into the analysis and that West European censuses are on the whole less concerned with ethnic issues. The evidence suggests that factor analyses have confirmed that there is a West European cultural variant to the social ecology of cities.

Robson (1969), in his study of Sunderland, takes up the issue of cultural factors in the explanation of human organisation and behaviour. He is in agreement with Jones's (1960) earlier analysis of Belfast which also confirmed that neither concentric zones nor sectoral organisation explained the patterns of residential segregation. In the case of Sunderland, the impact of vast areas of local authority housing which had been a part of British housing provision for more than half a century was most marked. In addition, the effect of planning controls could not be ignored. These particular interventions by local authorities were also shown to have distorted patterns in Exeter (Morgan, 1971). The subsequent principal components analysis of social, demographic, tenurial, household composition and housing condition data identified a first component closely associated with indices of social class, especially high social class areas, which explained 30 per cent of the variability in the data. The sectoral pattern of component scores for the socio-economic component mirrors that of other cities (Figure 3.6a). The two major areas with high component scores are both sectors. The second component was a measure of housing conditions and amenities which both distinguishes between areas of local authority urban renewal and private areas, and between the various peripheral local authority areas by age. The third component identifies those areas of subdivided housing, declining in value, which often house elderly single persons.

Table 3.1 Multivariate analyses of West European cities

Authors	Sweetser	Robson	Herbert	Schaffer	Mattheisen	Van Engelsdorp-Gastelaars and Beek
Date of publication	1965	1969	1970	1971	1972	1972
Census/data date	1960	1961	1961	1961	1965	1960–66
City	Helsinki	Sunderland	Cardiff	Penzburg	Copenhagen	Amsterdam
Number of variables	42	30	26	66	20	31
Dimensions and percentage of total variance						
I	Socio-economic status 35.2	Social class 30	Housing conditions 27.4	Social/employment status	Social rank	Social rank I 30.4
II	Progeniture 16.1	Housing conditions 29	Housing conditions 23.5	Age of population	Non-familism	Housing conditions 20.5
III	Homeownership 14.6	Housing 8	Ethnicity/high status 10.7	Migration and mobility potential	Housing	Family stage 12.9
IV	Established familism 13.8				Activity rates	Social rank II 5.3
V	Female careerism 7.6				Youth	

Table 3.1 Multivariate analyses of West European cities (continued)

Authors	Friedrichs	Schreifer	Vilsteren and Everaers	Kreth	Lando	Kesteloot	Glebe and O'Loughlin
Date of publication	1977	1977	1977	1977	1978	1980	1980
Census/data date	1970	1970	1970	1970–75	1971	1970	1970
City	City of Hamburg	Bremen	Arnhem	Mainz	Venice	Brussels	Düsseldorf
Number of variables	16	49	26	50	38	48	40
Dimensions and percentage of total variance I	Social rank 24	Social rank 18	Socio-economic status 31.4	Employment 36.7	Social rank	Social rank	Social rank
II	Urbanisation 20	Family status 13	Family stage 26.4	Age and household structure 36.7	Self-employment	Inner city	Religion
III	Building density 18	Urbanisation 11	Housing 18.4		Demographic vitality	Housing	No residential space
IV	Mobility 17				Housing tenure	Self-employment	Religion
V	Segregation 10				Recency	Age and feminisation	

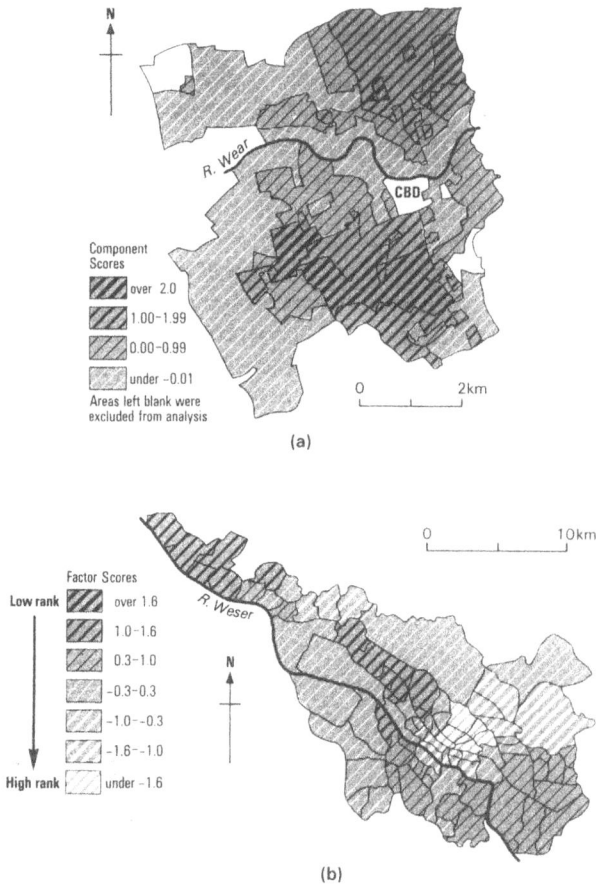

Figure 3.6 Factorial ecologies of cities: (a) Sunderland: component scores (after Robson) (reproduced by permission of Cambridge University Press); (b) Bremen (1970) factor scores: Factor I socioeconomic status (after Schriefer) (reproduced by permission of Bernd Schriefer, Bremen)

Friedrichs (1977) undertook an analysis of the city of Hamburg using sixteen variables. The results drew attention to the ranking of factors in the city and the region. The first five factors explain 69 per cent of the total variance in the region and 83 per cent in the city, thus emphasising the contrast that exists between the city and region in West Germany. In the city (Table 3.1) social rank was the most important factor and the urbanisation factor the second, whereas this position was reversed in the region. A similar reversal occurs with the third and fourth-ranked factors, building density and mobility. The second factor was highly loaded in favour of social democratic and communist voting, multi-family buildings and the retired, and was taken to be an urbanisation dimension. The third factor, 'the agglomeration component', identified the peripheral regions in a similar fashion to Robson's components and could well relate to areas of social housing on the fringes of the city. It would seem from these studies that, as a result of the interplay of culture and social history, the outer suburban areas of Sunderland and Hamburg are more similar than are the cities of West Europe and North America.

A further interesting facet to social analysis studies in West Germany has been provided by Schriefer's (1977) study of Bremen. His findings with regard to the 1970 data are tabulated in Table 3.1 and the familiar sectoral pattern of social rank scores has been plotted in Figure 3.6b. As in other studies, it is only the fourth factor, locational character, that contains high positive loadings on foreign-born populations. Schriefer was able to compare his findings in 1970 with those for 1961 although adjustments had to be made for changed areal units and a changed number of variables. He was able to show that in both 1961 and 1970 the ranking of the first four factors had not altered, thus suggesting a degree of permanence in the socio-spatial structure of Bremen. The areas of high factor scores for socio-economic status (Factor 1) show a distinct sectoral pattern characteristic of this factor's scores. All the studies based on 1970 data contain factors which consistently explain less of the total variance than the earlier studies using data from the previous decade.

Subsequent studies of the German speaking cities such as O'Loughlin and Glebe's (1980) study of Dusseldorf have further emphasised the importance of the social status variable. They cite fifteen factor analyses and principal components analyses and in eleven of these the social status variable is the first factor, while in the study of Wurzburg it was replaced by employment type. In the three other studies of Bonn, Munster and Mannheim family status and life cycle is the dominant factor; it is the second factor in six other studies although in the Dusseldorf study the second factor is religion. The religious dimension differentiated areas in Bonn and Amsterdam.

Both Schriefer and Vilsteren and Everaers (1978), who studied Arnhem, used Rees's distance grouping procedure in order to categorise the social areas of the city. In Bremen, the distance grouping identified two types of high-status region, the centre and the sector stretching east. The newest residential areas are identified as distinct and two predominantly rural areas are differentiated mainly by the age of buildings (Figure 3.7a). The identical procedure in Arnhem was adopted and mapped at three levels of grouping, the middle of which produced a similar number of categories although the data used differed. The upper-class regions, as in Arnhem and Bremen, are sectors on either side of valued open space (Figure 3.7a and b) whereas the working-class districts are a mixture of districts closest to the Rhine in Arnhem and the Weser in Bremen, and peripheral areas interspersed with middle-class residential areas. This patterning is recurrent throughout West Europe as Table 3.1 confirms. However, White's comment that the plethora of these studies 'add relatively little extra detail to the patterns of social class representation' is valid and may account for the decline of the method of analysis.

Studies of the degree of segregation between social groups have also taken

(a)

(b)

Figure 3.7   Social areas: Distance grouping maps: (a) Bremen (1970) social areas (after Schriefer) (reproduced by permission of Bernd Schriefer, Bremen); (b) Arnhem (1979) social areas (after Vilsteren) (reproduced by permission of Katholieke Universiteit Nijmegen)

their lead from North American literature. Gisser (1969), working on Vienna, and Belleville (1962 quoted by Gisser), in his study of Paris *intra-muros*, have both drawn attention to the increased segregation at the uppermost and lowest ends of the socio-economic spectrum. This theme was examined for a range of British towns and cities by Morgan (1975). His analysis showed that the degree of residential differentiation replicated the Viennese and Parisian findings more closely than it did for Chicago and Cleveland on which the study was based. The highest and lowest socio-economic groups were more segregated, and the indices of segregation in all three West European studies were higher for the uppermost socio-economic groups than in the two North American cases. The indices were approximately the same for other groups, with the exception of the lowest where the West European cities had lower segregation indices than the North American cities. Therefore, the limited evidence would suggest that residential segregation in West European urban areas is more pronounced for the upper social groups, a fact that is reinforced in the maps showing factor scores for the social rank factor (Figure 3.6). Morgan (1975) was also able to show that segregation did not increase significantly with city size although the proportion of males in the professional and managerial occupations, which was influenced by the functions of the urban area, did find expression in the degree of segregation. The exceptional case of segregation is that noted by Boal (1970 and 1972) in the rather uneasy political climate of Belfast where the religious-based communities have become increasingly segregated as a response to conflict and the need to find psychological support.

'Radical' studies

The dissatisfaction with both behavioural and positivist approaches of multivariate analysis stemmed partly from the fact that neither approach offered any all-encompassing explanation for the socio-spatial framework of cities. This explanation, as indicated in the two previous chapters, has been attempted in particular by the radical analyses of social processes within cities. The city cannot be analysed in the view of the structuralists apart from the web of structural interests which determine it. It might not be solely the structure of a given society at one point in time but rather the impact of a series of ideologies which have each created their own city. Thus, in its built form, the city becomes a palimpsest of past ideologies as Lichtenberger (1970) indicates in her studies.

Perhaps the most important modern exponent of the structuralist approach has been Castells, and in his analysis with Godard of Dunkirk (Castells and Godard, 1974) he has attempted to show how the particular approach offers an explanation for the complex processes of spatial segregation. Structuralism has a particularly strong tradition in France which dates back to Levi-Strauss. Castells' main theoretical stance was outlined in his earlier work *La Question Urbaine* (1972): that a society and the social forms of that society such as space are the product of the actions of individuals whose behaviour is determined by the individual's location within the structure of society and by the individual's economic power. The social-class conflicts that are the consequence of the structure of capitalist society are expressed through and in urban planning and urban social movements such as the protest movements opposing urban renewal in Paris (Castells, 1978) and those opposing motorways in Barcelona (Borja, 1978).

For his major empirical analysis (Castells and Godard, 1974) the city of Dunkirk was selected because the city was undergoing rapid expansion following the establishment of the USINOR steel plant, an enlarged port, oil refineries and shipyards. The new economy of the town produced advantages for industrial development, for the development of the centre and its office and retail function allied to the growth of a residential area for the managers, and the development of a new working-class housing area related to redevelopment of older housing areas and rural housing. Besides these groups, there is a small local middle class who are trying to manage the local community against the growing power of the former three social groups. These are represented politically by the Gaullists and apolitical groups in control of the city, the social democratic group controlling the *département* and the socialist communist union with its power base in the working-class areas. Not only are these groups socially and politically separate but they are also spatially segregated (Figure 3.8). Migration patterns are reinforcing segregation, as we have seen in other cities; 80 per cent of residential movement is to areas of similar social status and between 32 and 37 per cent of movement is within the same *commune*. High status has become concentrated by the sea while the rest of the housing is spread through the city with relatively weak segregation.

Castells goes on to illustrate how the varied social interests struggle for power and how the ideologies of each group find expression in the townscape. The official planning documents become tools of the struggle used by the dominant groups to aid continuing control. Infrastructural improvements are seen as ways by which it is hoped that the town will approve of the national proposals for new urban areas. The establishment of the Maison de la Culture is seen as an example of the bribery of the middle classes. The development plans are instruments of social control, and, as such, the *Livre Blanc* is seen as the product of the commercial middle classes of the town and the big industrial managers of the port because it argues for a town centred on the old core and the new port-industrial region. The social conflicts in Dunkirk can be crystallised into four groups:

(1) opposition between the old and the new;
(2) opposition between the middle classes and workers;
(3) opposition between Dunkirk/Flemish/port and the suburban/ foreign/industrial zones; and
(4) opposition between the town and the country – a dichotomy that has had no substance in England and Wales (Pahl, 1970).

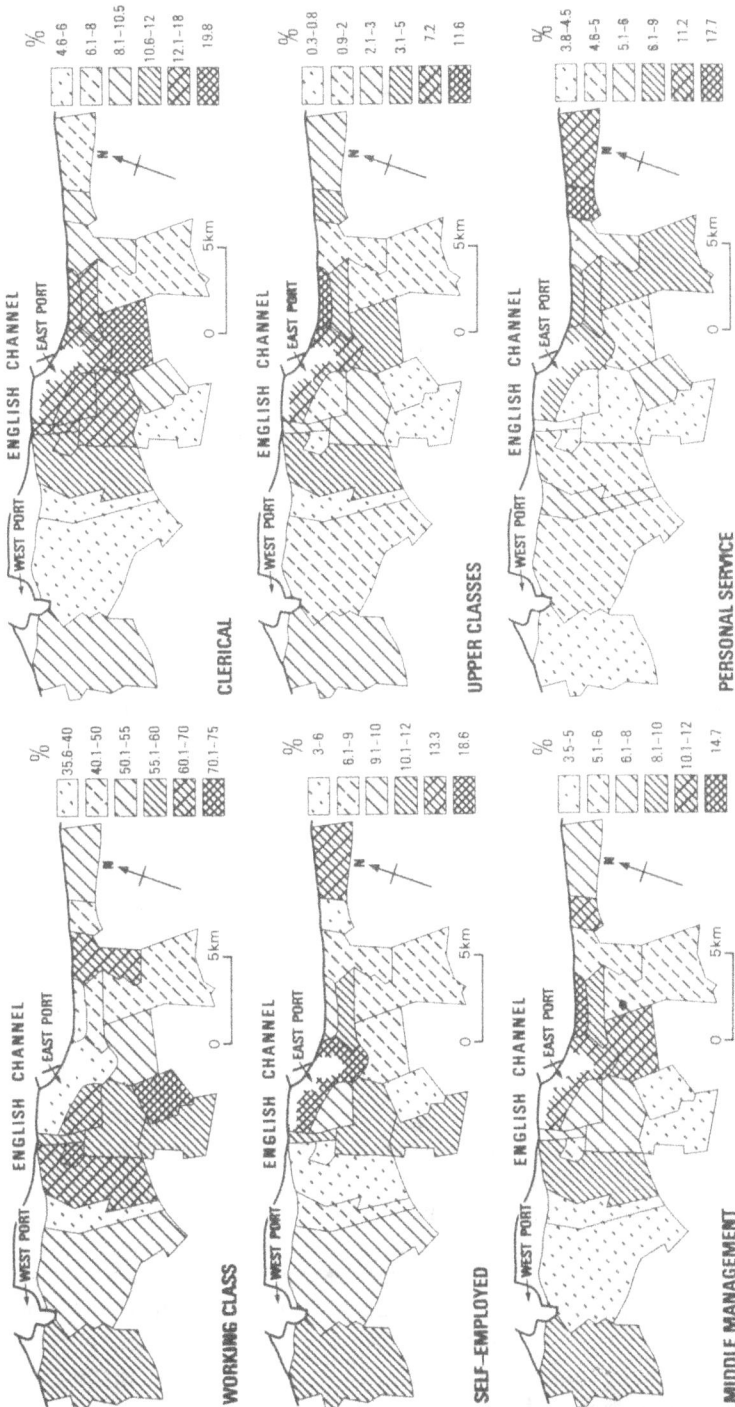

Figure 3.8 Social areas of Dunkirk (after Castells) (reproduced by permission of Mouton, Paris)

Table 3.2   Urban conflicts and their spatial expressions

		New developments		Old districts	
		Middle class	Workers	Middle class	Workers
Country	Local Fleming/Dunkirk	? Lacking substantiation (existing in places)	HLM, Reconstructed Dunkirk (canal side)	Centre of town, Malo les Bains	Docker *quartiers*, port, St Pol
	Newcomers	Residential area Malo les Bains	New areas and industrial areas	? Lacking substantiation (existing in places)	Shanty towns, hovels and chalets
Town	Dunkirk			Old rural towns and large landed properties to the east of town	Villages
	Newcomers	Suburban semi-rural area	Single-person homes, caravans		Rehabilitation of rural towns for immigrants

Source: (Castells and Godard, 1974)

In diagrammatic form (Table 3.2) these conflicts explain almost completely the form of the built-up areas of the city especially when they occur in combination. The combinations of conflicts are the basis of four ideologies. First, the ideology of consumerism which is characteristic of the dominant middle classes. This is the ideology of dominance of new over old, middle class over workers, the immigrant over the local and the town over the country. Second is the family ideology of the *petit-bourgeoisie* where the old dominates the new, the middle class the worker, the local the immigrant, and the country the town. In the third ideological type, the basis is that of the traditional community in which the old dominates the new, the worker dominates the middle class, the local dominates the immigrant, and the town dominates the country. Finally, there is the ideology of class conflict where the new and the worker are opposed to the old and the middle class, while the newcomer and the town are opposed by the local and the rural interests. Each ideology dominates in particular districts of Dunkirk; consumerism is especially strong in the residential zone of Malo les Bains; familism dominates the rural towns and the older suburban sprawl of the inter-war and post-war period; the traditional community ideology is strongest in the *quartiers* around the port; the class conflict produces opposition between the new residential areas and the centre of Dunkirk.

In order to obtain clarity from the variety of ideologies and conflicts that are present in the city, the planning process is utilised not as the product of the dominant ideology, consumerism, but as the outcome of a political balancing act in the interests of the dominant ideology. The *schema d'aménagement* is seen as a *scenario de compromis*, the product of a conciliatory effort by the state that attempts to retain the decisions of the dominant group so long as the various social groups or movements can coexist. In this way the planning process becomes a part of the urban political process, part of the way in which national ideology is translated into local urban development and by which the locally dominant ideology responds to the pressures from other ideologies. The conciliatory efforts and responses, however, are regarded as only involving those options or areas of the plan where they do not involve any fundamental contradiction of the dominant ideology. The neutrality of the plan, whether in a social or technical sense, is seen as a part of the method by which the dominant ideology is able to control the planning process.

At the level of the individual city the structuralist theories offer a new, more universal theory of socio-spatial patterns, although it is notable that the work relies on simple percentages for relatively large areas in contrast to the small area data of the multivariate techniques. The work can also be criticised on the grounds that it is rather retrospective in its view and cannot be utilised in a more predictive manner than the mere extrapolation of continuing political studies. *Monopolville* remains the sole study which examines the whole urban system. Other studies, including several by Castells himself (1973, 1978 and 1984), using the same theoretical position, focused on particular policies where the conflicts were reduced more easily to one or two dichotomous situations and a few areas of the city. Hence there has been a concentration on the urban protest movements by Borja, Pickvance and Lojkine (Harloe, 1977).

Castells (1973) has also taken this more restricted empirical material in an analysis of urban renewal and urban conflict in Paris. Urban renewal is seen as accentuating residential segregation and, in particular, the city of Paris is becoming a preserve of the higher social strata. This has been linked to Paris's growth as a tertiary centre (Chapter 4). Centrality has become the mainspring of change, so much so that the new RER lines were built to develop the central location. At the same time, new commercial centres were being developed in the areas of renewal to aid the ideology of consumerism in its dominance over the city. Renewal is a means by which the working-class political hold on France's capital city is broken by the increasingly middle-class inhabitants and voters. By handing over areas to private development the authorities are, in Castells' opinion, creating the conditions in which the capitalist system can continue. Thus, Algerian areas and those of the lowest social strata are subject to renewal whereas the areas of most insalubrious dwellings are ignored. By references to several renamed areas Castells illustrates the effect on the city's social geography. In one *quartier* the removal of insalubrious dwellings was soon replaced by an 'urban reconquest' conservation and renewal programme in an area formerly housing ethnic minorities, immigrants and the working class. Where social housing was promised the number of homes built was not enough to rehouse the pre-existing households. Protest movements were galvanised into action by the political left. Political and physical battles were lost, though, more importantly for the researcher, the theoretical stance of the Marxist was strengthened by the evidence collected showing the dominance of consumerism, centrality and the support of the organs of the state (Castells, 1978). Private affluence and public poverty are part of the same basic problem, the class struggle.

In his later studies of Sarcelles and the citizen movement in Madrid, Castells (1984) acknowledged that he developed 'the heritage of Marxist theory' by introducing class conflicts and urban social movements into a theory which regarded these forces for change as unthinkable because they could not succeed. He showed how the citizen movement in Sarcelles actually pressurised the Grands Ensembles' managers to change their development policy and thus make it unprofitable for the state to develop. The final nail in the coffin was the reversal of the policy in 1974 which legitimised the efforts based on the social movements on neighbourhood associations (*associaciones de vecinos*). In this way a group could work within the existing structures to alleviate some of the excesses of state-provided housing rather than overthrow the organs of the state. His subsequent study of Madrid's citizen movement illustrates the role of the movement in removing shanty settlements, rehabilitation at San Blas, the 1980 conservation plan for the centre, social housing developments in the centre and improved services to peripheral areas and producing urban reform. Similar movements also developed around housing shortages in Rome, Milan, Turin, Naples and Florence. The role of many of the social movements declined as the left's role in urban government, especially in Spain and Portugal, rose in the late seventies and early eighties (Gaspar, 1984).

A further perspective on the socio-spatial organisation of cities comes from the neo-Weberian approach. The emphasis and central concern shifts away from the holistic view of the economic and social system to the analysis of power within the social system. The emphasis is placed on both spatial and social constraints on access to urban resources that highlight, in particular, the role of bureaucracies, the urban managers and others who give access to life chances, such as the mortgage lenders (Ford, 1975; Williams, 1976 and 1978). On the whole, the main examples of this work, like that of the neo-Marxists, have not been developed in the context of a whole West European city but rather elements of the city. Where urban conflicts occur the neo-Weberian sees them as being independent from other social conflicts and able to be identified by the researcher as the degree of access to housing or housing class. Therefore, works such as Rex and Moore's study of Sparkbrook, Birmingham, have concentrated on urban managerialism and the conflict between housing classes rather than on those who control the city bureaucrat, or the origins of conflicts in the search for housing (Rex and Moore, 1967). Damer (1974) also looked at the way local authority managers helped to create Wine Alley, while others have looked at community action groups, residents' associations and their basis within housing classes (Dennis, 1972).

Both of the radical perspectives do help to exemplify the basic premise of this book, that the West European city is distinct. Although the theories both of the Marxists and the neo-Weberians begin from the premise that the socio-spatial structure of cities is the product of a particular mode of production and the variety of attitudes and patterns of behaviour among those who operate the capitalist system, both explanations rely, at least as far as the empirical evidence suggests, on the need to introduce explanations of the local structure of society. Castells, therefore, relies very heavily on the role of the multiplicity of planning and development agencies that are peculiarly French in their operation, powers and interactions with other agencies at work in the built environment. To understand the empirical evidence demands an understanding of France. To understand Labyrinth Stadt, as the authors indicate in their material, it becomes essential to understand the West German political and economic system (Andritsky *et al.*, 1975). To understand the evidence of urban managerialism or the property machine it is essential to have a knowledge of the political and economic system of Great Britain. There are many who might agree with Pfeiffer's (1973) view:

> It is possible that Marxist theory satisfies the psychological need for a general simple theory of our society but it does not help us solve urban change. It delivers no criteria about how to handle a complex system of homes and hospitals, streets and shopping centres, sewers and subways for different groups with different interests on how to balance the old and new parts of a city.

Castells (1978) did point out that 'Marxist theory... has no tradition of the treatment of an urban problem'. As yet it would seem that the holistic perspective provides a greater understanding, but that it depends on the descriptions of systems which are distinct as a result of national histories, and the nation's own variations of its dominant ideology. In addition, as the Dunkirk study has shown, cities can still have distinct ideologies such as the local 'Flemish' one which are almost unique; the holistic perspective still has to take into account the uniqueness of a city's or a region's own history and culture. The time dimension and evolution are undervalued. It is this continual interplay of the main currents of history, national and/or local identity and ideology that the structuralists cannot as yet obscure.

No matter whether one adopts the stance of description, positivism, reductionism or structuralism, distinct elements in the socio-spatial structure of West European cities are in evidence. Perhaps from the overwhelmingly Anglo-American view of the city, too much emphasis has been placed on the value of studies emanating from North America. It is only in recent years, with the translation of the works of Castells, that more English-speaking academics have come to recognise that West European social scientists have been aware of the socio-spatial processes at work in the city. The observations of these workers have suggested that socially the city in West Europe is a distinct entity, the product of the variety of cultural histories.

References

Andritsky, M., *et al.* (1975) *Labyrinth Stadt*, Du Mont Aktuel, Cologne.
Ashworth, G. J. (1978) *Gelderland Field Guide, 2nd Edition*, Geography Department, Portsmouth Polytechnic.
Boal, F. (1970) Social space in the Belfast urban area, in *Irish Geographical Studies in Honour of E. E. Evans*, (eds N. Stephens and R. Glasscock), pp. 373–93, University of Belfast.
Boal, F. (1972) The urban residential sub-community – a conflict interpretation, *Area*, **4**, pp. 164–168.
Boal, F. *et al.* (174) *The Spatial Distribution of some Social Problems in the Belfast Urban Area*, Northern Ireland Community Relations Commission, Belfast.
Borja, J. (1978) Urban movements in Spain, in *Captive Cities*, (ed. M. Harloe), pp. 187–211, Wiley, London.
Boustedt, O. (1967) Zum Beteulung des Dichtebriffes in der Raumordnung und Landesplanung, *Mitteilungen der Deutschen Akademie für Städtebau und Landesplanung, Heft*, 11.
Boustedt, O. (1975) *Grundriss der Empirischen Regionalforschung Teil III Siedlungsstruckturen*, Schoedel, Hannover.
Breheny, M. J. (1987) Return to the city, *Built Environment*, **13**, pp. 189–192.
Buursink, J. (1977) *De Hierarchie van Winkelcentra*, GIRUG, Groningen.
Castells, M. (1972) *La Question Urbaine*, Maspero, Paris; translated (1977) as *The Urban Question*, Edward Arnold, London.
Castells, M. (1973) *Luttes Urbaines*, Maspero, Paris.
Castells, M. (1978) *City, Class and Power*, Macmillan, Basingstoke.
Castells, M. (1984) *The City and the Grassroots*, Edward Arnold, London.
Castells, M. and Godard, F. (1974) *Monopolville: L'Entreprise, L'Etat et L'Urbain*, Mouton, Paris.
Damer, S. (1974) Wine Alley: the sociology of a dreadful enclosure, *Sociological Review* **22**, pp. 221–248.
De Jong, D. (1962) Images of urban areas: their structure and psychological foundations, *Journal Institute of American Planners*, **28**, pp. 276–286.
De Lannoy, W. (1978) Enkele aspecten van de residentiele differentiatie in de Brusselse Agglomeratie, *De Aadrijkskunde*, **3**, pp. 251–262.
Dennis, N. (1972) *Public Participation and Planners Blight*, Faber and Faber, London.
Elkins, T. H. (1975) Style in German cities. Urban development since World War 2, *Geographical Magazine*, **48**(1), pp. 17–23.
Eyles, J. (1969) *The Inhabitant's Images of Highgate Village*, Discussion Paper 1095, Department of Geography, London School of Economics, London.
Ford, J. (1975) The role of the building society manager in urban stratification, *Urban Studies*, **12**, pp. 295–302.
Friedrichs, J. (1977) *Stadtanalyse*, Rowohlt, Hamburg.
Gaspar, J. (1984) Urbanisation, Growth, Problems and Policies, in *Southern Europe Transformed*, (ed. A. Williams), Harper and Row.
Gisser, R. (1969) Okologische Segregation der Berufschichten in Grossstadten, in *Soziologische Forschung in Osterreich* (eds. L. Rosenmayer and S. Hollinger) pp. 199–220, Bollaus, Vienna.

Gittus, E. (1964) The structure of urban areas: a new approach, *Town Planning Review*, **35**, pp. 5-20.

Gordon, G. (1971) *Status areas in Edinburgh* (unpublished PhD), University of Edinburgh.

Hamm, B. (1975) *Untersuchungen zur Sozialen Wirtschaftlichen und Baulichen Struktur der Stadt Bern*, M. S. Bollingen.

Hammond, E. (1968) *London to Durham*, Rowntree Research Unit, University of Durham.

Harloe, M. (1977) *Captive Cities*, Wiley, London.

Herbert, D. T. (1967) Social area analysis: a British study, *Urban Studies*, **4**, pp. 41-60.

Herbert, D. T. (1970) Principal components analysis and urban social structure, in *Urban Essays: Studies in the geography of Wales* (eds. H. Carter and W. Davies), pp. 79-100, Longman, London.

Herbert, D. T. (1976) Social deviance in the city, in *Social Areas in Cities*, Vol. II (eds D. T. Herbert and R. J. Johnston), pp. 89-121, Wiley, London.

Hoyle, B. S., Pinder, D. A. and Husain, M. S. (eds) (1988) *Revitalising the Waterfront: International dimensions of dockland redevelopment*, Heinemann, London.

Johnston, R. J. (1978) Residential area characteristics: research methods for identifying urban sub-areas, social area analysis and factorial ecology, in *Social Areas in Cities* (eds. D. Herbert and R. J. Johnston), pp. 175-217, Wiley, London.

Jones, E. (1960) *A Social Geography of Belfast*, Oxford University Press, London.

Kestelot, C. (1980) De Ruimtelijke sociale struktuur van Brussel Hoofdstad, *Actu Geographisca Lovaniensia*, **19**.

Kreth, R. (1977) Socialraumliche gliederung von Mainz, *Geographische Rundschau*, **29**, pp. 142-149.

Lando, F. (1978) La struttura socio-economica veneziana: un tentativo d'analisi, *Rivista Veneta*, **12**, pp. 125-140.

Lichtenberger, E. (1970) The nature of European urbanism, *Geoforum*, **4**, pp. 45-62.

Lichtenberger, E. (1972) Die Europaische Stadt - Wesen modelle probleme, *Raumforsch. Raumord.*, **16**, pp. 3-25.

Lob, R. and Wehling, H. (1977) *Geographie und Umwelt*, Hain, Mesenheim.

Mann, B. (1965) *An Approach to Urban Sociology*, Routledge and Kegan Paul, London.

Matthiessen, C. W. (1972) Befolknings - og boligstrukturen i københavns kommune belyst ved en principal-component analysis, *Geografisk Tidsskrift*, **71**, pp. 1-19.

McElrath, D. (1962) The social areas of Rome: a comparative analysis, *American Sociological Review*, **27**, pp. 376-391.

Mischke, M. (1976) *Faktorenokologische Untersuchung zur Raumlichen Auspragung der Sozialstruktur in Pforzheim*, Karlsruhe Universität, Karlsruhe.

Morgan, B. (1971) The residential structure of Exeter, in *Exeter Essays in Geography* (eds W. Ravenhill and K. Gregory), pp. 219-236, Exeter University.

Morgan, B. (1975) The segregation of socio-economic groups in urban areas: a comparative approach, *Urban Studies*, **12**, pp. 47-60.

Moser, C. A. and Scott, W. (1961) *British Towns: A Statistical Study of Their Social and Economic Differences*, Oliver and Boyd, London.

Nellner, W. (1976) Die Innere Gliederung Stadtischer Siedlungssagglomeration, *Forschungs und Sitzungsberichte Akademie fur Raumforschung and Landesplanung*, **112**, pp. 35-74.

Niemeyer, G. (1969) Braunschweig - Soziale Schichtung und Sozialraumliche Gliederung Einer Grossstadt, *Raumforsch. Raumord.*, **27**, pp. 193-209.

O'Loughlin, J. and Glebe, J. (1980) Faktorokologie der Stadt Dusseldorf, *Dusseldorfer Geographische Schriften*, H. 16.

Pahl, R. E. (1970) *Whose City?*, Longman, London.

Pfeiffer, U. (1973) Market forces and urban change in Germany, in *The Management of Urban Change in Britain and Germany*, (ed. R. Rose), pp. 45–52, Sage, London.

Pocock, D. (1976) Some characteristics of mental maps: an empirical study, *Trans. Institute British Geographers, New Series*, 1(4), pp. 43–512.

Rees, P. H. (1968) *The Factorial Ecology of Metropolitan Chicago*, (unpublished masters thesis), University of Chicago.

Rex, J. and Moore, R. (1967) *Race, Community and Conflict*, Oxford University Press, London.

Richardson, H., Vipond, J. and Furbey, R. (1975) *Housing and Urban Spatial Structure: A Case Study*, Saxon House, Farnborough.

Robson, B. (1969) *Urban Analysis*, Cambridge University Press, London.

Robson, B. (1975) *Urban Social Areas*, Oxford University Press, London.

Schaffer, F. (1971) *Prozesstypen als socialgeographisches Gliederungsprinzip, Mitteilungen der Geographischen Gesellschaft München*, No. 56.

Schriefer, B. (1977) *Die Sozio-okonomische Struktur der Stadt Bremen: Eine Faktorokologische Untersuchungen* (mimeo), Bremenä Universitat, Bremen.

Sweetser, F. (1965) Factorial ecology: Helsinki 1960, *Demography*, 1, pp. 372–385.

van Engelsdorp-Gastelaars, R. and Beek, W. (1972) Ecologische Differentiatie Binnen Amsterdam, *Tijd. Econ. Soc. Geog.*, 63, pp. 62–78.

Vilsteren, G. J. and Everaers, P. C. J. (1978) Factor-Ecologie van Arnhem, *Nijmeegse Geografische Cahiers*, No. 12.

White, P. (1984) *The West European City: A Social Geography*, Longman.

Williams, P. (1976) The role of institutions in the inner London housing market, *Trans. Institute British Geographers, New Series*, 1(1), pp. 72–82.

Williams, P. (1978) Urban managerialism, *Area*, 10(3), pp. 236–240.

CHAPTER 4

Economic Activities in the City

All cities were established as specialised centres of activity, providing a location for those engaged in a number of civic functions, including trade, learning and crafts. In the Middle Ages, the economic *raison d'être* of cities continued to be trade and exchange, whilst they performed other functions – both secular and religious – and were centres of small scale industry. In many cities, it was the merchants or the skilled craftsmen who comprised the governing class, as befitted their economic importance. In early industrial societies, many cities grew enormously under the impetus of manufacturing industry, dependent on a large spatially concentrated labour force. The end-product of course was often an ill-planned confusion of activity – Lewis Mumford termed the nineteenth-century urban product of the Industrial Revolution 'Coketown', reflecting the image of urban centres in the industrial regions of Britain at that time. The legacy of that period to the contemporary city was not only the unsatisfactory urban fabric but, in time, an out-dated economic structure and a repellent image of poverty and pollution which would require major change.

Cities in Western Europe have in fact shown a remarkable ability to adapt their physical structures as a result of changing economic activity. In very broad terms, the decline of traditional large scale industry, whether it be steel-making in the Ruhr, textiles in Twente, or ship-building in Newcastle, has left a physical mark on the city. This is hardly surprising, since it has been previously argued that it was the growth of just these activities which fashioned much of the modern city, determining its physical and social characteristics. The industrial revolution and its adoption of large scale production and social and functional segregation were reflected in the tracts of industrial plant and workers' housing in many Western European cities, most notably in those located on the coalfields of the countries of the North-West of the continent.

Equally, smaller-scale industry has been the subject of major changes, with the decline of industrial workshops in so many cities, to be replaced by larger scale modern facilities. It has been argued that the process of urban re-development has forced the decline of the small industrial plant, or at the least made its survival difficult. But even in the absence of that agent of change, small scale industry has declined and, with it, the associated plant and facilities. The cutlery industry provides one of many examples, where in Sheffield, Solingen or Thiers in the French Auvergne, the old workshops often housing one or two workers have been replaced by a larger scale of operation, newer modern processes and plant. Traditional economies and the communities based on them have been transformed and the physical framework for the activity has become out-moded.

The changes in the manufacturing sector of the economy have been accompanied by a changing locational pattern of activity, in part an actual shift, but more often a change in locational emphasis, as new industry has sought the suburban fringe, leaving the inner areas, to take advantage of cheaper land, less congestion, or a locally available workforce (Watts 1990). In its wake are left the almost intractable problems of finding new economic functions to re-use the land and in some cases the buildings left behind.

The decline of manufacturing employment, and with it the major shift in locational pattern of employment, has been accompanied by very significant developments in the service sector. In short, the change has been a structural one with major inter-sectoral shifts in employment taking place. Not only has there been an increase in service sector jobs, however, but this sector also has seen a considerable intra-urban movement in its location. This shift is of itself of major importance to the physical form of the city, putting pressure on peripheral sites for commercial use, resulting in a series of policies of control throughout Western Europe.

These profound changes, as the city becomes a centre for the production of services rather than goods and, more broadly, a place of consumption rather than production, affect all the following chapters. In this chapter, however, the process of change will be briefly considered in terms of the sectoral shift of employment towards the service sector, before moving on to consider the planning problems presented by such changes, and the range of solutions investigated within European urban centres.

Economic activities – a brief overview

The traditional division of economic activities into primary, secondary and tertiary has proved to be inadequate for examining the economy of the Western European city at the end of the twentieth century. The first, including principally agriculture, is of minimal importance but the shift from the second, manufacturing, to the third, services, has been spectacular. The increasing range of services makes it sensible to distinguish between those engaged in the transport and selling of physical goods, and services dealing in such intangibles as knowledge, education, entertainment, health, or security. It is this last quarternary sector that has exhibited not only the strongest growth in recent decades but also in many respects proved often to be intrinsically urban in its locational preferences. Consequently the Western European cities can be decreasingly characterised as places of industrial production and increasingly as centres of control, interaction, creativity and enjoyment.

The process of change towards the service sector is not, of course, confined to

Table 4.1 Service sector employment as proportion of total employment in countries in Europe, 1983, with selected international comparisons

Austria	51.3	Greece	41.2	Portugal	40.7
Belgium	65.8	Ireland	53.2	Spain	48.4
Denmark	65.5	Italy	51.6	Sweden	64.7
France	58.0	Luxembourg	59.5	Switzerland	55.3
Finland	54.2	Netherlands	67.1	Turkey	23.5
Germany (FR)	82.4	Norway	64.3	UK	63.7
Canada	69.0	USA	68.5		
Japan	66.0	USSR	41.0		

Source: *Basic Statistics of the Community 1984*, Eurostat, Luxembourg, 1985

Table 4.2 Service sector change in countries and regions

	Country	Service % change	Metropolitan Centre	Service employment % change
1975–83	Belgium	12%	Brussels	5%
1979–80	Denmark	39%	Copenhagen	28%
1971–81	Greece	37%	Athens	36%
1975–83	Spain	11%	Madrid	8%
1974–83	France	20%	Paris	13%
1971–81	Ireland	28%	Dublin	33%
1976–83	Italy	15%	Milan	7%
1970–71	Norway	43%	Oslo	28%
1971–81	Austria	29%	Vienna	16%
1975–84	Sweden	26%	Stockholm	24%

Source: from Illeris, 1989

cities, but is reflected in national economies. The proportion of the working population as a whole engaged in service activities has increased dramatically so that, by the 1980s, the service sector employed the majority of the working population in every Western European country. This is well illustrated by the data in Table 4.1, where the service sector employment in European states is shown alongside some international comparisons.

Whilst national economies have seen absolute change, there is evidence to suggest that overall the process is one of the rest of the country catching up with its major metropolitan region. Table 4.2 indicates service sector employment change in Western European countries, in most cases involving significant growth. On the other hand, in the major metropolitan areas, the rate of growth has been slightly below that of the nation as a whole.

Industrial change

Whilst there are difficulties in describing precise models of industrial location to parallel those depicting residential patterns in cities, there is a degree of consistency in the dynamic processes of location which have affected cities in the second half of the century. The trends can be summed up as inner city decline, balanced by suburban and exurban growth. Such a summary is, however, a considerable over-simplification of the real process of change, since it is a complex one which involves far more than simply a locational shift of industry from one part of the city to another. The shift from manufacturing industry to the service sector – the process of de-industrialisation – has involved the death of some manufacturing plants and the birth of others. The locational implications of this are that inner city industry has frequently disappeared completely, rather than being relocated, whilst new firms involved in new manufacturing processes have chosen non-traditional locations, often on the periphery of the city.

In seeking an explanation for this phenomenon there is no doubt that a shortage of land for expansion was one important factor. A process of shift within the centre prior to a movement to the periphery led to the 'incubator hypothesis' of Hoover and Vernon, in their study of New York in 1962 (Hoover and Vernon, 1962). The thesis was that in their early years firms valued the advantages of the central site – accessibility; external economies afforded by the proximity of other firms and services; cheap, adaptable premises, even if they did not allow expansion opportunities; access to the labour market and to the markets which they served, etc. Once they were strong enough to move from the central area of the city, firms would move to the periphery to take advantage of other factors more appropriate to mature firms – economies of scale; access to the inner-city road network; access to a suburban workforce, especially a female workforce for some manufacturers, etc. That movement, if it were to take place at all, would come only after a number of moves within the central area. Whilst conceptually neat, there is only limited evidence to support this process, although elements of it are no doubt identifiable in many cities. In Glasgow, for instance, the inner conurbation had 66.8 per cent of the employment total in 1951, but only 58.2 per cent by 1981 (Keating, 1988), although little evidence is available of physical movement being the decisive agent of that changed balance.

A second major factor which has contributed to the process of industrial change is the process of urban renewal, which has affected many inner city areas in the post-war period. This process has been especially damaging to small companies, many of which may be in the throes of the incubation described above. Urban renewal has frequently swept aside small workshops and low rent premises to be replaced by new residential and commercial premises which cannot offer the same opportunities to small industrial companies. It has been estimated that the survival rate in some cases of urban renewal in the UK has been as low as 40 per cent of the firms originally located in a renewal area.

An example is provided by the eastern Paris suburb of Belleville, which was the focus of political protest whilst an urban renewal policy was implemented from 1970 onwards. Not only were the apartments of the original population swept away, but so too were their workplaces – the two were often one and the same. This meant that the traditional industries of the area – carpet-making, glass-blowing and leather-working – declined markedly, since they were small in scale and dependent on a ready supply of small premises. Somewhat late in the day, the planning authority of the city of Paris changed its urban renewal policy in order to preserve more of the original economic activity, but not until the industrial structure of the area had been changed quite fundamentally.

In the late 1970s, Lloyd and Mason (1979) documented the changes which were transforming Manchester and its industrial structure. Their work warrants close attention since many of the problems associated with manufacturing locational shift are particularly well illustrated by the experience of Manchester in the post-war period. The city saw its major industrial growth in the nineteenth century, resulting in the development of a commercial centre, encircled by an inner city area comprising a mixture of poor quality housing and industry – predominantly textiles, clothing and engineering. The City of Manchester Plan of 1945 led to large scale slum clearance in the following two decades, which meant that not only housing but also workshops and small industries were cleared from the inner areas of the city. By the mid-1960s, and the starting point of Lloyd and Mason's analysis, however, there were still significant numbers of industrial plants in the inner zone of the city.

Recognising that there are three components of industrial change, *viz* the contraction and expansion of employment, the transfers of firms into and out of the area in question, and the births and deaths of firms, they conclude that industrial migration was actually the least significant factor in the case of Greater Manchester. The actual change which took place between 1966 and 1972 was dramatic, with a 33.2 per cent decline in manufacturing employment, amounting to some 30,387 manual jobs, a loss of 604 industrial establishments, 25.6 per cent of the total, in the same period. A more detailed appraisal of this change showed that whilst plant closures contributed over 35,000 of the jobs lost, and there was a further net loss of 6,000 jobs when the shrinkage of companies was much greater than those which expanded *in situ*, 11,000 new jobs were created in new plants opening up. These statistics demonstrate that the process of industrial change is far from simple and the notion of simple suburbanisation needs considerable qualification.

Table 4.3 indicates that the small number of large plant closures, i.e. those employing over 50 people, actually accounted for some 65 per cent of the jobs lost, even though they accounted for only 13 per cent of the total plants lost.

Other studies of British cities support the general conclusions of Lloyd and Mason. Massey and Meagan (1978), for example, showed that only 11 per cent of the loss of jobs in British cities could be attributed to locational change, with industrial restructuring shifting the balance from declining traditional industries to new growth industries being a far more important factor in determining overall job loss. Dennis documented the loss of jobs in London from 1961 to 1975 during which period the city lost one third of its jobs in manufacturing. During this critical period, however, the proportion of jobs in manufacturing compared with that in services changed quite markedly, making it difficult to conclude that the resulting changes in the intra-urban pattern of

Table 4.3   Inner city closures (excluding transfers) of manufacturing plants in Manchester: size distribution

Size	Employment	Plants	Percentage of total employment	Percentage of total plants
1–9	3066	555	10.85	55.39
10–19	2252	166	7.97	16.57
20–49	4604	150	16.29	14.97
50–99	5143	79	18.20	7.88
100–249	5570	38	19.71	3.79
250–499	2781	7	9.84	0.70
500–999	4844	7	17.14	0.70
1000 +	–	–	–	–

Source: Lloyd and Mason, 1978

employment and economic activity are simply a question of locational shift. Indeed, Dennis concluded that industrial decline in London was not simply confined to the inner city but was a feature of the entire conurbation. Furthermore, addressing the impact of outward movement, he showed that in the case of London, outward movement of employment was a more important factor in the outer zones of the city than in the inner core.

The conclusions which can be drawn from these analyses are wide-ranging and profound in terms of their implications for European cities. The inner city has become an unattractive location for new industrial growth. Urban renewal has done little to assist this – indeed, Lloyd and Mason suggested that comprehensive redevelopment has passively, if not actively, discriminated against the inner city manufacturing plant (Lloyd and Mason, 1978, p. 69). Certainly, the post-war period has seen a major shift in the location of industrial growth, with suburban areas and small and medium-sized towns on the periphery of major metropolitan areas seeing an expansion of manufacturing and other employment, with a concomitant decline in the central city. What has been described, of course, is an integral part of the process of counter-urbanisation discussed elsewhere and, as following sections of this chapter will illustrate, it is a process which is by no means confined to manufacturing industrial change.

In Britain a succession of central government inner city initiatives have been implemented over the last 20 years. These in the 1970s were generally local authority managed and central government financed investment in area-based urban renewal projects, whilst in the 1980s a whole alphabet of market approaches combining public and private financing were attempted within purpose created management structures (Ashworth and Voogd, 1990). These included Urban Development Corporations, Enterprise Zones, Economic Development Corporations and many purely local public–private partnerships. Among the most successful have been those focused on London's Docklands.

Within London Docklands, which had seen the demise of the docks and their associated industry in the post-war period, the Isle of Dogs Enterprise Zone was designated and the Docklands Development Corporation established to foster its development. It quickly attracted a range of economic activity, as well as residential developments. Linked to central London by a new light rail system, Docklands has attracted only a limited amount of manufacturing, with the majority of the new economic functions being within the service sector, most notably the 1.12 million square metres development at Canary Wharf on the Isle of Dogs, to house a major financial centre which is in effect an extension of the City of London.

The control and direction of office development

The office was not fully recognised as a major urban function requiring specific direction and control until the 1960s. Even then, there was a wide variation in attitudes towards potential office development, from a permissive *laissez-faire* policy such as that prevailing in Brussels to comparative restraint, of which the most notable example is London. Between the two has been a broad spectrum of policy approaches to dealing with the growth of the office sector.

The policies may be divided into two groups. On the one hand there are those which have favoured deconcentration to subsidiary centres at some considerable distance from the central city, while control has been simultaneously exercised to a greater or lesser degree in the central city itself. This policy may indeed be seen on a national as much as a metropolitan scale. In contrast have been policies to encourage office concentration in locations other than the centre of the city, but often in a suburban rather than a more distant location. In practice, in most cities where office control policies have been found necessary, they have oscillated between elements of each of these two broad policies, but in the final analysis one can usually point to a mixed emphasis in either one or the other direction.

There is a distinction between countries with an active national policy for office location and those with no planning at this level. In the case of the United Kingdom, in 1964, as a control measure, an office development permit system was introduced and survived until 1979. Central government control at this level was further underlined by the proposals of the Hardman Report (1973) on the dispersal of civil servants from London to the provinces. Similarly, both the Swedish and Dutch governments took steps to control their own office development and to move jobs to provincial locations from the capitals. In other countries, however, notably West Germany, there has been no such national policy within which to fit a local plan for a particular urban centre, although in the case of West Germany the absence of such a policy is largely explained by the federal structure of that country.

At the scale of the medium-sized city, it would be true to say that German planners have been the most positive in offering solutions to the demand for office space experienced in the last thirty years. At a local level, a close examination of German cities exhibits five major responses to the demand for office space in the city. The five responses are:

(1) inner ring-road developments, often closely linked to the central area;
(2) overspill from the centre into desirable residential area;
(3) the establishment of inner suburban nodes of office development; and similarly
(4) nodes in the outer suburbs; and finally
(5) the use of major transport interchanges as office locations, frequently linked to the establishment of entirely new office parks.

The first response represented an early attempt to provide office expansion and can be identified in Cologne and in Hamburg. Overspill into desirable residential districts has been evident in both Hamburg and Frankfurt. In the latter case such growth into the city's Westend area met with considerable opposition. In Hamburg, the Harvestehüde and Rotherbaum districts to the north of the city (originally developed in the nineteenth century along the shores of the Alster lake) have provided other classic examples of this style of overspill development. A more restrictive policy is now operated in this area to prevent its further transformation into a totally office-dominated quarter. There can be little doubt that, in social terms, such office growth can be very damaging in its disruption of well-established urban communities. For instance, the Westend district of Frankfurt lost 6,729 persons between 1961 and 1970, or a loss of 22 per cent of its population, whilst the area had 17 per cent of the office space of the city, compared to 31 per cent in the centre itself before policies of restraint were introduced.

The establishment of specific nodes of development has been a part of a more coordinated planning policy, which in German cities has taken the form of comprehensive structure plans. Where inner suburban schemes have been tried they have been conceived not solely as an office project, but as part of a wider scheme to revitalise the inner suburb and enable it to share the advantages of the outer, often more affluent suburb. In this sense, they have their parallels in Paris where the establishment of new suburban growth poles in the suburbs, such as that at Créteil, in the south-east of the city, have included a large element of office employment, often in the public sector. A German example can be found in Hamburg, where the Hamburgerstrasse development, while situated on one of the eight designated axes of growth for the city is in an inner suburban area relatively deprived of modern facilities. The development, in common with others like it, included a retail centre and other services as well as a residential component.

Other suburban nodes have been planned, but are rather less frequent. At Laatzen, in southern Hannover, office development has been permitted associated with a new covered suburban shopping centre. Similarly to the east of the same city an office development was permitted at Roderbruck. Such solutions have obvious disadvantages, not least that of limited levels of accessibility by public transport since, as in the case of Laatzen, they are sited at the termini of public transport routes to the central city.

In a European context, the most distinctive planning contribution to the problem of the office has been the office park. Such suburban 'greenfield' locations have now become as commonplace in Europe as they are in North America, where the concept originated. There is a considerable variety of styles of office park, since some house other uses – research or retailing, for example – but they share the common characteristic of offering accessibility, especially for the private car, and an open environment, in marked contrast to a central city location.

Some developments are on a very large scale, such as that of Hamburg City Nord, which is examined below in more detail in a comparison with La Défense, the new suburban development to the west of Paris. Other examples include Burostadt Niederrad, midway between Frankfurt and the city's international airport. Such locations are by now classical locations for office parks in both North America and Europe. Hamburg City Nord met with initial problems, associated with the image of its location within the city, which led to problems of staff recruitment. In addition, its position in relation to the public transport system is not particularly good, despite being close to the airport and in a European context perhaps more than that of North America, this is still a major location requirement.

The various policies adopted in Germany and elsewhere may be summarised diagrammatically and are shown in Figure 4.1. From the description of the German policies given above, it will be apparent that the results of some though not all of these policies included in Figure 4.1 can be identified in German cities. The notable exception is the long-distance decentralisation solution. This policy has been attempted in Britain on a large scale both in the public and private sector; it is true, however, that here as elsewhere, such as Sweden, the long-distance decentralisation of offices is particularly important for offices in the public sector forming part of a broader government regional aid policy.

While medium-sized German cities provide a variety of approaches to office location, it is the major capital cities of other countries in Europe such as London, Paris, Vienna and Brussels which have had to face even greater problems associated with large-scale demand and supply of office space.

The special pressures for new office space in Brussels led to the so-called 'Manhattan plan' for the Noordwijk district. Here the 1967 development plan laid out an office complex covering 53 hectares, which provided office accommodation for 75,000 workers in around 1 million square metres in 40 tower blocks up to 160 metres in height, together with 15,000 parking spaces and 5,000,000 square metres of shops. The scale of this development compares with La Défense or Croydon, but it is located much closer to the traditional central business district than either the Parisian or London cases (van Hecke, 1977).

In London, the history of office development policy illustrates the growing awareness of the importance of office development since 1945. In the period immediately after the war, there was little apparent concern about offices other than a desire to reconstruct those lost through wartime destruction. The County of London Development Plan in 1951 acknowledged the fact that there was some congestion, but nevertheless increased the area zoned for offices from 322

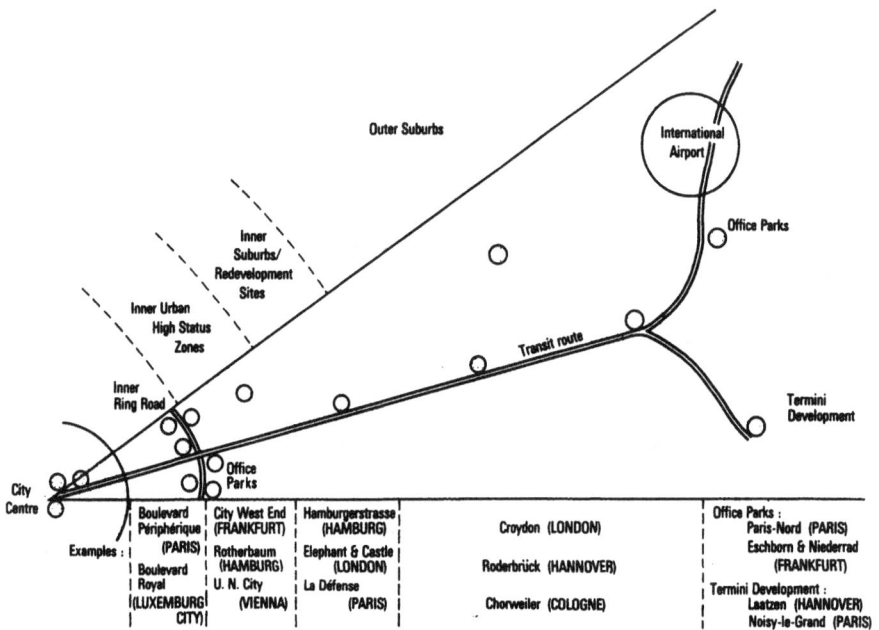

Figure 4.1   Spectrum of office locations in West European cities

to 374 hectares. However, the theme of deconcentration was already apparent, as indeed it had been as early as Abercrombie's Plan for Greater London in 1943. This theme was embodied in the Plan eventually approved in 1955. Active moves towards decentralisation took place on a limited scale, although the fact that administrative control was divided between the former London County Council (LCC) and the surrounding counties, prior to the creation of the wider Greater London Council (GLC) in 1964, made coordination difficult. Increasing commuting problems, brought about by an increase in office floor space from 7.7 million square metres in the central area in 1948 to 11.5 million square metres in 1962 (Daniels, 1975), emphasised further the need for major policy changes.

During the 1950s offices were developed at a rapid rate in London, successfully competing against other land-uses. By the early 1960s there was an increasing call for greater decentralisation and for offices to move to various towns and suburban centres outside London in the period after 1956, a trend already encouraged by some county councils such as Middlesex and Essex. But such moves were not without their own locally generated problems, and by 1964 most counties had reversed their policies. By that time, firms were being attracted to the peripheral areas not only from the LCC area but from elsewhere in the United Kingdom. Thus these concentrations outside London were not necessarily helping directly the problems presented by London itself. A major exception to the policy reversal of 1964, however, was Croydon, which continued to pursue an active policy of office development.

The Croydon Corporation Act passed in 1956 was the first instrument designed to encourage major office expansion in a medium-sized suburban centre. The local authority, eager to attract new offices, initiated the act to enable the purchase of one hectare of land in the central area of the town. There was no doubt a fair measure of civic pride as well as financial incentive as the motivating force behind this action. This area and adjoining land the local authority leased to private developers resulted in 19,000 square metres of offices and only 150 square metres of shops (Daniels, 1975). Fortuitously, this action by Croydon coincided with a general adoption of a deconcentration policy, resulting in the allocation of a further 18 hectares by the local authority for decentralising offices. In fact, development has spilled over the limits of this area and new office space has been provided in excess of 600,000 square metres, in which nearly 30,000 workers are employed. The success of Croydon has been dramatic in concentrating the dispersion of office functions in such a large-scale development.

The importance attached to the decentralisation policy was underlined in 1963 by the publication by the British government of a White Paper on London's office problems, including the recommendation to establish the Location of Offices Bureau (LOB) with a role to encourage decentralisation and to supply intending companies with information on potential moves. The White Paper looked towards suburban office centres as well as those further afield.

While the 1963 White Paper produced broad strategies, a White Paper in the following year introduced a major planning control. It was embodied in the Control of Offices and Industrial Development Act 1965, and took the form of a system of Office Development Permits (ODP) for new office buildings and for conversions from other uses. The ODP system was applied retrospectively from November 1964 to all new buildings for which permission was sought. During the 1970s, the ODP system was applied more generally than its original area of operation of the London metropolitan region, within which all office space of more than 300 square metres required an ODP prior to development, though by the 1970s this was raised to 1000 square metres. After 1979, however, with the adoption of a more *laissez-faire* policy, the system was dismantlèd, with the only remaining control being that of the local planning authority over the building development itself.

The policy of dispersal brought with it other problems. In some cases, companies with decentralised offices found it necessary to re-establish a stronger London base – the oil company BP being one such example. The exodus of many office workers to locations outside the capital left many inner London boroughs which were not important locations for offices without the means of achieving a balanced employment structure. As early as 1977, this stimulated central government to redefine the role of the Location of Offices Bureau to include the attraction of jobs to the inner city, although the LOB did not survive to actively carry out this role. This problem has not been fully resolved, however, and in the 1989 Policy Review by the London Planning Advisory Committee, the need to make the inner suburban nodes attractive for office development was acknowledged.

The situation in Paris has certain parallels to that of London, but also some important differences. On a national scale, the designation of eight *metropoles d'équilibre* in 1965 fostered the concept of deconcentration from the capital. The impact on the office sector, however, was minimal and indeed only Lyons could be seen as offering any viable alternative to Paris as an office centre, with a limited concentration of financial and regional functions.

Since the late 1950s, in Paris itself, a major part of the office location policy has revolved around the successful completion of La Défense, the major suburban node to the west of the city. The first operational regional plan for Paris suggested that offices should be encouraged in a number of new growth nodes, both in the inner suburb and in the new towns. Historically there has been a general and marked trend to office development moving towards the west of the city. The east–west imbalance of the central city itself was best illustrated by the fact that at the time of the Regional Plan, during the 1960s, the population was approximately equally divided either side of a central north–south dividing line, whilst job growth to the west in the period 1962–68 was much more significant than the east, the share being 68 per cent and 32 per cent respectively.

Until the mid-1980s the policy in Paris has therefore been to favour office development in certain parts of the city and to discriminate against such development elsewhere. In 1974, the government introduced an annual 'ceiling' of office development of 900,000 square metres per annum of which 20 per cent was to be in the five new towns, while only 70,000 square metres was permitted in the central *département* of Paris itself.

The major instruments of control lay in the granting of planning consents (*agréments*) which were required for any building in excess of 1000 square metres, and in a taxation system levied on new building which was locationally selective. In a general sense, the *agrément* system operated to favour the hitherto less-favoured eastern sectors at the expense of the west, although it also discriminated in favour of the new towns of Paris and its new suburban growth poles, whereas with certain exceptions it worked against Paris and its western department of Hauts-de-Seine. The exceptions lay in the eastern part of the Paris departments, especially the Bercy–Gare de Lyon zone which has been developed as an office quarter away from the more traditional western concentrations within the old *ville*.

The building development tax (*redevance*) was similarly selective, as Figure 4.2 indicates. Zone 1 covered the western part of the department of Paris, and the western department of Hauts-de-Seine. The second level of tax applied to Zone 2 to the west and south-west of Zone 1, again discriminating against the suburban areas which had proved more attractive in the 1960s and early 1970s. Zone 3 covered most of the rest of the continuously built-up urban area. There were, however, a large number of nodes within these broad zones, especially in

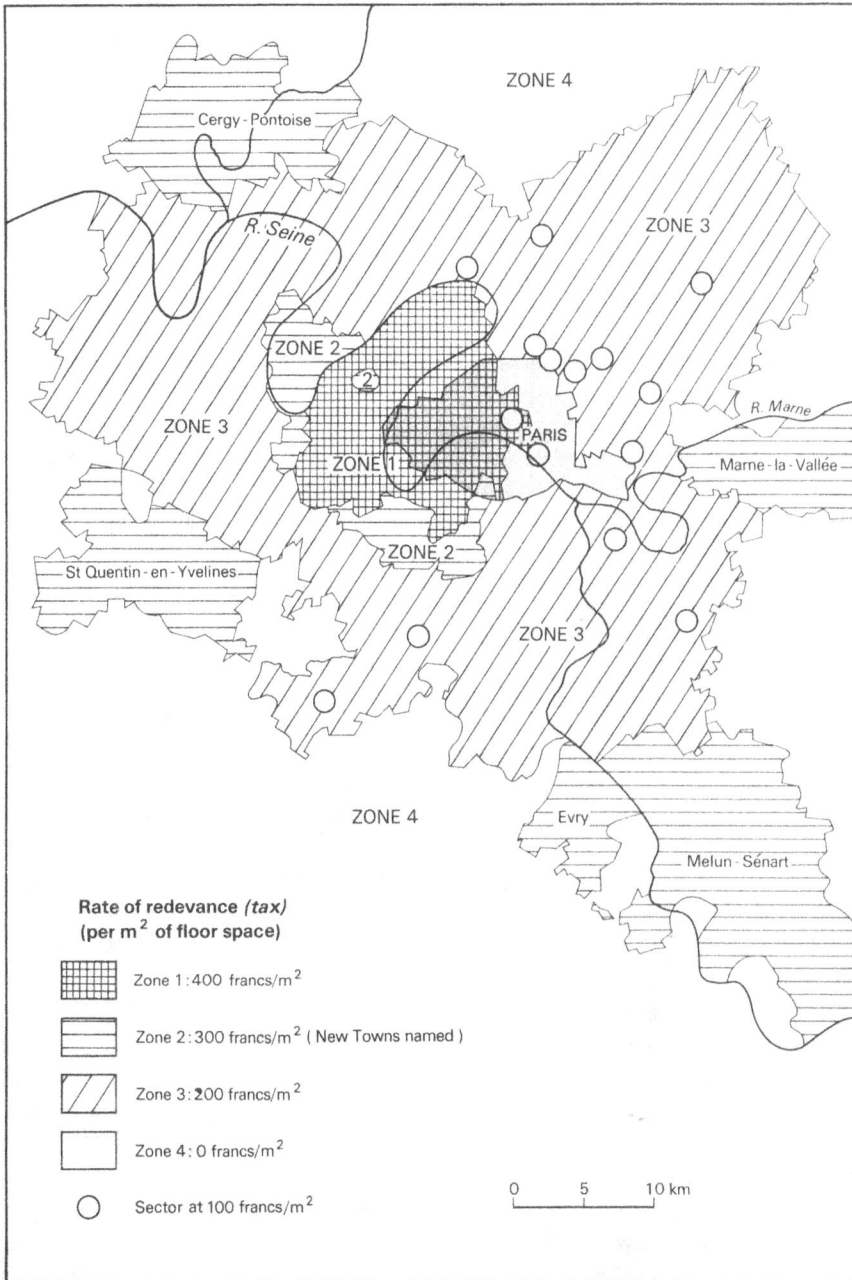

Figure 4.2 Office development tax zones in Paris 1971–1984

the case of the city, where office development was to be encouraged, and which therefore attracted a lower rate of *redevance*. For instance, in central Paris two areas, the controversial Les Halles site and the Bercy–Gare de Lyon site, were designated as exceptions to the area surrounding them. The tax was originally introduced in 1960 as a uniform tax on development, but it was altered in 1971 to be imposed differentially within the Paris regions, favouring the eastern and peripheral area and penalising the western and central areas.

The success of these measures was hard to assess but, in any event, the controls on development were lifted virtually completely in 1984, largely as a response to the growing competition from other European cities. Since that time, every effort had been made to secure the position of Paris as a major European business and cultural centre in direct competition to London (Datar, 1989).

The *Livre Blanc* published in 1990, to set the framework for the revising of the Regional Plan, identified Bercy and Tolbiac, both sites in eastern Paris, as major office centres. The encouragement of these foci for new service employment will do much to redress the long-standing imbalance between the east and west of the city.

Throughout the post-war years, La Défense has been pivotal in the office development policies of Paris, paralleling the similar development of City Nord in Hamburg.

La Défense, with a planned office floor space of 1.55 million square metres, is a high density concentration of tower blocks. Part of a much larger scheme of urban rehabilitation, including the suburb of Nanterre, it includes 6820 apartments in its plans. Its transport system is such that it could be carried out only on a completely cleared site, with multi-level transport systems affording links to the SNCF, the express metro (RER), as well as a recent extension of the metro itself, bus services and major road routes. It has an employment level of approximately 110,000 people, with a further 20,000 residents. It has not been without its developmental problems, such as that of vacant office floor space during the economic recession of the 1970s.

By the end of the 1980s, however, the original plans had been fulfilled and evidence of the success of La Défense lay in the large amount of office space in areas adjacent to the site itself, representing overspill demand. La Défense had attracted many major multinational concerns such as IBM, Dunlop, Esso and Rank-Xerox, together with major domestic office users such as Electricité de France (EDF) and many of the major banks and insurance companies. The completion of the 'grand projet' of La Tête de Défense – a spectacular open box-shaped office tower – in time for the 1989 Bicentennial celebrations, underlined the importance as well as the success of the project which at its conception may well have appeared to have been grandiose and perhaps over-ambitious. The development of a major office complex out of the city centre, however, offered the opportunity for innovatory office technologies (especially in the latter phases of its development), without unduly scarring the historic townscape of the capital.

Hamburg's City Nord, built from 1967 onwards, is an essentially similar though somewhat smaller concept. It is different in physical appearance, however, since it has lower-rise office blocks than La Défense. It occupies a 95.1 hectare site, with a total of 638,000 square metres of offices, 20,000 square metres of shops and 50,000 square metres of residential space. It has a central area of shops, restaurants and hotels and a workforce of 32,000. In both La Défense and City Nord, the plan existed to create a concentration of office activity sufficient to attract the city-centre office to a suburban location. In the case of City Nord, it has had some success, with for instance the oil company BP AG leaving behind 22,000 square metres in nine locations in the city-centre of Hamburg to take up 31,000 square metres in City Nord; and, similarly, Esso moving from 33,000 square metres in eighteen different locations to 36,000 square metres in the new development.

While in Britain, France and Germany, office development since the 1960s initially came under increasing control, in Belgium, and particularly in its capital, Brussels, there was far less restraint. Office development began on a large scale in 1971, mainly financed by overseas finance and powered by the growing European role of the city. British property companies were foremost in this process and the effect was a virtual transformation of l'Avenue des Arts, with modern speculatively built office blocks changing its skyline. The problems of oversupply were very evident in Brussels, with vacant office space being the subject of much public criticism. Brussels and its experience during the 1970s emphasised further the need for strong policies towards the control of office development. Since 1975, there has been a process of decentralisation especially towards the Flemish part of the country (Illeris, 1989).

The office has been recognised now as a major urban function, but it has taken time to evolve effective means of control; and even now the degree of effectiveness is difficult to assess, if only because control policies have rarely been sustained. In addition there is the fundamental problem of the spatial scale of assessment. Office location policies in the 1970s were designed to reallocate from overheated to underused cities within national systems, and thus tended to constraints and controls to attain regional policy objectives. The increasing internationalisation of economic activities, the reduction in the national barriers to location within the Community, and the re-emergence of market philosophies in public sector planning, all combined to redefine the scale as one of international competition between cities, and thus favoured policies of incentive, support and promotion. The international competition for companies operating on a European and world scale is such that most countries have attempted to adopt policies which would encourage the presence of such companies. But in itself the recognition of the office as the major twentieth-century employer, and as such a vital component in urban and regional planning, is important. Provision for office space in urban plans is now at least as important as that for industrial or manufacturing uses, which dominated early post-war planning ideas.

Control and development in the retail sector

In the field of retailing, there has been considerable variation in the extent to which European urban centres respond to recent changes and innovations in this sector of the economy. One reason for this is the wide variation in the structure of the retail trade itself (Burt and Dawson, 1988). At one extreme, in Italy, the continuing dominance of the independent retailer has hindered innovation in the form of large-scale, edge-of-town, or out-of-town retailing. Such innovation requires available capital which has usually been the preserve of large retailing organisations or property companies. The latter are unlikely to be involved when there is little demand for new retailing premises since the retailers are small in scale, operating from only one location. At the other end of the scale, in the United Kingdom a considerable dominance of the retail trade by a relatively small number of retailers has ensured a competitiveness often involving large-scale innovation, backed by considerable capital formation.

A second reason lies in the variation in control of the retail sector. In some cases, new retailing units have been built with practically no control. For instance, new hypermarket development in both Germany and France was originally subject to little restriction. Only later in the mid-1970s did planning authorities seek to direct this form of retailing. On the other hand, in the United Kingdom the new retailing typified by the hypermarket was the subject of the most careful control from the outset and only during the 1980s has that control slackened, although not to the point of total release.

Three major retailing developments can be distinguished in the West European city, each bringing with it a consequential series of responses in the commercial structure of the city. Firstly, there is the development of hypermarkets and superstores, often, but not always, in an edge-of-town or out-of-town location. There are problems of defining a hypermarket, but a useful early definition was put forward by MPC and Associates Limited (MPC, 1974). They defined a hypermarket on an initial count as a store (a) with a selling area of at least 2500 square metres on one level, specialising in food but with a wide range of other convenience goods; (b) adopting self-service methods with at least fifteen checkouts; (c) usually 3 to 6 km from the town centre; and (d) with its car parking area, adjacent to the store, in excess of three times its selling area. In fact other developments not fulfilling each of these criteria have been built, often termed superstores, sometimes smaller than a hypermarket, and more importantly varying somewhat in their retailing methods.

A second development has been the introduction of the regional shopping centre, similar to North American shopping 'malls'. Such centres situated outside established cities contain at least two department stores and around a hundred other shops together with other associated facilities. Car parking is provided on a large scale. The regional centre is planned to serve a much wider area and a larger population than the other 'out-of-town' facilities defined above. An intermediate stage of this type of development can be identified in West Germany where Ruhrpark outside Bochum and the Main-Taunus

Centrum near Frankfurt were each developed as major regional centres, but with open malls rather than the fully enclosed centres currently being developed.

The third development has been the redevelopment of the city-centre with major new retailing facilities, either in a pedestrianised precinct or in covered malls. In the recent past, many of these have been on the scale of the out-of-town regional shopping centre. The impact of these three changes in retailing is very varied in Europe and each is now examined in turn to illustrate this variation and to point to the consequential changes in the city's retailing structure.

Hypermarkets in Europe showed a very uneven pattern of distribution during their period of development. The widespread diffusion of the hypermarket in the French, Belgian and German city contrasts sharply with the situation in Britain and elsewhere. The extreme case is that of the Netherlands, where a combination of shopkeeper interests protecting existing city centre investment and urban planners committed to the 'compact city' effectively prevented any real hypermarkets being developed. Early studies of the French hypermarket suggested that its impact on pre-existing retailing within the town was not as great as at first feared. However, it was this fear which held back the development of hypermarkets in the United Kingdom. Belgium, however, showed far less reluctance and generally welcomed new forms of retailing more quickly than many of its neighbours. In France, the first supermarket was opened in 1958 and the first hypermarket as early as 1961. By the mid-1980s, the situation had been reached where a city such as Marseille was surrounded by a ring of no fewer than twelve hypermarkets (Burt and Dawson, 1990). The pattern of location has thus had both the time to develop, and a general absence of planning restraints designed to preserve the existing locations.

In Britain, however, more resistance was present. In 1972, British local planning authorities were requested to inform central government of any large-scale retailing proposals of over 5000 square metres, outside existing city, town or district centres. By so doing, if effectively removed this particular form of land-use from local control (Department of the Environment, 1972). The concern of central government essentially hinged around the possible problems for existing areas, the transport implications for large scale out-of-town developments and the feeling that not all members of the community could share equally in the use of this new form of retailing. There was, therefore, a desire to ensure the continuing prosperity of the existing retailing centres. It was felt particularly that the non-car owner would be disadvantaged by a lack of access to hypermarkets unless close attention was paid to public transport facilities. This policy by central government ensured a very slow diffusion of hypermarkets in Britain. The first was opened in Caerphilly in South Wales in 1972, but was rather small by standards of continental Europe, with a selling area of only 5500 square metres.

Meanwhile, elsewhere in the United Kingdom, and especially in many towns in northern England, superstores were being widely developed. These were defined, once more by MPC and Associates Limited (1974), as having a selling area between 2000 and 4000 square metres, with a more limited range of non-food items than a hypermarket. The superstore's method of merchandising was often somewhat crude, with goods sold direct from their cartons stacked in specially designed caging rather than being displayed on shelves. The number of checkouts varied between ten and twenty whilst the car-parking area was between one and three times that of the selling space. Superstores, so defined, became very popular, particularly where sites could be easily found within or on the edge of industrial cities. Often an inner city site could be found for such a store, while the local authority was pleased to see a new commercial use in a declining area. It was noticeable that there was a considerable delay before the introduction of superstores in cities in southern Britain, where sites in these less industrialised urban centres were more difficult to obtain. A market leader in this field was ASDA (Associated Dairies), but as the newer stores of that company illustrate, by the 1980s the difference between the superstore operators and true hypermarkets had all but disappeared.

Initially at least, however, the hypermarket was restricted everywhere. Central government did allow a development at Chandler's Ford in South Hampshire, which opened in 1977, but its progress was to be closely monitored particularly in terms of its impact on surrounding retailers. In the event, the impact study demonstrated that although some local shops had been affected, its influence had been spread very widely and was not felt by any one particular centre (Wood, 1976).

In December 1977, the British government's Department of the Environment (DOE) issued a further report on hypermarkets (Department of the Environment, 1977). It noted that the larger food stores already permitted had little impact on the independent food retailer within 15 minutes' drive time of the new developments. They had rather more impact on multiple food retailers' and cooperative stores' supermarkets, where there had been an average fall in turnover of 20 per cent for such stores within the catchment of hypermarkets, compared to an average fall of around three per cent. In general, the larger the supermarket, the greater had been its impact.

The development control policy note issued in 1977 highlighted the major planning issues once again, with the DOE stressing the importance of access for both pedestrians and public transport as well as for the motorist and seeing particular merit in accommodating new stores within existing urban areas. This location was seen as being particularly advantageous if it was within an expanding or developing town centre, or if the new retailing development could act as the nucleus for a district centre. The hypermarket 'issue' was also considered within the wider debate on the fate of the inner city. Indeed, in October 1978 the Greater London Council, in a policy statement, pointed to the acceptability of such new forms of retailing in the inner parts of the capital.

By the 1980s it was evident that there was a more relaxed response to the demand by retailers to be allowed to build new stores. Local planning authorities were still careful in terms of the siting of the stores, and often they

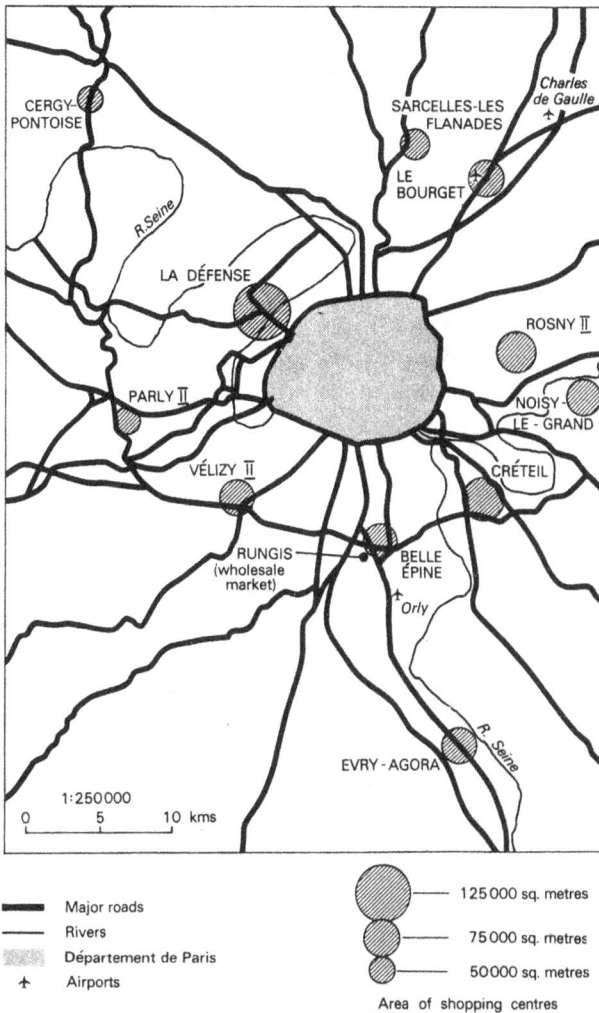

Figure 4.3 Regional shopping centres in the Paris region

could be seen as an integral part of a policy to assist inner city regeneration. Retail parks were established within or just outside most major cities, resulting in an explosion in terms of the number of out-of-door large-scale retailing facilities, often involving not only food retailing, but also furniture, DIY and electrical goods.

In a number of European centres, the second major development, the regional shopping centre, has been adopted. Nowhere, however, has it been as widely adopted as in the Paris region. The regional shopping centre represented an American concept introduced into the European city. In Paris, one of the first to open was Parly Deux in 1969. The promoting company sought to demonstrate that it was possible to provide a standard of retailing equal to that of the central city. Accordingly, Parly Deux was built to very high standards and fitted well into the affluent suburb of Le Chesnay near Versailles in southwest Paris. Comparatively speaking, Parly Deux was not a large centre; indeed, its 55,000 square metres is small compared with many of its successors. Its arrangement of retailing and other facilities is, however, typical, with two major department stores sited at either end of a covered shopping mall, which contains some 107 additional shops, together with restaurants and cinemas. In 1970, Parly Deux had a turnover of 288 million francs, in 1971, 387 million francs, and by 1972 420.85 million francs (Groupe Balkany, undated). Encouraged by this success, the same promoting company built other centres in the Paris region with Velizy Deux (75,000 square metres) opening for trading in 1972, Rosny Deux (83,000 square metres) in 1973, and Evry Deux (71,000 square metres) in 1975. These, together with other similar centres, now form a complete network of such centres around Paris, as shown in Figure 4.3.

The *centre commercial regional* was not, however, without its problems. On the credit side, they did much to give to the suburbs a much-needed structure and as such fitted into the first Schéma Directeur for the Paris Region, originally published in 1965. A number of the centres were situated either in the new towns or in new suburban growth poles, as indicated in Figure 4.3.

The centres were planned, however, in a very short period of time, with the possibility that true levels of demand had not been accurately ascertained. In these circumstances, the more marginal sites were somewhat exposed in economic terms. In any event, at least one of the regional centres, Créteil Soleil, had problems, and a further one at La Défense was delayed in its opening. It should be pointed out that the relative weakness of the major French retailers compared to similar concerns elsewhere made Créteil somewhat vulnerable. The centre had a branch of Printemps, which suffered major losses in 1975 and 1976. It countered by closure of branches, including in 1976 the furniture and domestic appliance sections of its store at Créteil, to be followed in October 1977 by its complete closure. This action shifted the focus of attention to the planned centre at La Défense, one of the largest of the *centres commerciaux*. Both Printemps and the other 'anchor store', La Samaritaine, sought to postpone the opening of this centre by three years, claiming that the 'climate' was not right for such a new development and that La Défense was running behind schedule.

Elsewhere the concept of the regional shopping centre was adopted rather more slowly. In the United Kingdom, the problems associated with land acquisition and the obtaining of planning permissions meant that only one centre was built in the 1970s – Brent Cross in north-west London. This project was initiated in 1956, with planning permission being granted in 1972, and opened for trading in March 1976. Unlike Paris, suburban London is extremely well served by a hierarchy of important suburban centres such as Richmond, Kingston-upon-Thames and Croydon. There was, however, a noticeable sector in the north-west of the city into which a major regional centre could fit and draw its trade from further out along the same sector. The strategic location of Brent Cross at the junction of the M1 motorway, the A41 and the North Circular Road was a major factor in its success.

The demand from developers for major out-of-centre regional developments was considerable and ultimately difficult to resist. Two very large centres were built in Enterprise Zones – the Metro Centre in Gateshead and Merry Hill in Dudley in the West Midlands. In a sense, the dam had been breached and by the end of the 1980s, a number of very large regional shopping centres, often with associated entertainment facilities, had received permissions, although there was still a presumption against new development if existing city-centre retailing was seen as being under threat.

While many of the regional shopping centres described above are in the existing built-up area, they are not a part of any existing central city retailing facility. The redevelopment of a part of a city-centre, or the addition of a major retailing development adjacent to it, is the third retailing development which may be identified. In this case, the examples are by no means confined to Britain and France, since the destruction of the 1939–45 war gave to many the opportunity to replan the retailing centres. In others, such as Stockholm, there has been a deliberate policy to redevelop the pre-existing central zone to produce a retailing centre which offers more modern retailing premises in traffic-free surroundings.

The initial developments in city-centres during the postwar period were often open-air pedestrian precincts (see Chapter 5). By the 1970s, however, the new shopping centres were generally covered shopping-areas. Some were on a very large scale, dominating the pre-existing retailing and often incorporating existing railway developments, thus providing both a national accessibility and a use for the space over railway facilities as at Hoog Catheryne (Utrecht) or Babylonia (The Hague). In the UK the Eldon Square development in Newcastle dominated the pre-existing city centre retailing facilities. Others attempted an integration of residential functions and retailing, such as the limited integration in Nottingham with its Victoria Centre, or the more general integration with a range of urban functions, including offices, such as La Part Dieu in Lyons.

There are problems attached to the building of major new covered centres which may dominate the retailing of a city simply by nature of their size. In their impact study of the Eldon Square development in Newcastle, Bennison and Davies (1977) analysed the movement of retailers from elsewhere in the city and showed a shift northwards of the centre of gravity of the retailing, leading to the considerable decrease in the importance of some of the southern zones of the city-centre. On the other hand, some similar developments are able to bring new retailing to a centre without having such adverse effects. The lack of such problems may well be a reflection of the increase in size of the centre, with the larger centres well able to sustain a major addition. One such example was the Arndale Centre in Manchester, consisting of 200 shop units, eight major stores, a multi-storey car park for 1800 cars and 20,000 square metres of office space. In total, the new lettable area, built in early 1980s, was in excess of 100,000

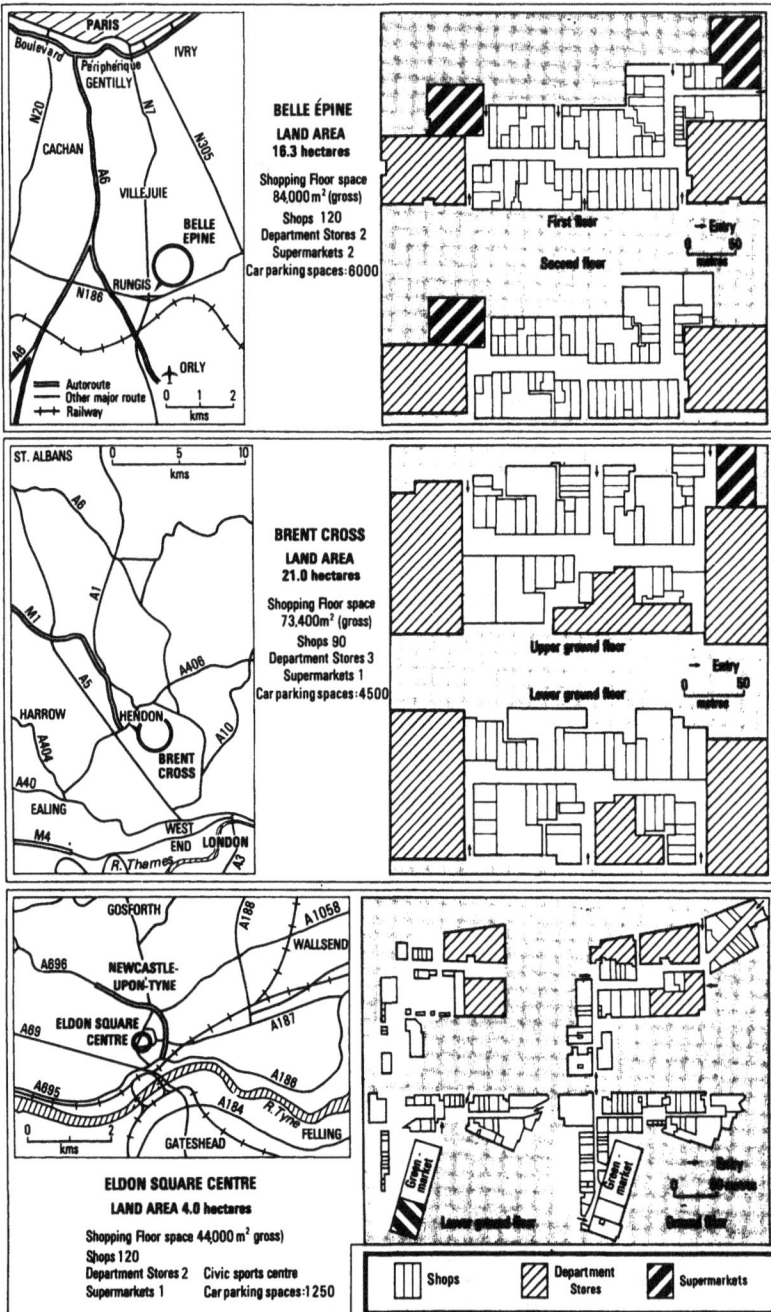

BELLE ÉPINE

LAND AREA
16.3 hectares

Shopping Floor space
84,000 m² (gross)
Shops 120
Department Stores 2
Supermarkets 2
Car parking spaces: 6000

First floor
Second floor

BRENT CROSS

LAND AREA
21.0 hectares

Shopping Floor space
73,400 m² (gross)
Shops 90
Department Stores 3
Supermarkets 1
Car parking spaces: 4500

Upper ground floor
Lower ground floor

ELDON SQUARE CENTRE

LAND AREA 4.0 hectares

Shopping Floor space 44,000 m² gross)
Shops 120
Department Stores 2   Civic sports centre
Supermarkets 1   Car parking spaces:1250

Figure 4.4   Regional shopping centres – the alternative approaches: (a) Belle Épine (Paris); (b) Brent Cross (London); (c) Eldon Square (Newcastle upon Tyne)

square metres, making it one of the largest covered shopping centres in Europe, but capable of being absorbed into the city centre without major adverse effects. In the case of Newcastle, however, 40,000 square metres of retailing in the Eldon Square centre was a very major addition to the retailing space of the city, and quite sufficient to change the whole balance of the centre.

Rivalling these British developments in size, is La Part Dieu development in Lyons, with its retailing component totalling 110,000 square metres, with two department stores, three other major stores and 250 shops. It differs, however, with its greater commitment to office development with 450,000 square metres of office space.

Figure 4.4 illustrates three of the major centres, discussed above to demonstrate their major differences. Each one may be seen as a new regional shopping centre, but they vary in their location from the outer suburban (Belle Epine, Paris), the inner suburban (Brent Cross, London) and the central city (Eldon Square, Newcastle upon Tyne).

The last 20 years have seen rapid changes in the retail sector. While it is tempting to suggest that those cities which redeveloped at a relatively early stage in the period missed the opportunities afforded for the construction of covered malls, this may not be supported by the facts of the situation. The fitting of major new centres into pre-existing shopping centres involves a delicate adjustment process and there is in Europe no consistent evidence to show that it is easily achieved, although it is by no means impossible. In a similar fashion, the out-of-town shopping mall has been shown to present problems unless it is very carefully controlled. It should also be said that the refurbishment of existing retailing areas, through the adoption of pedestrianisation schemes, for example, has often proved to be a very powerful strategy to combat the attraction of brand new retail developments. If coupled with conservation and environmental enhancement schemes, such projects often result in high quality specialist retailing districts (Ashworth and Tunbridge, 1990).

The theme of this chapter has been the persistent movement of centres of economic activity away from the centre of the city. Whilst this is true of traditional economic activities considered so far – manufacturing, offices and retailing – it is also true of newer services, such as specialist research and development. European cities have seen the growth of the science park, devoted to research and development and its commercial application. Usually, but not always, such parks have been associated with major research centres, and especially with universities and have been built on the edge of existing cities. Cambridge Science Park is one example and the largest in the UK, whilst Chilworth Research Centre on the edge of Southampton is another. In both cases, research from the respective university has given rise to commercial development of innovation.

In France, the somewhat looser concept of the *technopole* has been defined to include major research and development facilities in a number of cities. Montpellier, Nice, Rennes, Nancy/Metz, Caen and Grenoble have all developed *technopoles*, which in some cases include the university establishments themselves. In the case of Montpellier, for example, the *technopole* incorporates a number of establishments of higher education, with a total of 45,000 students and 6000 people employed in research and development establishments. A series of separate *poles* are designated, including a *pole informatique*, with an IBM facility employing 3000 people, and a *pole euromedicine*, with major research laboratories focused on biotechnology and pharmaceutical research (Brunet *et al.*, 1988).

Economic activities have indeed changed markedly in European cities of the post-war period. Few cities have adjusted to the new demands with complete success. The shift of location of economic activities has put additional pressure on the transport networks and has disrupted traditional relationships between place of work, home and shopping. Yet it is difficult to point to any one 'correct' solution, except that cities need to be adaptable and ready for change. It is by no means certain, however, that complete *laissez-faire* is the answer. The experience of Brussels in the late 1960s and early 1970s should not be repeated and neither should the results of the undirected market economy be permitted to take effect without heed of the consequences.

References

Ashworth, G. J. and Tunbridge, J. E. (1990) *The tourist-historic city*, Belhaven, London.

Ashworth, G. J. and Voogd, H. (1990) *Selling the city: market approaches in public sector urban planning*, Belhaven, London.

Bennison, D. and Davies, R. L. (1977) *The Local Effects of City Centre Shopping Schemes: A Case Study*. Paper presented to PTRC Summer Annual Meeting, University of Warwick.

Burt, S. L. and Dawson, J. A. (1988) *The evolution of European retailing*, ICL, Slough.

Burt, S. L. and Dawson, J. A. (1990) 'From small shop to hypermarket; the dynamics of retailing', in Pinder, D. (ed.) *Western Europe: challenge and change*, Belhaven, London.

Brunet, R. *et al.*, (1988), *Montpellier Europole*, Ed. GIP.

Daniels, P. W. (1975) *Office Location, An Urban and Regional Study*, Bell, London.

DATAR (1989), *Les Villes Européens*, la Documentation Française, Paris.

Dennis, R. (1978) The decline of manufacturing in greater London 1966-1974, *Urban Studies* 15(1).

Department of the Environment (1972) *Development Policy Control Note 11*, HMSO, London.

Department of the Environment (1977) *Development Policy Control Note 13*, HMSO, London.

Hardman Report (1973) *The Dispersal of Government Work from London*, Cmnd 5322, HMSO, London.

Hoover, E. M. and Vernon, R. (1962) *Anatomy of a Metropolis*, Harvard University Press, Cambridge, Mass.

Illeris, S. (1989), *Services and Regions in Europe,* Aveburg.

Keating, M. (1988) *The City that Refused to Die*, Aberdeen University Press, Aberdeen.

Lloyd, P. and Mason, C. (1978) Manufacturing Industry in the Inner City: a case study of Greater Manchester. *Trans. Institute of British Geographers* New Series 3(1), pp. 66-90.

Mason, C. M. and Harrison, R. T. (1990) Small firms: phoenix from the ashes, in Pinder, D. (ed.) *Western Europe: challenge and change*, Belhaven, London.

Massey, D. and Meagan, P. (1978) Industrial Restructuring versus the Cities, *Urban Studies*, **15**, pp. 273–88.

MPC and Associates Ltd (1974) *Retailing in Europe*, MPC, Worcester.

van Hecke, E. (1977) *De Lokalisatie von de Tertiare Sector in de Brusselse Agglomeration*, Universiteit Leuven, Leuven.

Watts, H. D. (1990) Manufacturing trends, corporate restructuring and spatial change, in Pinder, D. (ed.) *Western Europe: challenge and change*, Belhaven, London.

Wood, D. (1976) *The Eastleigh Carrefour, a Hypermarket and its Effects*, Department of the Environment Research Report, 16, HMSO, London.

CHAPTER 5

Urban Transport Planning (Planning for Movement)

Some of the earliest recorded European planning legislation was concerned with solving the urban transport problem, when in the first century AD wheeled vehicles were prohibited within the city of Rome during daylight hours. The nature of this problem and the difficulties inherent in most solutions to it, stem, at their simplest, from a dilemma that is as old as urban civilisation itself. Cities are established and flourish because agglomeration offers a reduction in the costs of transport, while the very concentration of people and activities into a small physical space results in congestion, which leads to competition for space, conflict and increased transport costs. If the reaction of urban governments is inertia, then the increasing congestion inhibits the accessibility that is the city's *raison d'être*, whereas if they act to regulate the use of transport they raise its costs with potentially the same effect.

It is possible to view the task of the urban transport planner as simply to reshape the morphology of the city to accommodate the changing demands of transport made upon it, with the overriding goal of maximising the efficiency of movement. Not so long ago in the major conurbations of Western Europe the context within which such planners worked could be described with little exaggeration as a 'large, rapidly changing urban area, a central concern for rising car ownership; an expectation of large investment funds for new and modified infrastructure; a belief that society should cater for all the demands for mobility which was anticipated by a future date' (Moseley, 1979). Although it is tempting to regard such a limited view as historic in that it would rarely be stated without qualification by responsible governments in the 1990s at least in North-Western Europe, it remains one of the competing approaches to planning urban transport, especially in many of the southern European metropolises, such as Barcelona, Athens and, in an extreme form, Istanbul, which experienced in the 1980s a growth in the use of the motor car similar to that undergone in Western Europe a decade earlier.

A number of basic difficulties were always inherent in such an approach. First, future demands were predicted on the basis of current transport provision for a limited group of transport users, and the supply and demand for transport facilities are in practice related, not independent, variables. Consequently it has become apparent from projects as distant and diverse in the cities they serve as London's M25 ring road to Istanbul's Bosphorus bridges that demand is supply-led. Second, the function of transport is not only to satisfy current and predicted demands for movement. It is a land-use in its own right, occupying as much as 30 per cent of the land area of cities and thereby being a major determinant of the physical structure of cities. It acts therefore as both a creator and a reflection of the values, priorities and choices of lifestyle of citizens. Quite simply, 'in defining the functions of transport, one is drawn back to the question of what sort of city one wants' (Thomson, 1977), and to the consequent political questions of who is the 'one' that is to benefit, pay and decide. The Economic Commission for Europe (1973) identified from among many, three families of models that illustrated the differing aims of transport planning in European cities which have continued to compete for the following decades. Apart from the 'economic priority models', where the purpose of transport is to maximise the efficiency, in general or particular, of the city's economic functions, there are 'welfare priority models', which stress the accessibility of citizens to the services of the city and the distribution of the benefits of that accessibility to groups in society; and 'environmental priority models', where undesirable impacts on the quality of the urban environment, whether in terms of visual cityscapes, or the quality of air and sound, are minimised.

Clearly once urban transport planning is considered in these terms then it confronts the more general planning dilemmas which underlie many chapters throughout this book, such as that of welfare against development, or the conservation of the built or natural environment against functional change. Consequently a study of transport planning becomes subsumed in a study of urban planning goals, instruments and effects in general. It will therefore appear in different guises in all chapters leaving this chapter only to sketch the main continental dimensions and attempt an international comparative approach rather than review the copious literature on particular national or city transport policies.

The potential value of an international comparative approach, where common solutions to common problems can be designed, has long been apparent. An international perspective is intrinsic to many aspects of transport, which involve movements across boundaries and use easily exchangeable technologies and methods of operation, and has been strongly reinforced by the realisation of the European Community that a harmonisation of transport policies and practices is a prerequisite for the free movement of goods and people. The Economic Commission for Europe called in 1974 for a definition of 'those problems which are of equal concern to all cities and states which require for their solution the closest kind of cooperation, joint research and the exchange of data and experience'. Such international forums have aided the diffusion of ideas on this topic across international boundaries more rapidly than has been the case for many of the urban functions discussed in this section. Distinctive national policies are less easily discerned. In continental terms, however, the distinctive history of the European city, the importance of its relict building forms, and the attitudes of Europeans towards accessibility, renders the experience of North American or Australian cities less relevant. If the Western European city in particular has distinct planning goals, as is a central assertion of this book, then transport planning, as a major instrument of attaining these goals, will be distinctly West European.

The variety of proposed solutions makes a rigid classification of examples difficult, as there are few cities that have not at some time or another considered some application of most of the planning and management ideas discussed, and equally few cities that can stand as archetypes for particular policies. Throughout Western Europe, a common appreciation of the nature of the problem and a common determination that a solution must be found have not led to any consensus on either the goals to be achieved or the methods of attaining them. There is, however, a common sense of urgency that pervades the plans of city after city regardless of size or national identity. Few would be so dramatic as the Parisian planners in 1972, when they claimed to be faced 'with two alternatives; either the traffic problem will be resolved in the future, or the economic, social and cultural life of the region, already becoming day by day more sluggish will be completely paralysed (Préfecture de la Région Parisienne, 1972). Neither resolution nor paralysis has occurred either here or in other cities, but both the difficulties of finding solutions and the high costs of not finding them remain.

The demand for transport in cities

The typical journey taken within the urban region is short, generally around 1.5 km. To this characteristic must be added two other peculiarities: the wide spatial variety of journey origins and destinations that produces a complex and dispersed network of flows throughout the city, and a distinctive and extreme peaking of demand over time which produces regular predictable daily and weekly rhythms.

The main exceptions to the dispersed pattern of demand are the through traffic that results from the traditional 'gateway' function of cities as route centres and transhipment points, and the long and medium distance commuter flows that typically result in high density flows along narrow, usually radial, routes. In London, a typical 'strong centred city' (Thomson, 1977) possessing a high concentration of economic activities in a limited central area, around 1.5 million people are transported from the suburbs into a few square kilometres, within 1.5 hours each working day. The concentration of attention on these remarkably concentrated and regular commuter flows should not conceal the even more numerous, but less dramatically concentrated or easily predictable, flows that criss-cross the city. Goods deliveries for example account for around a third of all vehicles on city streets. The journey to work rarely accounts for more than 35 per cent of transport demands, and many of these trips are both short and not radial to the city centre, encouraged by changes in both the location of economic activities (Chapter 4), and urban residential preferences (Chapter 6). Similarly, journeys to shopping, education, recreation and entertainment are likely to have quite different attributes. In addition urban transport, when defined as a planning problem, has frequently been influenced by the experience of the major metropolitan urban regions whose characteristic transport demands and capacity to accommodate them may be quite different from those in smaller cities.

The requirement is therefore for a transport mode capable of handling economically a large number of short journeys over a wide variety of routes, with a capacity to accommodate a short but intense peaking of demand. Neither speed nor comfort are high priority attributes but use of limited and expensive space and minimising the negative external effects on other urban users are. No single transport system has been devised that meets these criteria. Collective transit systems satisfy the requirements for high capacity and the conservation of space but for the majority of intra-urban journeys the flexibility of routing and the instant availability of individual transport, especially the private car, taxi, van or lorry have proved ideal.

This response of urban travellers to their movement needs has created the problem which the solutions described in the rest of this chapter are designed to solve. The motor vehicle, especially when used by an owner-driver, has a number of fundamental disadvantages as a means of transport in cities. Above all it is profligate in its use of space, precisely where such space is scarcest, both when in use and also for the 98 per cent of its existence when it is stationary. Estimates of the number of cars parked for periods of up to eight hours on working days in the central areas of cities such as Paris or London are around a million, and whether on or off-street, legal or not, the result is an enormous consumption of the scarcest of urban resources, space. Most European cities devote between 10 and 20 per cent of their total land area to accommodating cars. In Cologne, for example, road and parking space took up twice the area of the city's parks and almost half as much land as was devoted to buildings of all types. This disadvantage has become more apparent with the post-war growth of cities in extent, the substantial increase in the real disposable incomes of citizens and their increasing capacity to realise their desires for both more living space in lower density residential areas simultaneously with more personal mobility.

The benefits of the motor car to the owner were more obvious, tangible and measurable than were the costs incurred by the city as a whole. Awareness of the costs of congestion, accidents, visual intrusion and damage to the environment, and social inequities was delayed by distortions in the consumer choice between collective and private transport modes. The low direct marginal costs of car use eroded the competitive position of alternative public transport, thus creating the need for social accounting which in turn brought urban transport onto the political arena and endowed the policy alternatives with particularly strong ideological overtones.

The range of possible solutions

Few towns, and none of the major European conurbations, have committed themselves, except in utopian planning visions (as described in Chapter 2) to a single transport system. It is therefore not realistic to assign labels such as 'motor car cities', 'bus cities' or even 'bicycle/pedestrian cities'. It is possible to arrange European cities along a spectrum which ranges from individual to collective transport according to the dominant mix of the selected policies. Almost all cities have endeavoured to maintain, and usually enhance, some public transport while also making some effort to accommodate the motor car.

Such a spectrum of strategies can be related to two others, namely a concentration-dispersion of economic activities and of building density. The important point is that concentration and high densities can be related to collective transport, while dispersal of activities and low building densities can be most effectively served by individual transport, especially the motor car. Thomson (1977) calibrated such spectra, suggesting that 120,000 city centre jobs was a maximum capable of being maintained in a motorised 'weak centred' city, while at the other extreme high density transit systems can support well over a million such jobs in 'strong centres' such as London or Paris. Such a comparison of morphological and transport archetypes begs the question whether the transport policies were devised as the only possible response to the existing urban structure, or whether favoured transport strategies are being used as an instrument for the creation of a desired urban form, such as the Dutch declared preference for 'the compact city' (Rijksplanologische Dienst, 1988). Certainly 'new town' planning (Chapter 12) can provide some of the best examples in Europe of the deliberate adoption of both 'weak centred' (as in Milton Keynes) and 'strong centred' (as in Cumbernauld or Zoetermeer) models. In Copenhagen, on the other hand, residential suburbanisation and the decentralisation of employment which has left only about 250,000 jobs in the inner city, had occurred without a deliberate central design, and the task of the transport planners has been to attempt to serve the transport needs of what they regard as an undesirable functional and morphological structure. At the other morphological extreme cities such as Madrid, Amsterdam or Brussels have responded by developing high density transport systems to serve a structure that has evolved while also endeavouring to decentralise functions which in turn will create new transport requirements.

'A large city is mechanically different from a small city, just as a horse is mechanically different from a dog' (Thomson, 1977). The size of the city, in both population and physical extent, is an important determinant of the chosen point on the spectrum of solutions. Size not only increases the intensity of the problem but also alters its nature. Small towns will satisfy a large portion of their transport demand by walking and cycling, and the use of the motor car poses fewer problems. As size increases beyond a threshold, which may range from fewer than 50,000 to more than 150,000 inhabitants, the flows become large enough to support economically viable bus services and even possibly

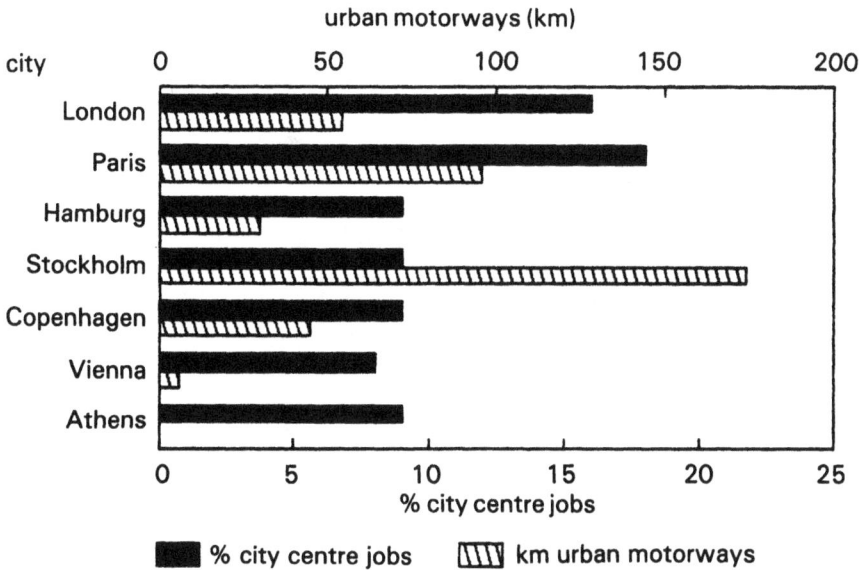

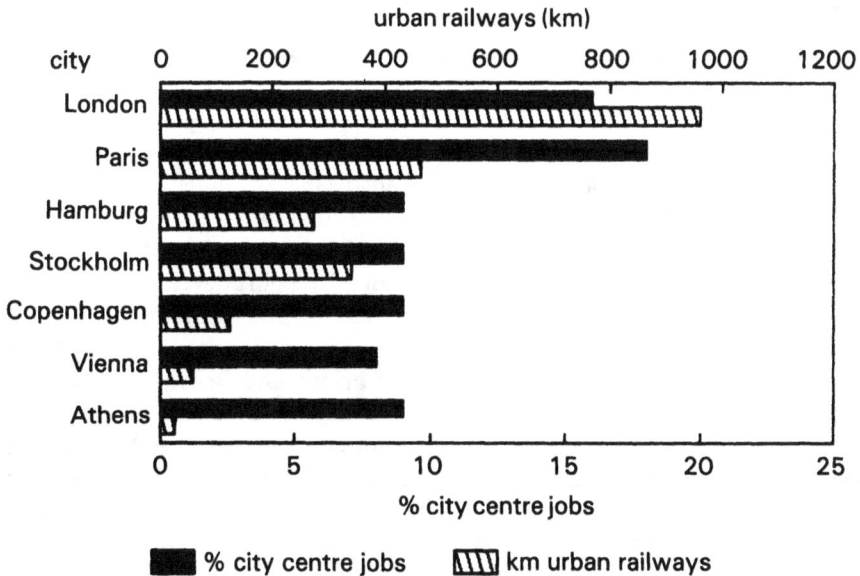

Figure 5.1 Urban motorways and urban railways in selected cities

justify policies of restraint on car use. Surface and underground railways generally need populations of well over a million to provide flows of sufficient density to justify their construction costs, although this threshold can be much lower for tram (as in cities such as Utrecht) or 'mini-metro' technologies (as in a number of German cities).

Bus/tram passengers (million)

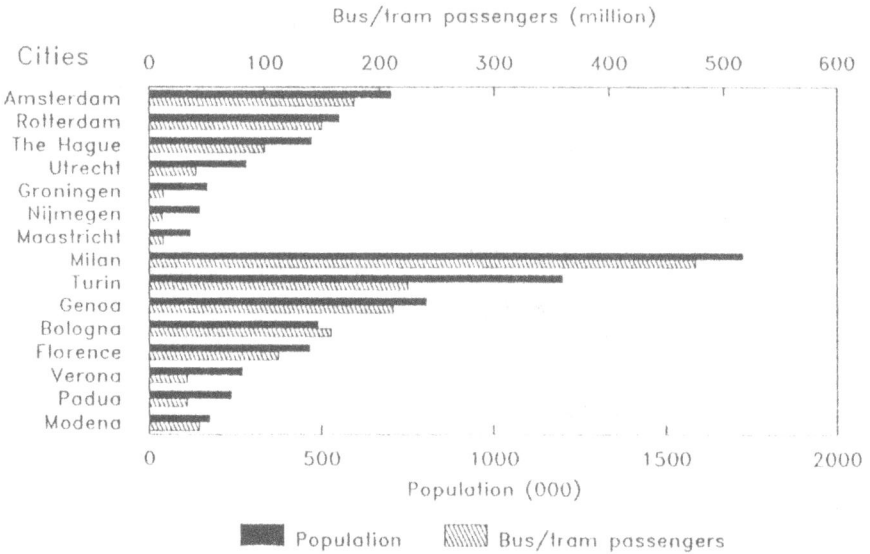

Figure 5.2   Public transport use in selected Dutch and Italian cities

The physical setting will exacerbate particular problems and favour particular solutions. Istanbul's dependence on the frequent high density ferries over the Bosporus between its European and Asian parts is perhaps the most dramatic example of such a physically divided city. Even a much smaller city such as Rotterdam would have difficulty justifying an underground railway system on size alone. However, the first two bridges built across the Nieuwe Maas in 1878 were adequate until the 1930s and were supplemented by the first trans-Maas tunnel in 1942. Further expansion of the port developments on the south bank and the southern residential suburbs, together with increasing road use for both freight and passengers, caused traffic through the tunnel to increase threefold in the 1950s. The new Brienoord bridge and Benelux tunnel were constructed but were used to capacity almost immediately. Demand predictions of ever increasing magnitude in the next decade could be met by either at least a further six bridges or tunnels or, as was preferred and implemented, by building a short metro line under the river. This Rotterdam river crossing problem finds an echo in many other cities, such as in Stockholm where surface railways provide most of the inter-island links. Coastal and lakeside sites, such as Geneva and Zurich, and 'offshore' island sites such as Portsmouth, similarly distort the simple relationship between population size and traffic intensity. The tendency of political boundaries to be drawn along rivers is a further complication that may hinder the development of an effective policy for handling the cross-river flows, as for example in Mainz where the river constitutes the *Land* boundary. The extreme case is perhaps found in Venice where the single causeway between the mainland and historic lagoon city could not begin to handle the traffic that would normally be generated by a city centre. Instead its inadequacy provides the opportunity for the complete exclusion of the motorcar and its replacement by the public transport of gondola and *vaporetti*.

Transport policies have to be imposed upon an existing city with a given morphology and transport infrastructure. The presence of a valued historical legacy has often, in practice, favoured the adoption of policies of motor car restraint (see Chapter 7) and the encouragement of forms of transport considered to be less environmentally damaging. Similarly, the survival of transport installations, such as city centre railway stations, engineering works and rights of way, may render viable transport systems whose construction could not now be justified. Again in practice it has often been public transport systems, especially railways and trams, that have been favoured by such inertia.

The social character of a town, in terms of age, income and class should be a determinant of the mix of transport policies adopted, and even the viability of different forms of transport in particular districts of the city, but this is difficult to demonstrate empirically. A higher than average reliance on public bus services and profitability of bus undertakings can be found for example in Bournemouth and Bradford, and this can be related to the elderly age structure of the former and income structure of the latter; but quite contradictory examples can be as easily produced from various countries. Logically these social variables should favour the adoption of public transport solutions but the profitability of undertakings clearly depends on many other factors.

Attitudes towards transport, and particularly the use of transport planning to attain other desired planning objectives, clearly depend on the overall ideology of the city's administrators. Transport policies evolve in practice as an oscillating balance between economic and welfare motives, between interventionist and market approaches and between the various lobbies that represent individual transport systems which influence the decision makers at local, national and increasingly European Community levels.

At a local level, for example, it is to be expected that a city with a longstanding left-wing government, as in Bologna, would operate rigorous car restraint policies and experiment with partially free public transport, while a right-wing city government, as in Milan, has more often favoured new road construction. In most cities elements of both change and stability are present. There are examples to be found where change in local political control is either a cause or an effect of change in urban transport policy. Public opposition to urban motorway proposals, as in London during the early 1970s, reflected a more general struggle in the town halls in many European countries over the last two decades. The struggle was largely between the transport specialists implementing a received conventional wisdom from the post-war period, and the representatives of an articulate public opinion which had shifted radically against attempts to uncritically accommodate growing motorcar use. The timing of this dissent varied with the national planning laws and customs, with differences in the pace of change in transport use, and also, as in other aspects of urban planning, with what can only be termed national character. The result is that, broadly speaking, these changes were evident a decade earlier in Scandinavian and Dutch towns than in the cities of Britain, Belgium, France or Italy, with the towns of Spain, Greece, and Portugal trailing a further decade behind.

Finally, the choice of solution cannot be determined by reference to the needs of the individual city alone. Regional, national and even international demands for movement impose further constraints on the range of choices available. Cities act as gateways for sea, air or land movement. A typical illustration of this would be the long debated dilemma of Geneva, where growing international traffic on the E2 and E15 main roads could only be accommodated by costly new construction through the southern suburbs or under the lake, in a city with the highest population density of all Swiss cantons (see République et Canton de Genève, 1977).

It is apparent that many of these constraints on the choice of transport policy were more effective in the cities of Western Europe than in North America. This can be related to both the valuation placed upon existing urban structures (Chapter 7) and a consequent reluctance to accept too rapid change, but also to the wider sets of values reflected in choice of residential amenity and lifestyle (Chapter 6). In addition there is to a varying degree in the countries of West Europe a longer standing welfare tradition and public acceptance of restraint on motor car use that has developed in response to a quite different experience of the availability of urban space. Thus in this aspect of planning, as in those regulating commercial activities, retailing and residential choice, divergence from the North American pattern and the emergence of a distinctive West European set of strategies is apparent.

Solutions: planning for private transport

(i) Policies of accommodation

The steady increase in the popularity of the motor car for intra-urban travel in the last 20 years has been well documented and explained in all the countries of West Europe. It is not really surprising that the initial response of planners to this growth in public demand was to endeavour to accommodate it by providing the road and parking facilities that would enable its advantages to be realised. Not only was the growth gradual but the benefits to the user were apparent from the first, while the costs to a wider society only became obvious when the number of vehicles in use had increased beyond a certain level. By the time that threshold had been reached, a substantial portion of citizens had invested in cars and adjusted their patterns of life appropriately, while many others aspired to do so. 'The goodlife of the early 1960s consisted of ceaseless mobility in search of an ever widening range of choice in jobs, education, entertainment and social life' (Hall, 1980, p. 86). To deny citizens this right then carried a considerable political risk in democratic Europe.

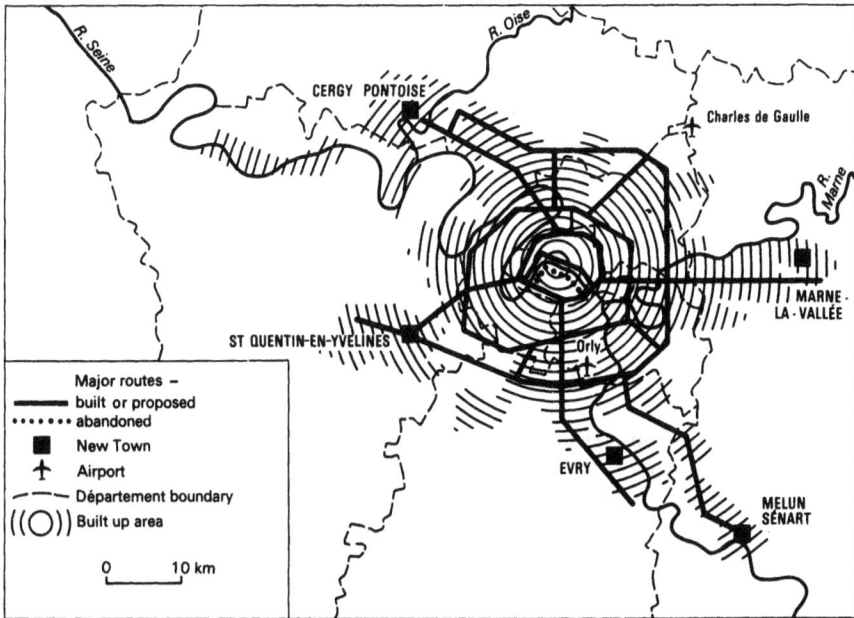

Figure 5.3 Major road planning proposals for the Paris region (reproduced by permission of I.A.U.R.I.F.)

By the end of the 1960s cities such as Munich, Birmingham and Turin were taking the lead in demonstrating their modernity through the construction of complete systems of ringways and radial routes of motorway standards, combining local authority initiatives with national government financial assistance. The influential *Traffic in towns* report (Ministry of Transport, 1963) was interpreted by some as a warning of the costs of adapting old cities to new transport demands but more immediately many practising planners saw it rather as propounding the development of 'transport corridors' which would remove traffic from interstitial neighbourhoods and thus justify the building of high capacity urban motorways. Buchanan himself at a GLC enquiry in 1971 commented that 'the failure to provide an adequate road network was producing devastating environmental results besides congestion'.

Parisian experience has been regarded initially as an archetype to be imitated, and less than a decade later as a warning to be heeded. Paris, in common with many other cities, responded to the growth in car ownership by an expensive attempt to accommodate it. The PADOG plan of 1962 and the *Schéma Directeur* of 1965 produced proposals for some 800 km of new urban expressway and two and a half ringroads (namely the Boulevard Périphérique at about 5 km from the city centre, the ARISO at about 10 km, and an outer Rocade de Banlieu). The flow of traffic through the inner city was to be maintained by instituting the most extensive one-way circulation system in

Europe, and a new west to east expressway through the centre, including tunnelling underneath the Place de la Concorde and the Louvre. A parallel east to west expressway along the *quais* of the left bank was also proposed. In addition, the eight major radial routes that linked Paris to the provincial cities were to be completed to motorway standards and linked to the ringway systems (Figure 5.3). Although there was to be a simultaneous substantial, but not equal, investment in new public transport facilities, there was no formal policy of restraint on car use, and even proposals to restrict on-street parking by metering were rejected on four occasions during the 1960s.

A distinct shift in emphasis occurred in the 1970s, reflecting changes in public opinion, in the conventional wisdom of planners (Chapter 11), and political control. The Boulevard Périphérique has proved inadequate and dangerous; the inner city expressway proposals were amended and in some cases abandoned; and even parking controls were introduced and enforced in the second half of the decade.

The other large scale archetype, London, had a tradition of high public transport use, especially for the journey to work, together with an interwar and immediate postwar history of very little major road construction. Feeling within the Greater London Council by the beginning of the 1970s was that the city had fallen behind its continental counterparts and a major road building programme was long overdue. The result was the London Ringway plan (Figure 5.4) which proposed no fewer than four circular routes totalling almost 600 km around the city region. Neither the plans, nor the roads themselves, were completely original, since successive London plans, including Abercrombie's 1945 plan, had included ringway proposals, and the incompletely implemented results of previous plans, such as the North Circular Road of the 1930s, were incorporated. The scale of the enterprise – which more than doubled the available road space in terms of vehicle kilometres – and the determination of the council to execute it were, however, original and provided a radical solution to the problem of filtering traffic from the nine radial motorways built since 1959 into the city.

Opposition was slow to develop and focused on escalating construction costs, the loss of existing housing and local environmental damage. Local objections successfully pruned the proposed routes and downgraded the road standards. It was, however, a combination of local protestors directly threatened by the plans and a more general mobilisation of public opinion in favour of public transport and a diffuse but increasingly popular conservation ethic (Chapter 7) that brought the issue into the party political arena and aided a change in political control. The London situation thus paralleled that in Paris, and even a later reversion of political control could not resurrect either the plans or the spirit in which they were conceived. Ever since Abercrombie, transport planning was in practice conceived as a process of demand modelling and prediction; it flourished where and when local planning was at its weakest and by the 1980s it became clear to many that, however accurate the answers given by transport modelling, it was just not posing the questions most relevant to

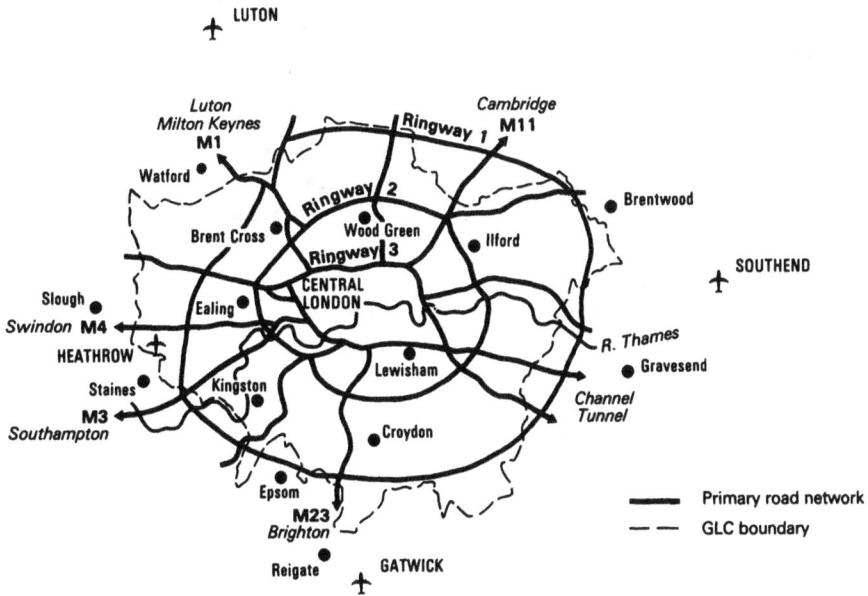

Figure 5.4 London's Ringway proposals

cities (Black, 1981). However, the defeat of uncritical accommodation policies and their consignment to the 'great planning disasters of our time' (Hall, 1980), was not in either city a solution in itself to the initial problem. This problem became more, rather than less, pressing in the course of the next decade while the political room to manoeuvre had been narrowed by the experience of the 1970s and other approaches had to be sought.

The accommodating policies of London and Paris, and the reactions to them, were reflected in differing degrees in most European cities, with the timing dependent upon both a diffusion spatially from North-Western Europe to Central and Southern Europe and down the hierarchy of population size. Athens was still relying on road building as a solution to its growing motorisation in the beginning of the 1980s and Istanbul completed a double ringway, quayside downtown throughway, and two new Bosporus bridges, at the end of the 1980s. Similarly many medium sized cities were still completing ambitious ringroad and urban motorway schemes in the course of the 1980s. However by 1990 there was a near universal consensus that 'no amount of ingenious juggling by engineers and architects is capable of adapting long established centres to accommodate large flows of traffic and vast highways, except over a period and at a cost that makes it an irrelevancy' (Bendixson, 1974).

(ii) Policies of restraint

Many methods of restraining motor vehicle use and making a higher proportion of the total costs of their use payable by the user have been canvassed and been the subject of experiment in cities through Europe. Even a policy of non-action, so that the resulting congestion discourages continued use, has, by default or design, been attempted (Roberts, 1988). The most ubiquitous policy, however, has been to physically prevent access for categories of vehicles to limited areas at defined times, and thus the dedication of these areas to pedestrian use. London Street, Norwich, completed in 1962 is generally acknowledged as the pioneer of pedestrianisation, closely followed by a host of largely medium-sized, often historic cities in West Germany, Scandinavia, the Low Countries and Britain, By 1974 the OECD could comment on a wave of 'creeping pedestrianisation', which increasingly included large cities and those with few pretensions to an historic fabric; Rotterdam's Lijnbaan was generally taken as the archetype for the restructuring of general comparative shopping streets in this way in the centres of major cities, such as Copenhagen's Stroget and Amsterdam's Kalverstraat. By the end of the decade pedestrianisation was effectively universal. It also took a wide variety of forms (see for example Kuhn's (1979) exhaustive cataloguing of the types and structure of pedestrian areas in West German cities). Two questions now arose: specifically, what are the spatial limits to pedestrianisation and concomitant vehicle exclusion policies; and more generally what is the relationship between such traffic circulation policies and land use change? Answers are being sought to these questions empirically and through often politically tendentious experiment.

As German cities had been in the forefront of pedestrianisation it is not surprising that it was there also that its limits were first tested. West Berlin (with 120 ha of inner city pedestrianisation) and Mainz (with a staggering 610 ha) began to be regarded as a threshold beyond which there would be an unacceptable loss of accessibility and thus commercial viability. Within these areas high amenity specialist shopping, often in conjunction with conservation policy (Chapter 7) shaped tourist-historic districts (Ashworth and Tunbridge, 1990); Bremen's Bottscherstrasse was an early, much imitated exemplar.

Outside the central areas two new types of traffic restraint policy were evolved. The *Woonerf* ('Residential Threshold') was originally a Dutch concept applied to largely housing areas in the inner residential ring within which motorised traffic was filtered out and slowed down but not completely excluded. The idea, which is cheap and arouses little controversy, has found wide acceptance in neighbouring countries. The second idea is the German *Verkehrsberuhigung* ('traffic calming'), which again can be widely applied to the older districts around the central area, and again involves slowing and selecting traffic rather than excluding essential users (Hass-Klau, 1988 and 1989).

Pedestrianisation is however only one facet of such policies which generally include some version of a collar around the central area. Examples of such rigorous prohibitions or delays on categories of vehicles entering central areas were notable *causes celèbres* during the 1970s, as in Norwich, Nottingham,

Leicester or Southampton, but they became so commonplace in the 1980s as to arouse little comment.

Restraints on vehicle parking can be as effective as those on their movement and simultaneously release road space for circulation. Again, differences within Europe have been largely only in timing, often explainable through the climate of local opinion. Cities such as London introduced metering as early as 1958 and more than halving the amount of on-street parking in the next 15 years (Netter, 1977), with others such as Vienna or Paris being a decade behind. Parking policies however can be much more than simple restraints on overall use; through pricing strategies, integration with public transport and park-and-ride schemes, and even signposting policies, they can be subtle instruments differentiating between types of users in pursuit of more general urban management strategies. Such a use depends on control not only of street parking but off-street, often privately managed, car parks, and also private parking spaces and garages. Few Europen cities would yet go as far as those of the Netherlands, Denmark or Sweden in using planning consent as a means of restricting parking spaces in public and private office developments.

(iii) Policies for alternative private transport

The dominance of the owner driven motor car as a method of individual transport has led planners to underestimate the importance of other forms of private transport and, with the exception of a handful of cities, neglect their potential. Travel by foot, for example, is an essential part of every journey and the main mode of transport for as many as one-sixth of journeys to work in a city such as London, and much higher in smaller towns. Pedestrianisation as discussed above can be more than a means of traffic restraint and an instrument in the revitalisation of downtown areas, it is also a transport facility. Networks of segregated pedestrian routes have been created in the central areas of many West German cities such as Bremen, Essen, and Munich. Although the idea of weatherproofing such routes is well established and has long existed in for example Exhibition Road in London's South Kensington or Kropke, Hannover, complete 'underground cities' linking transport terminals, shopping malls and entertainment facilities, as in Montreal, have as yet to be constructed on any comparable scale in Europe. Even further, the introduction of pedestrian conveyors is generally currently limited to transport interchanges where the density of traffic justifies the investment.

The bicycle similarly provides cheap, pollution free, space-saving transport at a speed not far short of the average for inner city journeys. Its vulnerability to other traffic requires separate routing systems which are more easily inserted into new towns such as Stevenage or Zoetermeer, or in major new housing development such as Stockholm's Vasteras, than in existing towns. Nevertheless experiments, some shortlived, in priority systems and marked if not segregated track, have been attempted in cities as diverse as Cambridge,

Peterborough and Portsmouth. However the Netherlands remains the only West European country where the bicycle, and its motorised equivalents, is a major means of transport for all social groups in all types of city; in the rest of Europe it is an exceptional phenomenon of particular special towns. This has been achieved, and is maintained in The Netherlands, by high levels of investment in segregated track, circulation priorities over other traffic and simultaneous restraint policies on motor car use.

Hired cars and taxis combine some of the flexibility of the private car with many of the space saving attributes of public transport; a small taxi fleet, such as London's 10,000 cabs can handle a large volume of traffic, in this case about a third of all West End passenger traffic (Bendixson, 1974). Licensing anomalies often restrict their availability, most notably in Paris, in addition planners have been slow in many cities to recognise taxis as a 'public automobile service', and cost excludes some users. Nevertheless increasingly the privileged use of road space alongside buses, exemption from restraint schemes and a host of varied experiments in communal taxis, dial-a-ride schemes as in Amsterdam, Montpellier or Goteborg, or the Dutch 'trein-taxi' combination, are all an indication of growing recognition.

Solutions: planning for public transport

Planning solutions that rely on public transport playing a major role were motivated by four main trends.

(1) It has become evident in cities throughout Europe that public transport undertakings were in increasing financial difficulty. A cycle of declining patronage and contraction of services was identified in the 1960s and intensified through to the 1980s. Shrinking revenues of about four per cent annually in most West European conurbations provided little room for new investment. By the beginning of the 1980s neither operating costs nor capital development could be met by the undertakings themselves. This presented cities with the stark alternatives of either large scale subsidy from local or central government sources, at a time of ideological resistance to government expenditure in a number of North-West European countries, or extinction. By the middle of the 1980s the level of public subsidy on operating costs alone varied from around 25 per cent in London, 50 per cent in Frankfurt to 72 per cent in Rotterdam (Adam, 1986).

(2) The reaction to the alternative policies of unrestrained motor car use, described above, combined with the growing conservation consciousness to produce a powerful public transport lobby.

(3) Unrelated to the above, public transport use in many West European cities, and especially the major conurbations, began to rise in the course of the 1980s, reversing a generation-long decline, and rendering obvious and politically embarrassing the deficiencies in the system in terms of capacity and quality resulting from an equally long neglect.

(4) Finally, and least effective, was an equity argument that appealed on behalf of those 'captive to public transport' by virtue of age, infirmity or poverty, and linked public transport investment to more general inner urban revitalisation.

As a result there occurred a renaissance of interest in public transport. New systems were constructed and many existing ones often received their first new investment since they had been built in the nineteenth century. New organisational structures and authorities were created which often blurred the distinction between private and public with new relationships to the passenger market.

The extent and nature of this reaction varied widely across the continent and even within individual cities different forms of collective transport were supported often in a confusion of complementarity and competition. A traditional distinction is between rail, with its exclusive rights of way, and more flexible, if lower capacity, bus systems. The cities that make a proportionately high use of urban railways, especially for long distance radial commuter journeys, such as London, Liege, Liverpool or Hannover, use in almost every case an historical endowment of routes that would be prohibitively expensive to construct today. Most transport policies have confined themselves to encouraging the use of existing facilities and extending their scope, rather than building new lines. The exceptions tend either to be a tidying up of a Victorian heritage with new links, as in Liverpool (Halsall, 1978) or, more spectacularly, new lines often reflecting new technologies and substantial government finance. The new town of Zoetermeer near The Hague (Chapter 12), for example, was designed to include a purpose-built rail system linking the residential districts with frequent services (known as 'the sprinter') into the city. The quite different situation in the existing large conurbations was tackled in Paris by the Réseau Express Regional (RER) which provides the cross-city region routes linking the outer suburbs and airports to the inner city metro lines.

Surface railways were built to carry dense regular flows over the relatively long suburb–city line haul, while metros or subways were designed mainly as inner city distributors, although these distinct functions can become merged in the larger systems such as London and, more recently, Paris. Given the high financial and disruption costs of metro construction, and the very high volume of traffic needed to justify their use, it is understandable that their construction has come to symbolise the revival of interest in urban public transport. After more than 50 years of little or no investment, no fewer than eleven West European cities have built new underground rail systems in the last two decades. Of the fifteen pre-existing systems many have been extended, such as London's new Victoria and Jubilee lines and the extensions to lines 3, 8, and 12 in Paris. Many antiquated systems have been refurbished and a number of cities including Hannover, Cologne, Essen, Antwerp and Brussels have built underground sections to their existing tramways so that the distinction between tram, 'pre-metro' and metro becomes increasingly blurred. Even more remarkable has been the building of new metros in cities previously considered too small to support such high capacity transport. The case of Rotterdam is explicable in terms of the Maas crossing, but developments in Amsterdam, Marseilles, Seville, and Lyons owe more to a political commitment to public transport of which the metro is a symbol than accurate appraisals of financial viability.

Trams may either sustain the speed and capacity advantages of exclusive track like railways, or the convenience and flexibility in meeting demand through circulation on the public road. There are a wide, and in recent years growing, variety of light street railway systems running on, above and occasionally under, the surface. Some, such as the Copenhagen S-bane or the S-bahnen in Berlin or Hamburg, are little different from suburban railways serving longer distance radial flows; others such as the 'pre-metros' of Brussels, Oslo, Vienna, Utrecht, Stuttgart or even London's Docklands Light Railway, are high capacity short stage distributors within the central area; while some cities, especially in the Netherlands, Belgium and Germany, have retained the more traditional *street* railways, or even closely related trolley-bus systems, as in Arnhem, which operate with much of the flexibility of high capacity bus services.

Buses have carried the majority of public transport passengers in European cities for two generations. In small and medium sized towns they are commonly the only public transport available, while in the major conurbations they act as collectors and distributors for rapid transit and handle most short journeys. Despite their operational flexibility and absence of costly fixed investment in track a decline in the number of passengers carried has been evident in almost every city in West Europe over the last 20 years, presenting urban managements with a choice between ever larger public subsidies or sharp reductions in the quality of services; or, as in many Northern Italian cities (Wyles, 1988), both simultaneously.

The resulting strategies are broadly of two types, the welfare approach and the market approach, although in any given city elements of both are frequently simultaneously evident. Almost all local and national governments subsidise their bus undertakings through tax concessions, investment grants or operational subsidies so that passengers pay only a proportion, and in some experiments – for example Rome and Bologna – none of the incurred costs of transport. This is justified either as an aspect of social welfare for the carless residuum or an essential prerequisite of car restraint policies undertaken for other urban management goals. An alternative approach attempts to improve speed, reliability, comfort and convenience in competition with individual transport rather than just a safety net for the carless. The bus only lane, taken to its logical extreme in the exclusive bus roads of the new towns of Runcorn, Evry and Almere (Chapter 12), is a low cost near ubiquitous strategy. Combined with promotion, improved passenger information, and new timetabling such routing priority schemes have stemmed and even reversed the decline in bus passengers in cities as varied as Goteborg, Turin, Reading and Breda, while new technologies, working practices and ticketing systems can reduce costs. However such competitive approaches, even when combined with elements of deregulation and competition, are rarely successful unless combined with physical or financial car restraint policies which attempt to decisively tilt the market in favour of collective transport. Success (as in Freiburg) or failure (as in Vienna) depends upon the nature of the policy mix and appropriateness to purely local conditions (Roberts, 1988).

Solutions: planning for integration

The compartmentalisation of urban planning and management, and the similar technological division of transport by mode, characterised discussions of urban transport in most European cities until well into the 1970s. The last 20 years, however, have witnessed a steady movement towards integration both within transport and between transport and the management of those other urban functions that generate the demand for it and respond to its supply.

Even an ostensibly dominant car restraint plan such as that of Norwich, in the early 1960s included a wide mix of measures for different transport modes. The Norwich 'model' (Figure 5.5) was widely imitated and adapted, notably in Goteborg, Lyons, and Bonn in the 1970s and even in Groningen as late as 1977, with a restatement and further refinements in 1988 (Figure 5.6). All contain not only 'collars' around the central area for motorised vehicles, but also 'loops'

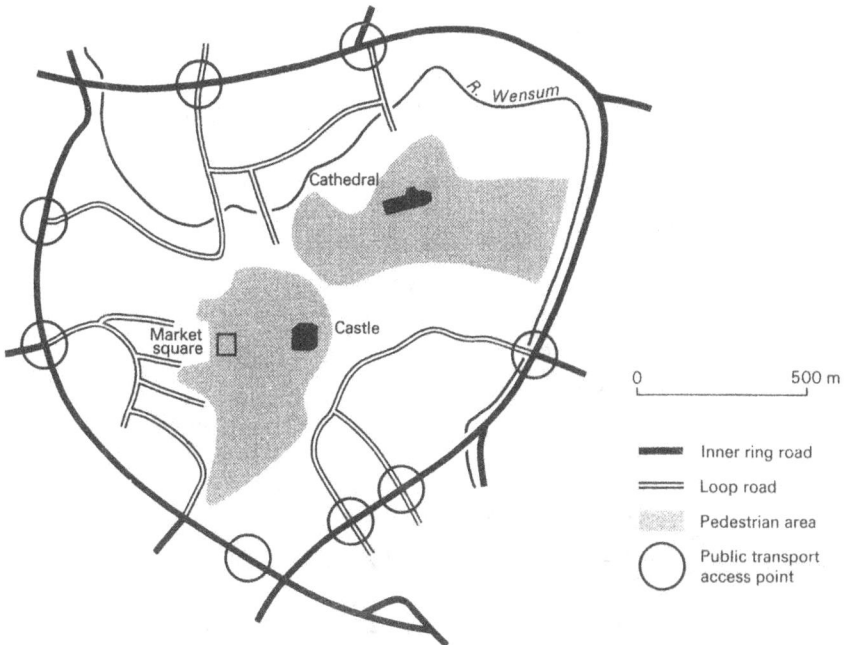

Figure 5.5 Norwich ring-loop traffic management scheme (reproduced by permission of Norwich City Council)

within the centre, especially for exempted traffic, including buses, taxis, essential delivery and emergency vehicles, and bicycles, and positive provision of peripheral car parking. Success, or the lack of it, appears to depend on the degree of the inter-modal integration and local compromise.

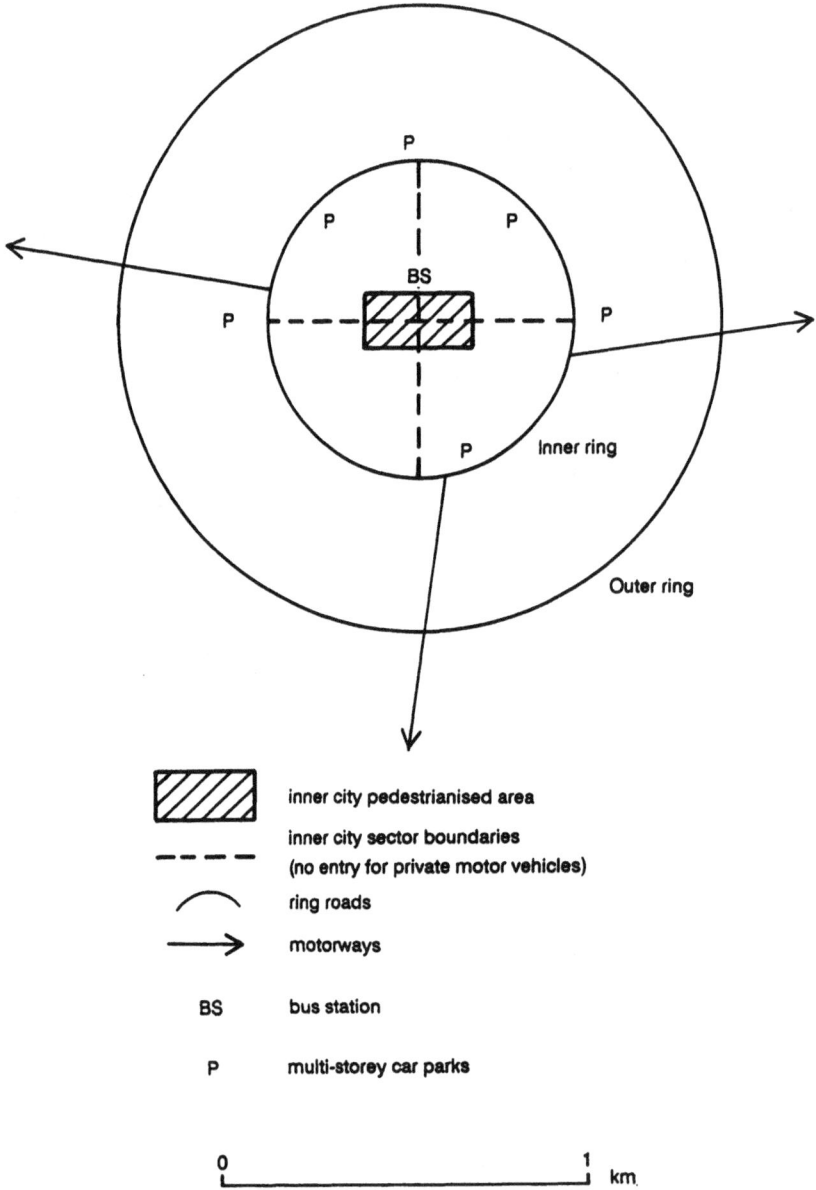

Figure 5.6   Gröningen's traffic management scheme

The largely piecemeal growth of public transport facilities, and the policies for their support, left a confused legacy of often competing installations, ownership and management. The lack of integration both in the physical sense of passenger interchange between different modes, and administratively in the shaping of coordinating the planning and operation of services, weakened the financial position of the undertakings and detracted from the convenience for the customer.

The realisation that almost all journeys are multimodal, and that different modes could specialise in separate parts of the journey led to more attention being devoted to interchanges. At its simplest this involved the siting together of bus, train, tram and taxi facilities most usually at the railway station. In the larger cities new central interchanges were the major landscape manifestations of more far reaching transport strategies. In Amsterdam for example buses are used as collectors in the less densely peopled suburbs, feeding, through a ring of outer interchanges, the tram and metro routes that handle the denser central area flows, which in turn focus on the main inner interchanges that include the railway stations. On a larger and more complex scale in Paris, a similar relationship between the SNCF rail lines, the RER, the metro and the bus system, has been imposed with varying completeness upon the very diverse legacy of public transport.

The flexibility of the private car can also be combined with the high speed and capacity of rapid transit in various ways. Park-and-ride facilities have been provided outside the inner city as a part of car restraint policies, and subsidised parking encourages a change of transport to railway, metro, tram or bus services. A pioneer in this was the Copenhagen S-bane stations. Long standing Dutch policy has been to reduce inner city public parking in the western cities and concentrate it at suburban rail stations; and secure, weatherproof bicycle storage has given a new dimension to park-and-ride. Even in a city such as Paris, where restrictions on parking have traditionally been lax and public opinion reluctant to accept change, a hierarchy of parking in association with public transport terminals with charges increasing with nearness to the centre has been brought into existence.

The problems of integrating transport management have generally proved to be more intractable. The example of the *Berliner Verkehrs Betriebe* established in the 1920s had few imitators, even in Germany until well into the 1960s: even then the creation of the *Verkehrsverband* for managing all transport within the Hamburg city-region was so exceptional as to be hailed a decade later as 'the outstanding institutional innovation of our time in the field' (Thomson, 1977).

The British Passenger Transport Authorities of the late 1960s and early 1970s tried to assume responsibility for 'the totality of transport', an ambitious task which at its best, as in the SELNEC region of greater Manchester, produced coherent management strategies for public transport, but all too often was little more than a forum for the exchange of information.

The Economic Commission for Europe suggested as early as 1973 that integration within city regions was necessary on three levels, namely long term strategic planning and investment, operational management, and coordination of services, fares and ticketing systems. No authority has as yet overcome the fragmentation of political control between neighbouring responsibilities, in particular the political separation of cities from suburbs, nor the inherited differences in financing, cross-subsidies, management philosophy and working practices between various forms of transport undertaking (Buchanan *et al.*, 1980). Both the Parisian RATP and Stockholm's SL come close to this ideal.

The apparent dilemma in many city regions in Europe by the late 1980s was that urban transport authorities could either be so incomplete in their spatial coverage and range of responsibilities for different forms of transport as to be fundamentally ineffective in the coordinating task they were set; or, conversely, so all embracing in their assumed responsibilities for movement within extensive city regions as to constitute a complete tier of urban-regional government, often distant from the detailed problems of management, and, as proved in the case of such regional authorities as the Greater London Council, or the Rijnmond authority for Greater Rotterdam, politically vulnerable.

Futures

Each section of this chapter has thus ended with the questions posed by urban transport being answerable only by reference to other aspects of urban planning and government. It is easy to point to the existence of an inextricable relationship between changes in transport and changes in the functioning of cities. The difficulty as far as planning the West European city is concerned is determining which is cause and which effect; or more precisely whether the roles of active and passive variables can be deliberately reversed. The increasing use of individual motor transport for people and freight has encouraged the suburbanisation, and even deurbanisation of housing, jobs and services which had been initiated by the train, tram and bus. This locational shift has been motivated by, and in turn has shaped, particular styles of living which have been achieved by many and are desired by many more. The movement demands occasioned by this way of using cities can in most instances only be adequately met by private transport. The transport planner has two equally unattractive alternatives: to respond to these demands and incur the wider costs to the city that most Europeans, with the spectre of the North American motorised city before them, are not prepared to pay; or to mount an assault upon the achievements and aspirations of citizens by frustrating their desire for movement and manipulating the life styles that support them. It is not surprising then that European cities have neither the means, will nor political invulnerability in a democratic society to develop fundamental solutions. However, the 20 years of short-term palliatives and local opportunist compromises described have at least postponed the arrival of the two conflicting visions of Armageddon: of total paralysis through increased demand, or the reproduction of Los Angeles throughout the continent in an attempt to meet it.

Radical solutions have generally relied either on salvation through new technology or the creation of new urban structures on the basis of a new valuation of cities. New mixed-mode vehicles, guidance, routing and propulsion systems have been announced at regular intervals but these generally contribute little to the economic, social and political dimensions of the problem – unlike many simple but effective innovations such as the 'flexibus' or automated ticket issuing and cancelling machine. 'Until a new approach is designed for the analysis of cities, to discuss how well they meet human needs, rather than how well they function as machines, there will be no reliable way of assessing the adequacy of these transport systems or the best way of improving them' (Thomson, 1977). Similarly, attempts to restructure the city to create patterns of demand more easily served by public transport have had some success, and certainly this factor is now routinely taken into account in most cities when planning the location of housing, medical and educational facilities. A cautious incrementalism (O'Sullivan, 1978; Adam, 1988) has become the conventional wisdom for the last decade of the century, if only by default. The modification of largely existing technologies and the encouragement of only those adjustments in lifestyle acceptable to the majority of citizens in democratic cities appears the pattern for the near future as for the present. While spectacular success is unlikely, equally those prophets who have regularly predicted the collapse of western urban civilisation beneath the unsolved urban transport problem are likely to be disappointed.

References

Adam, J. (1986) Transport Dilemmas, in Clout, H. and Wood, P. (eds) *London: Problems of Change*.

Adam, J. (1988) Cock-up or conspiracy, *Town and Country Planning*, 57(10), pp. 227–8.

Ashworth, G. J. and Tunbridge, J. E. (1990) *The tourist-historic city*, Belhaven, London.

Bendixson, T. (1974) *Instead of cars*, Temple Smith, London.

Black, J. (1981) *Urban Transport Planning*, Croom Helm, Beckenham.

Buchanan, M., Bursey, N., Lewis, K. and Mullen, P. (1980) *Transport Planning*, Saxon House, Farnborough.

Economic Commission for Europe (1973) *Role of transportation in urban planning development and environment*, Brussels.

Hall, P. (1980) *Great Planning Disasters of our time*, Weidenfeld and Nicolson, London.

Halsall, D. (1978) Rapid transit in Merseyside, *Area*, 10(3), pp. 212–216.

Hass-Klau, C. (1988) Trying to calm the urban motor car, *Town and Country Planning*, 54(2), pp. 51–53.

Hass-Klau, C. (1989) *New Life for City Centres* Anglo-German Foundation for the Study of Industrial Society.

Kuhn, W. (1979) Geschaftsstraasen als Freizetraum, *Münchener Geog.*, 42, Regensburg.

Metton, A. (1984) L'Expansion des éspaces piètonniers en France, in Metton, A. (ed.) *Le Commerce Urbain Français*, pp. 61–76, Presse Universitaires de France, Paris.

Ministry of Transport (1963) *Traffic in Towns*, HMSO, London.

Moseley, M. J. (1979) *Accessibility: the rural challenge*, Methuen, London.

Netter, J. (1977) Transports, circulation, stationnement, une politique pour les villes, *Urbanisme*, **160**, pp. 3–7.

O'Sullivan, P. (1978) Issues in transportation, in Davies, R. L. and Hall, P. (eds) *Issues in Urban Society*, pp. 106–131, Penguin, Harmondsworth.

Préfecture de la Région Parisienne (1972) Plan global transports, *Bull. Inform. de la Région Parisienne*.

République de Canton de Genève (1977) *Some aspects of town planning in Geneva*, Department of Public Works, Geneva.

Roberts, J. (1988) Where's downtown? *Town and Country Planning*, 57(10), pp. 139–41.

Rijksplanologische Dienst (1988) *Vierde nota over de ruimtelijke ordening*, Vol. 2, Staatsuitgeverij, The Hague.

Thomson, J. (1977) Verdict on Rome's free buses, *Financial Times*, 11 January.

Wyles, J. (1988) Italy's cities choke on traffic jams, *Financial Times*, 3 November.

CHAPTER 6

Housing

'Residential patterns largely expose the cultural values that people hold, the consumption profile they strive for, their views of themselves and the ways in which they judge others' (Daun, 1976).

To the Swedish ethnologist Åke Daun, housing policy, more than any other policy for the built environment in the cities of West Europe, is a mirror of the society that builds the housing. The results of successive attempts by societies to solve the need for shelter, and the mismatch between past provision and present-day needs and wants, constitute much of what is defined loosely as 'the housing problem'. Housing policies in all the countries of West Europe are formulated at a national level because each society perceives the problem as a national one. Therefore, in implementing housing policies, governments have translated ideology and social values into bricks and mortar with the assistance of city governments who have planned for residential developments within the guidelines of national policy.

Policy determinants in European urban history

Until the nineteenth century the provision of housing in cities was slow. The accretion of housing around the periphery of towns or the reconstructions within the urban area were constructed relatively slowly and, only in exceptional circumstances, did large-scale rebuilding or large extensions to the built-up area occur. One such case was the rapid spread of London beyond the limits of the walls in the years following the Great Fire of 1666; another was Gävle after the fire of 1888. In many cities in central Europe the Altstadt (the historic city centre) was still a recognisable residential area until the twentieth century and has remained so despite the growth in non-residential areas.

With industrialisation, a marked change took place in the provision of housing. Housing was no longer a part of the shelter provided by a feudal society but an integral part of the entrepreneurial philosophy underpinning industrialisation. To the industrialist, anxious to profit from new technologies, housing for workers was essential but the costs of this provision were to be kept as low as possible. The concept of housing markets with economic rents was introduced into housing in order that the industrialist/developer could gain as much profit as possible from the workers' shelter as from their labour (Vance, 1977). The result of entrepreneurial initiatives was the 'generalisation' of the housing stock in the new industrial centres of Europe. Vast areas of housing, almost identical in both outward appearance and internal provision, grew around the mills and mines of the Ruhrgebiet, the Nord coal-field, the South Wales valleys and the Borinage.

Bochum in the Ruhrgebiet was developed as a centre of coal-mining by several mine owners, who each built their distinctive 'colonies' adjacent to the thirty or more mines scattered through the city territory (Thomas and Tuppen, 1977). Colonies were also attached to other industrial sites such as the Krupp foundry. In the Emscher zone of the Ruhrgebiet, the settlements that sprang up were almost entirely residential colonies for the large mines, and visitors to Duisburg-Hamborn, Bottrop, Buer, Herne and Castrop-Rauxel today cannot but be struck by the generalised building forms.

Similar vast relatively uniform extensions to the built-up area occurred in northern France. The built-up area of Lille expanded fivefold in the period 1850–1919 while neighbouring Roubaix and Tourcoing merged into one another, tripling their surface area between 1830 and 1880 and engulfing other villages. The new factories of firms such as Agache and Kuhlmann each had a surrounding *quartier ouvrier* or *cité* with its monotonous housing so that tentacles of uniform, insalubrious housing extended out from the towns. By 1990 one-third of the population of Roubaix–Tourcoing were living in uniform cramped workers' settlements (Notes et Etudes Documentaires, 1976). Industrial growth had brought with it generalisation of housing, a tradition which has continued ever since. However, the moral responsibility that society possessed for the worst evils of the new housing was already being stressed by Dickens, Engels, Ruskin and Zola. The pressure for society to take responsibility for shelter had begun to force governments into action.

Thus, by the turn of the century, the foundations for public involvement in housing provision had been laid. The protective legislation which originated in the form of the health acts in Britain spread to other countries. The philanthropists such as Owen, Salt and Buckingham, and the new housing trusts such as the Guinness Trust, showed the value of improved standards, lower densities and a non-profit-making philosophy (see Chapter 2). The co-operative housing movement was founded in 1869 in Denmark, 1870 in Sweden and 1889 in Germany as an alternative means of providing housing, so establishing another strong tradition in European house building. In Italy the Luzzati Law of 1903 laid the foundations for the involvement of the banks in the provision of low income housing.

The problems faced by European society in the wake of the First World War inevitably resulted in policy changes which affected housing provision. The concept of the community as a provider of modern housing, which was a part of the British tradition of providing shelter expressed since the seventeenth century in the form of almshousing (and even earlier in the European monastic traditions), was strengthened after the war with increasing local authority provision of homes for rent. Rent controls were introduced in the private sector in many countries in the wake of the depression and have come to form an essential part of policies. In Germany the trade unions' involvement in housing building began in the period 1922–26 with the founding in Hamburg of a labour building association to cater for housing need – DEWOG, an advisory house building group, and GEKABE, a city and trade unions combine for house building – the forerunner of *Neue Heimat*, the trade unions backed housing authority (Fuerst, 1974).

By the end of the Second World War the historical precedents for present policies had been laid in the majority of countries. However, the destruction wrought on cities throughout Europe by military activity (including destruction in Spain during the Civil War) forced the authorities to devise policies which catered for the enormous shortages of housing. The Netherlands were short of 300,000 units in 1950 (Smith, 1973). Various Länder passed reconstruction laws, which were implemented by the cities in rapidly rebuilding housing with some success in Hannover and Kiel but ignored in Cologne by the city planning director with unfortunate consequences for the pace of reconstruction (Green, 1959 and 1964). In contrast, in Britain all initial investment in housing was channelled into the local authority sector for almost a decade. The most severe shortages (four to five-and-a-half million units) were found in the German cities where up to 60 per cent of dwellings were destroyed or damaged in, for example, Berlin, Hamburg, Hannover and Essen. Many felt it was easier to rebuild on new sites than to clear the rubble (Blacksell, 1968). Two-and-a-half million of the ten million homes had been destroyed in the new West Germany, whilst in Italy more than one million dwellings were damaged.

Social determinants of policies

There are a series of demographic trends which have influenced housing policies. Much of post-war policy was influenced by the rising birth-rate in the first fifteen years after the war. Also, household formation was expected to grow, with the result that the future task for most cities was one of accommodating increased numbers on a diminishing land area. The policy outcome was to concentrate both on adding to the housing stock on the periphery and also renewing the stock in the centre, two policies governed as much by the cost-benefit yardsticks of the economists as by the fulfilment of political targets. However, trends in fertility have altered since the mid-1960s with several countries reaching (West Germany, Switzerland) or approaching (Britain) 'zero population growth'. Inevitably changes in policy result. In countries where housing supply matches demand, more steps are being taken to improve the older housing stock including the poorer dwellings built in the immediate post-war period.

The structure of the West European population had altered with delayed and smaller families enabling women to work sooner and provide continuing additional disposable income. The resultant trends were expressed as demands for improved space standards in the home for their own sake or to accommodate a wider range of labour-saving domestic equipment. The most recent trends in the birth-rate suggest society will become older. Combined with increases in longevity and the retention of unitary family dwellings, partly reinforced by welfare state provision and the role of life insurance in providing, in particular, for widows, the number of single person households is rising and household formation is unlikely to decelerate at the same rate as the birth-rate.

In almost all countries the size of the average family has declined. Although lagging behind the North European cities, Naples illustrates this trend towards smaller families; the average size was 4.3 in 1961, 3.9 in 1971 and 3.4 in 1981, while in the affluent outer suburbs such as San Pietro there were 2.1 persons per household (Laino, 1988).

The growing trends of retirement migration affect certain city regions more than others. Early studies note the specific urban destinations for migrants aged over 65 years (Koch, 1976; Cribier, 1970; and Law and Warnes, 1973). In France, Nice and other cities of the Cote d'Azur, the new resorts of the Languedoc coast such as La Grande Motte, and the Atlantic resorts of Biarritz and Royan become focal points of retirement migration as did the rural fringe of almost all cities. In Germany, Koch noted a similar southward drift to the small towns of the alpine region, south and east of Munich, besides a flight to the urban fringes of Rhine-Main, Hannover and Hamburg (Koch 1976). By 1986 West Germany had more citizens aged over 60 years than under 20 and therefore, in terms of housing provision, the wishes of the older generation will influence the provision of new housing. The older generation have continued to express a preference for southern small towns. For some there is a back-to-the-city migration. In addition it is the single elderly who have partly contributed to the growth of single person households which form 33 per cent of the forecast 24 million households in 1990, a figure which will be lowered by the events of 1989–90 (Friedrichs, 1985).

Migration, as we have noted above, is an important variable in calculating the level of housing provision at a city level. Large scale migrations in central Europe did put extreme pressure on authorities in West Germany and Austria in the post-war period. In Sweden between 1940 and 1970 the proportion living in urban areas rose from 55 per cent to 80 per cent. This represented a fulfilled demand for housing an extra three million people in the towns and cities. In Spain, King has shown how the authorities there underestimated requirements caused by internal migration which provided a shortfall of approximately 850,000 dwellings in urban areas during the 1960s (King, 1971). Urbanisation in the Mediterranean states continued to increase the pressures for housing in the cities well into the last quarter of the twentieth century. In Italy the combination of migration to the northern cities from the south and urbanisation by emigration, which became domestic urbanisation when the migrants returned, helped to fuel the pressures upon Rome, Milan, Turin, Naples and Florence.

Urbanisation also resulted in increased pressures of population on the primate cities of the Mediterranean. In 1950 Athens contained 18 per cent of the national population and by 1980 this had risen to 57 per cent; 83 per cent of the Greek urban population lived in Athens and Thessaloniki. Forty per cent of Portuguese in 1980 lived in Lisbon and Oporto (Gaspar, 1984). Illegal settlements such as Quarto and Pianvra outside Naples become one response where official policies were inadequate to cope with the resultant flow of inter-regional migrants. In Thessaloniki during the first six months of 1976 6,000 illegal residences were constructed by migrants (Kouvelas, 1976).

Within cities the migration currents are far from even, despite the general levelling of the urban density gradients resulting from the decentralisation of people. The pressure of migration on eastern Lyon is greater than on the west, migration to the eastern suburbs of Dusseldorf and to south-east Geneva is stronger than in other directions.

Other cultural and historical determinants of housing density must not be ignored although here the evidence is more suspect. The French tradition of high urban densities, epitomised in the work of Le Corbusier (see Chapter 2) has been maintained in the concept of the *grands ensembles*. In Scandinavia the ideas of Sitte have been adopted and Gropius's influence on German housing is marked as well. In the latter case the emphasis on *existence-minimum*, the most concise spatial economy of rooms possible, imposed an overwhelming architectural rationality on the living environment. The low density anti-urban preference of British citizens is enshrined in most housing schemes in England and Wales, although not Scotland.

Aspirations for specific types of housing are important. At a local level perceptions of housing need, quality and desirability reflect current social trends and indicate potential patterns of demand, while at a national level perceptions concerning the housing stock and environment of a city can and do influence the rate of urban growth in different cities.

Wilmott and Young in a survey of London families showed that 50 per cent wanted a larger home and only 22 per cent a smaller one, and many in this latter group were elderly. Of those who had gardens, 44 per cent wanted larger gardens and 35 per cent the same size. The implications for the space demands of households are very clear in this case (Wilmott and Young, 1973). These figures confirm the overall pattern of demand in the United Kingdom which is for detached or semi-detached housing and with a small minority in favour of flat dwelling; only one per cent in a 1960s survey wished to live in high-rise flats and 11 per cent in low-rise flats. Racine and Creutz also have evidence to show that 68 per cent of French people want to live in a private house at a density of 20–25 houses per hectare; yet until recently the authorities governed by a philosophy of combating shortages have yet to respond completely to these new values (Racine and Creutz, 1975). Swedish authorities have also noted the fact that the numbers of houses in cities such as Stockholm rose rapidly while the population stagnated, a reflection of improved space expectations. Housing production has changed to accommodate the changed tastes of the Swedes. In the 1960s 70 per cent of dwellings constructed were apartments whereas by 1977 70 per cent were houses (Wohlin, 1977).

Another study originating from the Netherlands has attempted to relate the most desired settlement type to household size. The desirability of town depends on the stage in the life cycle although the urban village and village are generally the most popular locations. However, city life in the smaller urban agglomerations of 64,000 to 256,000 is desirable for a significant proportion of the families with small children. A similar survey comparing income groups and preferred cities did show that the lower socio-economic groups preferred the large cities, whereas the higher social groups preferred the middle sized centres or the villages. Otherwise almost a half of the middle socio-economic groups desired to live in settlements with under 16,000 inhabitants.

A further Dutch study relating desired residence to present residence, at the simplest level, indicates that people are happiest with a similar sized settlement to the one in which they already live. City residents opt for village life as would those in the large towns. The newly built areas within the urban area are preferred to residence either in the city/town centre or on the periphery. In a country with an overall population density as high as the Netherlands the implications of fulfilling such a set of preferences is serious and can be used both to call into question existing policies for housing and to justify policies that could result in altered preferences favouring the status quo.

Studies in the 1970s by Monheim of the attractions of German cities to the employees of 4,000 firms drew attention to the variable perceptions of a sample population of office managers (Monheim, 1972). Respondents were asked which city they would like to live in. The results show in Figure 6.1a indicate that 31 per cent wanted to live in Munich and 20 per cent in Hamburg. Only one in five wanted to live in non-metropolitan regions. Munich's role as the most desirable place of residence has grown since 1964 when it displaced West Berlin. Munich was also the most preferred location for industrial plant followed by Dusseldorf and Hamburg (Figure 6.1b) and the most preferred location for retirement (Figure 6.1c), despite the selection of more non-metropolitan centres in Bavaria. Munich elicited more sympathetic statements in all places of survey except Hamburg (Figure 6.1d) and was named almost twice as frequently as Hamburg and five times as frequently as Cologne. The implications for housing provision in and around Munich were obvious and Munich's policies have had to cater for one of the most rapidly growing populations in what was the Federal Republic. On the other hand the cities of the Ruhr, which received 22 per cent of the derogatory statements (Figure 6.1e), obviously must regard the nineteenth-century built environment as a liability.

Housing policy endeavours to accommodate known preferences, given the ability of the individual urban centres to respond to them. On the other hand opinions are given to the policy-makers which they can follow or counteract, as in the case of the Netherlands. The more direct social indicators derived from traditional demographic predictions and migration are still accepted by policy-

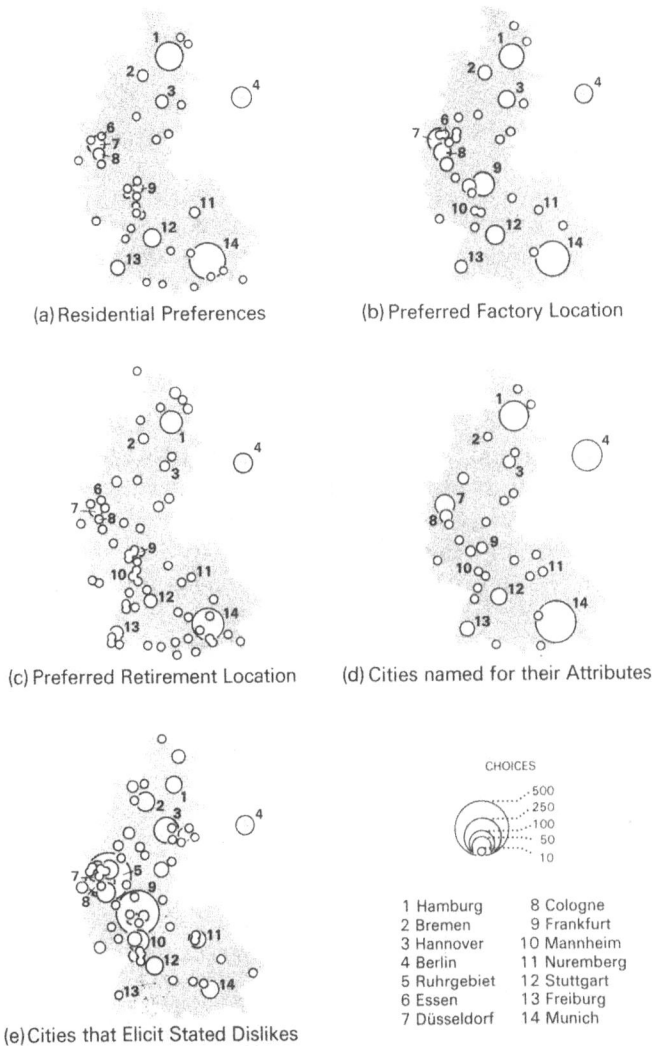

(a) Residential Preferences

(b) Preferred Factory Location

(c) Preferred Retirement Location

(d) Cities named for their Attributes

(e) Cities that Elicit Stated Dislikes

CHOICES

500
250
100
50
10

1 Hamburg	8 Cologne
2 Bremen	9 Frankfurt
3 Hannover	10 Mannheim
4 Berlin	11 Nuremberg
5 Ruhrgebiet	12 Stuttgart
6 Essen	13 Freiburg
7 Düsseldorf	14 Munich

Figure 6.1 City preferences in West Germany (after Monheim)

makers and must be seen as a foundation for present policies together with the experience gained from the examination of urban history.

This picture of metropolitan desirability altered somewhat in the 1980s. Between 1980 and 1984 all the major urban areas, with the exception of Munich, lost population. When the data are disaggregated they show that the urban cores averaged a 3.6 per cent loss of population whereas the peri-urban fringes gained population by 3.6 per cent in the case of Munich, 2.6 per cent in Nürnberg and 1.8 per cent in Hamburg. Nevertheless, even the peri-urban areas of the Ruhrgebiet, Frankfurt, Stuttgart and Hannover were losing people. Data for 1985 show that it was cities of less than 100,000 which were gaining population by migration (Riquet, 1988).

Policy shifts in the post-1945 period

Housing policies have shown similar shifts in emphasis which mirror the overall changes in urban policy since 1945. While the changes do not occur simultaneously, it is possible to detect phases which apply to a greater or lesser extent throughout West Europe.

The goals of housing policy were generally established in the brave new world of the immediate post-war period. The policies had to cope with the forecast population boom and the return to the cities by those who fled bombing or those fleeing East European socialism. They had to counteract the neglect of house building of the war years or, in the case of Italy, the neglect of low income housing. They were a response to the alarming findings of the post-war censuses. In the case of France the 1954 census revealed that 9 per cent of Parisians had no regular home, 17 per cent shared housing and 55 per cent had no toilet; the consequence was the rise of the *grand ensembles* (Castells, 1984). Industrialisation of Mediterranean cities drew many to work in the burgeoning factories and, in the case of Madrid, resulted in 1956 in 20 per cent of the population being housed in shanties. The policy response was to encourage the speculative building of low cost housing on the city fringe.

(i) Technocratic policies

In order to house people the politicians had turned to the expertise of architects and the infant profession of planners, mainly populated by architects and civil engineers, to solve the housing crisis. The main architectural influences on the residential planners were Gropius and his rationalistic view of housing and Le Corbusier's technocratic ideals. The second influence was that of the government economists who wanted as much housing as possible for the lowest possible outlay. In those countries where public housing provision depended on the local tax base, as in Italy, then wealthy cities such as Milan were able to invest more in public housing (Sbragia, 1979).

The built product of the application of technocratic planning principles was the provision of public sector rented accommodation on the outskirts of towns, together with the comprehensive redevelopment of insalubrious inner city areas. There was a proliferation of social housing schemes on the periphery of cities such as Sarcelles (Paris), Le Mirail (Toulouse), Osterholz–Tenever (Bremen), Neu Perlach (Munich), Råstätt (Jönköping), Wester Hailes (Edinburgh), Gran San Blas (Madrid) and Bijlmermeer (Amsterdam). Figure 6.2 illustrates the location of several such schemes with especial reference to German cities.

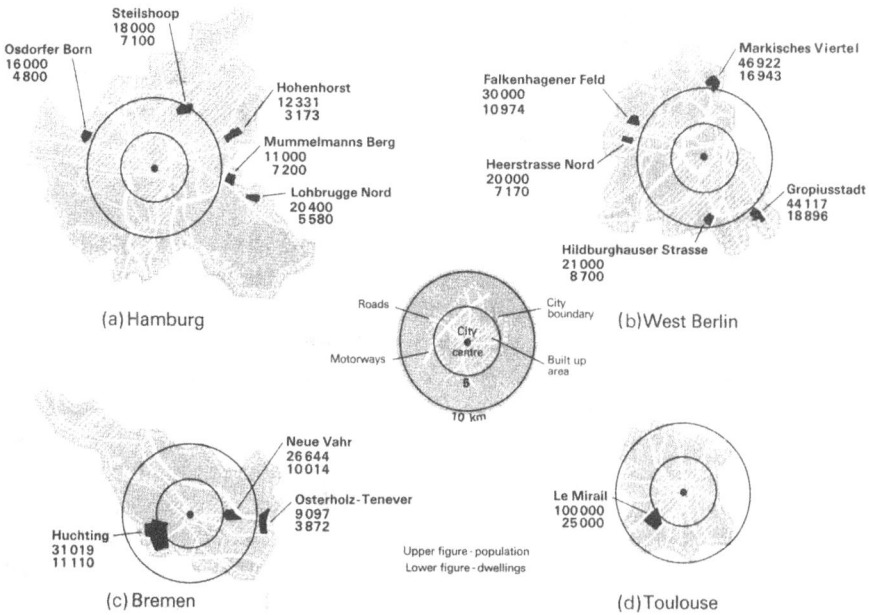

Steilshoop
18 000
7 100

Osdorfer Born
16 000
4 800

Hohenhorst
12 331
3 173

Mummelmanns Berg
11 000
7 200

Lohbrugge Nord
20 400
5 580

(a) Hamburg

Markisches Viertel
46 922
16 943

Falkenhagener Feld
30 000
10 974

Heerstrasse Nord
20 000
7 170

Gropiusstadt
44 117
18 896

Hildburghauser Strasse
21 000
8 700

(b) West Berlin

Roads

Motorways

City centre

City boundary

Built up area

10 km

Neue Vahr
26 644
10 014

Osterholz-Tenever
9 097
3 872

Huchting
31 019
11 110

Upper figure - population
Lower figure - dwellings

(c) Bremen

Le Mirail
100 000
25 000

(d) Toulouse

Figure 6.2   The location of peripheral estates in selected cities

Osterholz-Tenever, Bremen

Osterholz–Tenever was begun in 1970–71 as a development to house overspill population from the centre of Bremen. It succeeded an earlier scheme which was highly praised, Neue Vahr (Figure 6.2a). It was developed by the city and *Neue Heimat*, the former trade unions housing corporation, with funds from the provisions of the *Städtbauförderungsgesetz*. On completion 10–12,000 people live there at very high density in 7000 dwellings.

One of the last of its kind to be built, Osterholz–Tenever is situated, rather unusually, in a city which did not have a history of high density dwelling, although its Social Democratic government has an impressive record of housing provision, building 110,000 dwellings between 1953 and 1974. Compared with other schemes Osterholz–Tenever has attempted to meet the criticisms of other high density high-rise schemes. Owner-occupation has been encouraged in one-seventh of the dwellings but there are problems in attracting people to live in the 'alternative town of tomorrow'. The 59 hectare site is 12 kilometres from central Bremen and the transport links were initially inadequate because it was beyond the tram terminus. However, it is close to a new industrial zone which has attracted considerable investment. The physical form is varied with the units ranging from three to 23 storeys. Open space is found around the scheme but the open space within the scheme is rather formal in design and, despite discreetly placed vegetation, mainly concrete. Despite these attempts to alleviate problems, Osterholz–Tenever has one of the highest proportions of children under 15 of all the West German city schemes, 33 per cent, and less than 10 per cent of residents are elderly.

Gropiusstadt, West Berlin

This development, designed by Gropius and built over the period 1962–1975 on 264 hectares on the fringes of West Berlin–Neu Kölln, planned to house 60,000 people in 19,000 units (Figure 6.2b). Its history of development again illustrates the problems with which city housing authorities are faced. In the original plans of 1960 the 16,400 units were in low-rise blocks (74 per cent 4 storeys or less) but by 1974 only 12 per cent were low-rise, 58 per cent were 5–19 storey blocks and 6 per cent 19–31 storey blocks! In the final outcome 70 per cent of the 18,000 homes are in flat blocks (Becker *et al.*, 1976). Play spaces for small children in this scheme are inadequate for a teenage population and private open space is generally absent except on balconies. Employment is sited elsewhere in the city and this makes it difficult for young mothers seeking employment, unless they can be employed in shops on the estate itself. More recently the lurid confessions of a teenage drug addict sensationalised in the press have drawn attention to the area's social problems. The social services in the Bezirk of Neu Kölln were more prepared than other cities for the problems of an estate where 75 per cent of the population were under 45 years old in 1975. Offices of the youth service, social psychiatry section and school health authority were placed in the development. With 95.7 per cent social housing being constructed and only 4.6 per cent for sale (including housing financed from the social housing programme) Gropiusstadt continues to be very much a single social group housing development.

Le Mirail, Toulouse (Grand Ensemble)

The development of Le Mirail, Toulouse (Figure 6.2c) was planned in 1961 with the realisation that the city could grow into a centre for the electronics and aircraft industry. Seven hundred hectares were set aside for a programme to build 25,000 homes for 100,000 people, five kilometres from the centre. A high density housing area with pedestrian circulation, a university, a commercial centre and an industrial zone were planned but constrained by financial, legal and administrative problems throughout construction. The priority was to house people but the constraints did result in cost cutting exercises whose social consequences mirror those of other *grands ensembles*.

Sarcelles (Grand Ensemble)

Sarcelles on the northern outskirts of Paris was built between 1954 and 1974 to alleviate the Parisian housing crisis revealed by the 1954 census. The scheme comprised three phases of development: 1954–1960 when all the properties were

HLM (*Habitations à Loyer Modéré*) social housing in medium rise blocks; 1960-1966 when the Gaullist authorities built better quality homes to obtain more profit from rental; and 1966-1974 when condominiums were constructed for more middle class housing in order to change the image of the scheme, which had suffered from the bad publicity given to the *grands ensembles*.

The design of the apartments did not encourage ease of social mixing so that many women at home felt isolated during the day, a not uncommon phenomenon in any new social housing project, a complaint the French called 'Sarcellitis' after this *grand ensemble* (Lagarde, 1968). Despite subsidies the cost of housing was regarded as expensive, particularly as husbands were now commuting back to Paris. These costs were high partly because the owner/developer held monopoly power and residents were powerless until the Association Sarcelloise developed an effective citizen movement. However, the monotony of the blocks, poor maintenance, the lack of privacy caused by high density of blocks and poor noise insulation, the problems of public staircases and inadequate lifts, and fears for safety, besides the lack of life around, were the major criticisms that were advanced. It was particularly noticeable that actual play areas for children within the scheme were not always those designed for play. Cultural facilities only arrived when the Left was in power. The need for parking had been underestimated and there were problems of on-street parking and noise for ground floor residents. Private outside space, whether it was balconies or gardens, was valued and felt to be too little. The housing lacks the intimate small scale spaces and multiple use so characteristic of older French towns. By 1974 the policy on *grand ensembles* was reversed partly because they were less profitable than anticipated, a result attributed by Castells (1984) to the residents' associations such as Association Sarcelloise forcing better conditions out of the developers. Others would attribute the shift to a broader shift in social attitudes and attitudes to environmental quality ushered in during the seventies.

Easterhouse: Glasgow

A similar train of events can be seen throughout West Europe and the case of Easterhouse will serve to illustrate the parallels (Keating, 1988). Easterhouse was a high density, Glasgow Corporation-owned development of medium rise tenements built between 1950-1972 during the period of greatest housing shortage in the city. People arrived from low-income inner city areas to a townscape of blocks which lacked privacy, school buildings and shopping centre. The latter was opened in 1972 for an estate of 56,000 people. Despite the increase in amenities, Easterhouse remained an overcrowded development with 30 per cent of households in overcrowded conditions in 1981. Low income households (40 per cent), high unemployment (40 per cent in 1986), poor health and a lack of suitable jobs nearby have been difficult to counteract. The population may have fallen (to 46,000 in 1981) and attempts may have been made to assist the inhabitants (Frankie Vaughan in the 1960s, the Easterhouse Festival Society in the 1970s, and The Body Shop initiative in the 1990s to name but three), but the estate still is a classic example of multiple deprivation.

Most cities during the 1970s decided to stop the construction of large peripheral housing schemes. In 1976 West Berlin curbed the size of a smaller scheme, Ruhwalk, by 500 dwellings out of a planned 4000. The declining population, the problems of high-rise living, the costs of building which were not cheaper and the high maintenance costs have brought about a new look at social housing provision on the city fringes and a fresh approach to the older housing stock. It is a return to what Rappoport terms the vernacular tradition from the grand technocratic design. The policy shifts have taken housing back to reinstating the vernacular tradition reflecting the life and activities of people (Rappoport, 1968). By 1985 the West German press used headlines such as 'The Giants are in a Coma' to highlight concern over these estates. The solutions canvassed ranged from refurbishment, *Nachbesserungsmassnahmen*, to calls for demolition (Heineberg, 1988).

From the dawn of the seventies there was a radical shift away from the large scale residential extensions to cities, as the schemes of the technocratic planners were seen to be utopian dreams beset by economic, social, construction and design problems. The opposition to the housing schemes came from public pressure groups, urban social movements, citizen movements, residents' associations, *Burgerinitiativen*, who were concerned by the lack of provision in new housing schemes. These concerns were expressed about both the peripheral housing developments and the increasing number of comprehensive development proposals for the areas of nineteenth century housing.

Therefore, after the 1968 civil disturbances in France and West Germany and post-1975 in Spain, the expressions of people power combined with the economic and political realities of each nation initiated a shift in policy towards rehabilitation and rejuvenation. Conservation replaced clearance as the key word in housing provision. In West Germany over-provision of housing and new legislation resulted in an emphasis on conservation in the development plans of the seventies. In Spain the death of Franco in 1975 enabled the citizen movements (*associaciones de vecinos*) to focus upon poor urban facilities in the sixties public housing areas around Madrid, such as San Blas, and the similar lack of facilities on private developments, such as Mortalaza. San Blas was rehabilitated and conservation was stressed in the city plan revisions of 1978 and 1980 (Castells, 1984). In Italy urban social movements continued to focus upon housing shortages but they drew the attention of the electorate to the housing crisis. In some cases, such as Naples, it needed other events such as the 1980 earthquake to ensure a programme of both new peripheral housing (for example, Ponticelli) and rehabilitation programmes (for example, Barra S. Giovanni). In this case the Italian experience illustrates the time-lag of the shift away from peripheral expansion schemes. In the United Kingdom spatial targeting in the form of General Improvement Areas and Housing Action Areas, together with the Civic Amenities Act, were the policy responses to the growing concern over the destruction of our urban residential heritage.

(ii) Urban Renovation

Ostertor-Remberti, Bremen
The renovation of this area of 6.8 hectares just to the east of the city centre was undertaken by the Bremische Gesellschaft fur Stadterneuerung, Stadtentwicklung and Wohnungsbau mbH within the provisions of the federal Städtebauförderungsgesetze. The area was originally developed in the nineteenth century with distinctive housing called *Bremer Haus*. However, all the early attempts to provide an overall traffic plan for the city saw the area as lying astride an eastern inner ring road around the central area. From 1967 onwards, the east tangent became an accepted part of the planning strategy and planning blight occurred along the course of the planned road. However, in 1974 an urban renovation area (Sanierungsgebiete) was delimited on either side of the route; but by 1976 the road plan was abandoned and the renovation area enlarged as a result. The first attempts were made in one area in 1974, an area characterised by a mixture of residences and industry. Since then a whole series of studies has been undertaken (Bremische Gesellschaft, 1977) on the condition of the buildings, the environment and the social environment.

As one would expect in a blighted area, the proportion of immigrants rose rapidly from 6 per cent in 1970 to 18 per cent in 1976. At the same time the age structure was altering with fewer children and 45–65s but more old people. The number of single person households increased with an influx of students and the increase in the elderly to form two-thirds of all households, but there was also an increase in large households of the foreigners. Mobility in the area is twice that of the city as a whole.

As a result of the research programme by the interdisciplinary team a six part programme for the area phased from 1978–1980 to 1981–1983 was prepared, costing 39 million DM. The plans for the phases, comprising dwellings and industrial workshops, offices and retail outlets, were in the form of a social plan reinforced by federal government legislation. Most of the property (75 per cent) was rented by absentee landlords and the higher rents that renovation could bring were attractive to them. However, the financial assistance enabled the existing population to remain and new groups to be encouraged to enter the area.

The changes were concentrated in small areas, often by house, and conversion costs were paid. Loans were granted for modernisation and repairs at 2 per cent below the existing savings rate. In the first ten years there was no interest payment. The personal contribution as a rule amounted to 15 per cent of the modernisation costs and these were reclaimable against tax. For industries there was assistance to relocate. Throughout all stages the renters were made aware of their rights. Thus it was possible to build new residential apartments and sub-divide large single family residences into three family apartments. The policy has been to encourage the letting to young families who work in the area rather than encourage ghettos of the old (*Altenghettos*). The *Generationshaus* (a house with an apartment for elderly relatives) was encouraged, besides the development of an old people's day centre. The size of the student population of the area was to be maintained. The social outcasts are also considered in the plan, which hopes that no more will be attracted to the area, which is most unlikely once the rents begin to rise.

Ostertor-Remberti was one of 19 special study areas for renovation (others in similar locations included zones in Karlsruhe-Alstadt, Hamburg-St Pauli, Berlin-Wedding and Osnabruck), funded jointly by the Federal and Land governments. In all, federal assistance was given to 473 renewal schemes. The progress in the area to date is visible in the greatly improved external appearance of the houses. The social structure has been stabilised. Time has shown that the danger of gentrification of the area's housing market was a justifiable one, for in this area, as in other similar schemes, there has been a slow upward shift in the social composition of the area with the departure of the immigrant, student and lower income groups.

Ottensen, Hamburg

Ottensen was one of six inner urban areas that had been given priority treatment by Hamburg, the others being St Georg, Neustadt, St Pauli, Altona-Allstadt and Elmsbuttel Sud. It is under six kilometres from the centre of Hamburg but in terms of the city's history is a nineteenth century inner suburb of Altona, which became part of Hamburg in 1937. It is an area where 60 per cent of the housing was built before 1918 and where the mixture of industry and housing was being altered by the flight of both employment and workers and the influx of foreigners. The area of Ottensen is much larger (207.7 hectares) than Ostertor-Remberti. Also a nearby federal motorway had increased traffic flows through the area, which lacked open space. Ottensen was the subject of an intensive study which led to the acceptance of a development plan for the district (Ottensen, 1977).

The local aims of the plan were to renovate housing and to construct new apartments that partly compensated for dwelling losses to new industrial sites and transport improvements. In the sphere of the physical environment this demanded the creation of 'residentially oriented spaces', or open areas, to serve the housing including the preservation and refurbishing of the urban landscape, especially the small squares. Commercial enterprises and traffic that detracted from an area's attraction for residence were to be moved from the area. Yet, at the same time, the diversity of work and small scale factories had to be retained which could conflict with good environmental conditions. Similarly, conflict was foreseen between the preservation of dwellings and the provision of improved community facilities, and the dangers to other residential areas of traffic control measures in Ottensen.

Modernisation began in 1975 with twenty-one apartments and then a second, larger, area was tackled. In this area it was necessary to have some comprehensive redevelopment which was agreed with the participation of tenants and owners. Modernised homes in reorganised courtyards are a key point of the plan. New flats were provided on old factory sites. Grants similar to those in

Bremen were obtainable. The area for housing increased following the construction of dwellings, but the share of dwellings in all floor space fell due to increases in the intensity of use of the area for offices and businesses. Much of the housing that has been renovated was assisted by SAGA, the city's public housing company which encourages house by house renovation so that displacement is minimised. Over 500 homes were repaired and 900 modernised, in addition to the construction of 400 new apartments. Factories were converted, on local initiative, into a kindergarten and another became the old people's day centre. These improvements in the social infrastructure together with the policy of 'renewal in steps' will, it is hoped, improve the housing conditions in this area. The environment of deprivation has been slowly changing to one of conservation and renovation.

Hamburg was able to contemplate the renovation of Ottensen because the city could issue decrees requiring construction and modernisation besides stipulating sites where buildings may not be demolished. Similarly, they are able to designate sites where only social housing or housing for those with special needs may be constructed. In addition, as we have seen in both the German cases, importance is attached to social plans which must be prepared and discussed. In the case of Hamburg, the discussions included intepreters to represent the foreign worker interests. Hamburg also has been in receipt of generous federal funding for Sanierung; 39m DM in 1983 for its nineteen schemes involving 28,000 people and 12,000 dwellings.

In the further areas of SIKS (Urban Renewal Areas in Small Steps) 45 hectares containing 5700 dwellings have been improved at a cost of 10m DM. Further funds are available for modernisation of residences and between 1973 and 1985 126,000 residences were modernised (Moller, 1985).

Old housing in the Netherlands
In 1968 it was estimated that there were 375,000 slum dwellings, 250,000 poor dwellings in need of improvement, and 50,000 oversize dwellings for which there is no demand in the Netherlands. On an age basis it has been calculated that 20,000 new buildings enter the above categories each year. The overall increase in slums in the sixties was 10,000 dwellings a year. Over one in five slum dwellings and a quarter of those capable of improvement were located in the Rotterdam–Hague area of South Holland. Present policies represent a radical change in emphasis from expensive clearance and construction to a national and local policy of rehabilitation.

Of the homes that can be considered in need of improvement, 200,000 are already owned by housing corporations and municipalities and it is proposed to increase the number of those being improved from 6–8000 to 10,000 per annum. To achieve increased numbers of renovated houses it was suggested that the municipalities utilise their legal rights to make dwelling owners carry out the necessary improvements, for which government aid is forthcoming. Private individuals receive aid from The Premium Scheme for Improvement and Splitting of Housing, and this aid can be boosted further in areas of high unemployment. Funds are available for the worst slum clearance up to 80 per cent of the cost so that the municipality may redevelop the site. As a result of these policies renewal expenditure by the government has risen from 7.1 per cent of housing expenditure in 1971 to 11.9 per cent in 1976. In Rotterdam, eleven urban renewal districts have been designated within the provision of the 1965 Physical Planning Act and the 1976 Urban Renewal Act, containing a quarter of the city's dwellings, and in renewing the areas great emphasis is placed on local involvement and maintaining the existing population. It is intended to rehabilitate 1000 units a year and build a further 1500 units in the renewal areas, giving priority to existing residents. In Amsterdam 13,000 dwellings are included within the city's major rehabilitation and conservation scheme.

French urban renovation
In 1973 it was estimated that 30 per cent of the population living in pre-1949 housing was living in unsatisfactory conditions: sixteen million people in France would therefore appreciate improved housing and over the years since the 1958 decree on renovation several acts affecting housing renovation have been passed. The 1958 decree on urban renovation permitted the destruction of the housing stock of a *quartier*, alteration of landholdings and rebuilding as in Belleville, Paris. The Malraux Law of 1962 permitted *restauration immobilière*, exterior renovation with interior renewal, a form of conservation that is best seen in the Marais district of Paris. In 1967 the Loi Amélioration de l'Habitat permitted the improvement of hygiene, sanitary conditions, comfort and security of dwellings. The Vivien Law of 1970 was designed to tackle the worst insalubrious dwellings by relocation and destruction. Funds for renovation are obtained from Fonds d'Aménagement Urbaine, which in 1977 funded renovations to the sum of 177 million francs. In addition, 170 million francs were set aside for implementing the removal of slums and a further 570 million francs for housing improvements (Urbanisme, 1978).

It is claimed that the renovation of old areas has three main objectives: those of improving the value of the nation's real estate, maintaining social balance, and reinforcing the old urban symbolism and distinctiveness of districts. Thus the schemes for public restoration began with areas in the more historical towns such as Avignon, Sarlat and Tours, but some came to include areas in Lyons and Paris. However, no distinction is drawn between the true conservation of historic buildings and the renovation of dwellings. Many of the problems remain in Paris, where 620,000 households still did not have exclusive use of toilet facilities in 1977, and these were inhabited by 237,000 pensioner households and 194,000 households from the lower socio-economic groups. The social characteristics for the 35 per cent of poor dwellings in large city areas resemble Ottensen or Ostertor but the pace of improvement does not match that of Hamburg or Bremen.

(iii) Housing and urban regeneration in the eighties

In Manchester/Salford, the inner area had been subject to vast schemes of slum clearance and replacement by social housing schemes in the 1980s (Figure 6.3). Of the remaining undemolished housing stock about two-thirds was at risk because it was both pre-1919 and in private ownership in areas characterised by 50 per cent private renting where incomes of tenants and the many small landowners continue to be low. Building societies were unwilling or reluctant to lend to council nominated purchasers of pre-1919 housing, central government aid was cut back, improvement grants were not being taken – all contributing to continued deterioration. Sadly only 9400 houses were within the Housing Action Areas and General Improvement Areas in Manchester and Salford. Local authority housing had not been without problems. Inter-war council housing in, for instance, the Lloyd Street area of Manchester, was improved. 'Walk-up' flats of the immediate post-war period remained difficult to let and in need of modernisation while the deck access maisonettes such as the concentration of 2960 units in Hulme and Moss Side achieved national infamy for vandalism, rioting in 1981, structural faults, and problems for families with children. In all, 10 per cent of the social housing stock was difficult to let in 1978

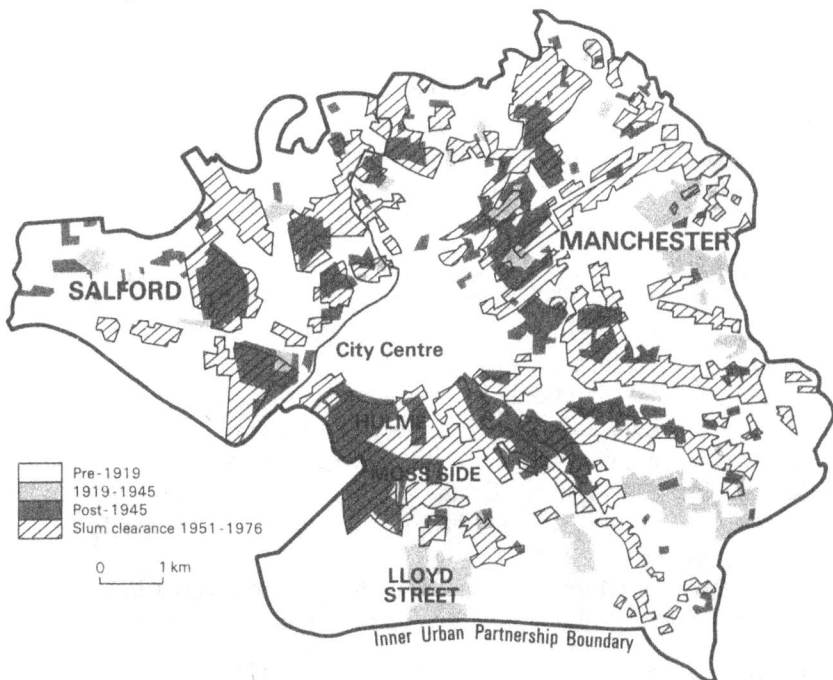

Figure 6.3 Slum clearance and age of buildings in the Manchester–Salford partnership area (reproduced by permission of Manchester and Salford Inner Area Study)

(Manchester/Salford IA Study, 1978) and it was no surprise that 48 per cent of Manchester's housing waiting list comprised inner city residents. Obviously the housing stress noted in inner Greater Manchester is part of a much broader all-pervading deprivation that characterises the inner areas of British cities and some outer areas as well. This picture was of Manchester in 1980; what has changed?

Gone today is the broad consensus that interventionist planning should be undertaken for the public good. Social desirability has been replaced by financial feasibility. Politically the New Right has enforced national policies upon the city which have brought about marked changes in housing policy which have relied increasingly upon the private sector and multi-national finance capital to dictate the nature of housing policy in the last decades of the century.

The sale of council housing to tenants has increased the correspondence between poverty and council housing because either only the least desirable public housing or only the housing occupied by those with least power in the marketplace is left unsold. In addition only 20 per cent of the profits from sales could be reinvested in house building with the result that new public housing could not replace the losses from sales. The third national imprint was a 54 per cent reduction between 1979 and 1985 in the housing resources budget (housing revenue allocation) whose effect on Manchester was the equivalent of approximately £50m being pruned from the housing budget in 1985 (Robson, 1988).

The policies which have been introduced all involve partnerships between the public and private sectors in the provision of housing, especially in the inner city because the outer city and beyond is the preserve of the private developer. At least four initiatives have aided the provision of housing in parts of inner Salford/Manchester:

(1) The Trafford–Salford Urban Development Corporation 1981 and the Mini UDC in Manchester 1987. Some of the worst 'sink' inter-war estates have been emptied and sold for refurbishing to the private sector. In this case the former council tenants have been relocated to 'better' estates and the cost of rehabilitation is being borne by the private sector.
(2) The Derelict Land Grant does provide private developers with funds to attract them into run-down areas such as Salford Quays.
(3) The Urban Development Grant, where for every £1 of government investment it is hoped to raise £4 for new housing and £2.5 for refurbishing, has been used since 1982.
(4) Urban Renewal Grants introduced in 1987 are being used in the Ordsall district for refurbishing houses by two major construction companies.

In the case of the Salford Quays development the local authority took the lead in partnership with central government, private developers and the dock owners to develop 300 houses as part of the scheme for waterfront generation to be completed by 1991. In this case the success of the scheme stems from the multiplicity of uses on the site – multi-screen cinema, business park, hotel, office, park and restaurants, and not just the housing – which is aimed at the upwardly mobile, professional, two-income, childless households in search of a 'designer' community (Shurmer-Smith and Burtenshaw, 1990). Those housed as a consequence of New Right ideology applied to housing are the 'children of Thatcherism' and not its abandoned waifs and strays.

Waterfront housing

As we saw in the case of Salford Quays, derelict docklands in seaport cities and inland harbours, the consequence of port decline and industrial abandonment of waterfronts, have offered many cities the opportunity to provide housing normally in conjunction with new business and commercial development symbolising rebirth and revitalisation (Hoyle *et al.*, 1988).

Within the London Docklands Development Corporation (LDDC), 2025 hectares are being redeveloped and refurbished. Housing forms a major part of the area but its provision has not been without problems. The activities of speculators raising land values from £60,000 to £4 million/acre in Wapping between 1982 and 1987 soon put much housing land beyond the reach of the local tenants. Therefore the affordable housing policy was developed by the LDDC to ensure that a proportion of the homes built on its land could be offered at a low price to local tenants. Between 1981 and 1987 12,000 new homes were being built on the LDDC area, especially on the Surrey Docks and at Beckton. In contrast, the conversion of old riverfront warehouses to luxury apartments, besides the building of new apartments on vacant river and dock frontages have both gentrified the area and brought the local low income population and the local authorities into conflict with the developers and the LDDC, whom they see as destroying the social structure of the area.

The use of docklands for housing is replicated in large and small sites throughout the United Kingdom in particular. Camber Docks, Portsmouth, Ocean Village, Southampton and Princes Dock, Glasgow contribute little to solving the housing problems of nearby inner city areas. However, in the case of Waterstad, Rotterdam the city has maintained a clear planning policy with an overt social dimension. A heavy commitment has been made to urban revitalisation policies aimed at disadvantaged communities. Waterstad has accommodated high income groups but, at the same time, the revitalisation policies elsewhere in the city have been predominantly for the original city population. This dimension is certainly lacking in the London case where national policies constrain the local authorities' ability to revitalise areas of public housing as was noted in the case of Salford. Hamburg is also developing policies which ensure that its dockland revitalisation does not destroy the social mix which exists in the area.

Housing as a public good

A similar socially oriented policy of housing rehabilitation has been pursued in

Marseilles (Baudouin, 1988). The Belsunce district had drastically changed its social composition between 1968 and 1982. Between 1968 and 1975 this district, associated with the wholesale clothing industry, lost 20 per cent of its French population as it was replaced by a predominantly Arab and Turkish population. The 'French' population comprised earlier generations of immigrants from Italy, Spain and Armenia, many of whom are now elderly. The current housing is of poor quality – 90 per cent built before 1914 and 60 per cent lacking in basic facilities of water supply and toilets – and is home to a low wage, mainly unemployed (50 per cent, 1988) population.

It is the task of SOMICA, a mixed private–public enterprise, to rehabilitate the area for the benefit of its 13,000 population. They began by getting the 'actors' together, not least the elderly who are those with interests in continuity. As a result, some of the hostel provision (the area's foreign born population is 71 per cent single men aged 20–45) has been dispersed to the city fringe, a park has been built on a cleared block, the Maghrebian market has been improved and the properties are being refurbished. Inner city refurbishing and greening of the residential environment are essential partners in this plan.

Bologna has provided housing within an overall plan for what the whole city needs. Bologna's theoretical basis of housing provision has been that the public good should be prevalent given the limitations imposed by the Italian economic and legal structure. The programme has attempted to stop speculation, to improve services and recreation space in the centre, to finance housing rehabilitation and to maintain a stable socially mixed population. By 1972 a plan for the five areas of the historic centre had been proposed which involved rehabilitation of residences in the centre where the costs of rehabilitation per square metre were 8 per cent lower than those of building in the periphery. The *per capita* costs of rehabilitation were one-third lower than those of developing the periphery. While renovation goes on the inhabitants are housed in special blocks of 33 apartments and 7 houses (*casa parcheggio* or parking houses) until their dwelling is completed, thus permitting a rolling programme of renewal. Private house construction has slowed because the best land has been zoned for co-operative and public housing. An unsuccessful attempt was made to enable public funds for land expropriation for new building to be utilised to expropriate private owners in areas in need of restoration and renovation.

Renewal in privately rented dwellings now takes the form of a contract over 20 years in which rents are restricted to the level of those in social housing (i.e. 12 per cent of average family income). Other conditions in the contract make it extremely probable that the property will be sold to the city or that the city will acquire more of the units. Corporate ownership is excluded. Not only is housing for low income groups guaranteed in this way but also allied plans for social services and transport reinforce the renewal programme's comprehensiveness. Bologna's approach to the total urban environment, rather than to sections of the urban environment, is unique in Western Europe although similar proposals are being utilised in Ferrara, Bergamo and Ancona.

Housing policies in Western Europe are full of contrasts but certain distinct trends do emerge. First, the national and local municipal involvement in housing is very strong, although the exact nature of involvement has been subject to the changing political ideologies of those in power during particular phases of city development. Second, there is a general movement towards the encouragement of owner-occupied housing in most cities although the socio-spatial dichotomy that this causes is of increasing concern. Third, there has been an acceptance that social housing schemes of the post-war period which were based on the genuine desire to assist the lower socio-economic groups did not have the sympathy of design or an appreciation of the social bonds that tied people to the older housing stock. Fourth, public housing policies are still governed by cost–benefit analysis rather than the shelter needs of society. Fifth, the need to reinstate housing townscapes that are more in keeping with a city's history and embody the physical elements of a nation's past ideologies, has been appreciated initially for aesthetic and social reasons and, all too often, for purely economic reasons by the housing managers. Finally, although we have tried to examine housing policy on its own, it is obvious that policies towards this sector of city planning cannot be isolated from other economic and social policies in the city, the nation and even the continent as a whole.

References

Barlow Report (1940) *Royal Commission on the Distribution of the Industrial Population*, Cmd. 6153, HMSO, London.

Baudouin, J. C. (1988) Crise urbaine et aménagement du centre-ville: l'exemple du quartier Bebunce à Marseille, *Peuples Mediterranéens*, **43**.

Becker, H. and Kein, K. (1977) *Gropiusstadt: Soziale Verhaltnisse am Stadrand*, Kohlhammer, Stuttgart.

Bengston, S. (1974) Housing and planning in Sweden, in *Public Housing in Europe and America*, (ed. J. Fuerst), pp. 99–109, Croom Helm, London.

Blacksell, M. (1968) Recent changes in the morphology of West German townscapes, in *Urbanization and its Problems* (eds M. Beckinsale and J. Houston), pp. 199–217, Blackwell, Oxford.

Boddy, M. (1976) The structure of mortgage finance: building societies and the British social formation, *Trans. Inst. Brit. Geog., New Series*, **1**(1), 58–71.

Bremische Gesellschaft (1977) *Sanierung Ostertor*, Bremische Gesellschaft für Stadterneuerung, Stadtentwicklung und Wohnungsbau mbh.

Castells, M. (1984) *The City and the Grassroots*, Edward Arnold, London.

Cribier, F. (1970) La migration de rétraite des fonctionnaires Parisiens, *Bull. Assoc. Geog. Français*, **381**, 119–122.

Daun, A. (1976) 'The ideological dilemma of housing policy', in *Plan International – Habitat 76* (ed. G. Corlestan), pp. 21–30, Swedish Society for Town and Country Planning, Stockholm.

Department of the Environment (1974) *Urban Guidelines Studies, Sunderland, Oldham and Rotherham*, HMSO, London.

Duclaud-Williams, R. (1978) *The Politics of Housing in Britain and France*, Heinemann, London.

Friedrichs, J. (1985) *Die Städte in den 80er Jahren*, Westdeutscher Verlag.

Fuerst, J. (ed.) (1974) *Public Housing in Europe and America*, Croom Helm, London.

Gaspar, J. (1984) Urbanisation: growth, problems and policies, in *Southern Europe Transformed* (ed. A. Williams), Harper and Row, London.

Gatzweiler, H. (1975) Zur Selektivitat Interregionaler Wanderungen, *Forsch. zur Raumentwick.*, Band 1.

Greater London Council (GLC) (1974) *A Strategic Housing Plan for London*, GLC, London.

Green, E. (1959) West German city reconstruction, *Sociol. Rev.*, **7**, 231-244.

Green, E. (1964) Politics and planning for reconstruction, *Urban Studies*, **1**, 71-78.

Hallett, G. (1977) *Housing and Land Policies in West Germany and Britain*, Macmillan, London.

Heineberg, H. (1988) The City in West Germany, *Geographische Rundschau*, Special Edition, pp. 20-28.

Hoyle, B., Pinder, D. A. and Husain (eds) (1988) *Revitalising the Waterfront: international dimensions of dockland redevelopment*, Heinemann, London.

Info Service (no date) *The Future of the Old Housing Stock in the Netherlands*, Ministry of Housing and Physical Planning Information Service, The Hague.

Keating, M. (1988) *The City that Refused to Die*, Aberdeen University Press, Aberdeen.

King, J. (1971) Housing in Spain, *Town Planning Review*, **42**, pp. 381-403.

Koch, R. (1976) Altenwanderung und Raumliche Konzentration Alter Menschen, *Forsch. zur Raumentwickl.*, Heft 4.

Kouvelas, S. (1976) Public discussion on Salonica's illegal settlements, *Technica Chronica*, **46**, pp. 41-42.

Lagarde, J. (1968) Les Grands Ensembles douze ans après, *Urbanisme*, **106**, pp. 30-34.

Laino, G. (1988) Naples à la fin des années quatre-vingt, *Peuples Mediterranéens*, **43**(2).

Law, C. and Warnes, A. (1973) The movement of retired people to seaside resorts. A study of Morecambe and Llandudno, *Town Planning Review*, **44**(4), pp. 373-390.

Lawson, R. and Stevens, C. (1974) Housing allowances in West Germany and France, *J. Soc. Policy*, **3**(3), 213-234.

Manchester and Salford Inner City Partnership Research Group (1978) *Manchester and Salford Inner Area Study*, Manchester and Salford Partnership.

Ministry of Local Government and Labour (1976) *Current Trends and Policies in the Field of Housing, Building and Planning* (mimeo), H 1024 Oslo.

Möller, I. (1985) *Hamburg*, Klett Landerprofile, E. Klett Verlag.

Monheim, H. (1972) Zur attraktivitat Deutscher stadte, *Berichte zur Regionalforsch.*, Heft 8.

Monheim, R. (1979) Wohnungsversorgung and Wohnungswechsel, *Geog. Rund.*, **31**(1), 17-28.

Notes et Études Documentaire (1976) Lille et sa communauté urbaine, *Notes et Études Documentaire*, pp. 4297-4299.

Ottensen (1977) *Urban District Development in Hamburg Illustrated with Ottensen as an Example*, International Union of Local Authorities World Congress, 1977.

Power, A. (1976) France, Holland, Belgium and Germany, a look at their housing problems and policies, *Habitat*, **1**(1), pp. 81-103.

Racine, E. and Creutz, Y. (1975) Planning and housing in France, *The Planner*, **61**, 83-85.

Rappoport, A. (1968) Housing and housing densities in France, *Town Planning Review*, **39**(4), pp. 341-354.

Riquet, P. (1988) 'Mobilité residentielle, marché immobilier et conjoncture dans les grandes villes allemandes', *Annales de Géographie*, **98**, pp. 1-39.

Robson, B. (1988) *Those Inner Cities*, Oxford University Press, London.

Sbragia, A. M. (1979) Milan and public housing policy, in *Western European Cities in Crisis*, (ed. M. C. Romanos), Lexington Books, Lexington, USA.

Shurmer-Smith, L. and Burtenshaw, D. (1990) Urban decay and rejuvenation, in *Western Europe: Challenge and Change* (ed. D. A. Pinder), Belhaven, London.

Smith, M. (1973) Housing in Amsterdam, *Housing*, May, pp. 9–15.

Thomas, W. S. G. and Tuppen, J. (1977) Readjustment in the Ruhr – the case of Bochum, *Geography*, **62**, pp. 168–175.

United Nations (1973) *Urban Land Policies and Land Use Control Measures. Volume 3, West Europe*, United Nations, New York.

United Nations (1974) *Compendium of Housing Statistics 1971*, United Nations Department of Economic and Social Affairs Statistical Office, New York.

Urbanisme (1978) Renovation urbaine, *Urbanisme*, **46**, pp. 162–163, Thematic Issue.

Vance, J. (1977) *This Scene of Man*, Harper and Row, London.

Watson, C. (1971) *Social Housing Policy in Belgium*, Occasional Paper 9, CURS, Birmingham.

Wendt, P. (1962) 'Post-World War 2 housing policies in Italy', *Land Economics*, **38**(2), pp. 113–133.

Wilmott, P. and Young, M. (1973) *The Symmetrical Family*, Routledge and Kegan Paul, London.

Wilson, H. and Wommersley, L. *et al.* (1977) *Change or Decay: Final Report of the Liverpool Inner Area Study*, HMSO, London.

Wohlin, H. (1977) *Framework for Urban Development in Sweden* (mimeo), Department of Planning and Building Control, Stockholm.

CHAPTER 7

Urban Conservation

Although there is an obvious relationship between the physical form of a city, its buildings and patterns of streets and spaces, which are brought into being to serve the functions that a city performs, the physical fabric commonly outlives the functions for which it was created. Houses often stand for two or three generations, public buildings frequently longer and the patterning of roads and squares can survive for many centuries. Much of the distinctiveness of the West European City stems from the antiquity of its surviving morphological forms. The type, location and space demands of the functions undertaken by the city, however, change much more rapidly. This discontinuity between form and function leads to a basic planning dilemma: should the existing urban forms be adapted or reconstructed to accommodate the new functional demands, or should these demands be constrained to fit the existing forms? The first alternative assumes that higher functional efficiency outweighs the financial and social costs of rapid and violent morphological change, while the second assumes that the benefits from preservation outweigh the costs of constraint.

The traditional models which attempt to explain the internal structuring of cities largely ignore the existence of this dilemma. The North American cities from which most of these models were derived have had short histories, dominated by rapid change, have consequently a smaller architectural inheritance, and until recently are characterised by attitudes that equate the value of a building inversely with its age. In Europe, however, the character and civic pride of individual cities, the quality of life of its inhabitants and, increasingly, the major economic activity of tourism, have depended to a large degree on surviving relict urban forms. 'If there is a "European" culture then it is first and foremost an urban culture' (Knox, 1985). Conservation planning is therefore a central part of the whole process of planning the European city, not merely a special constraint to be suffered in a few specific districts in a few specially endowed towns. A conservation ethic, in the sense of a recognition of this relationship of urban form and function, will thus be found pervading many of the chapters of this book, and the topic is only synthesised here in isolation from other urban activities for the sake of convenience.

The argument of this chapter is that the European city has experienced a period of particularly rapid change over the past 40 years which has accentuated the discrepancy between form and function, focused public and professional attention upon the consequent planning dilemma, and led in reaction to a revaluation of the surviving morphological forms. The conservation movement is to a large extent a peculiar phenomenon of this century and this continent, and its success in striking a new balance in urban planning between form and function has in turn created a set of new planning problems for the West European city.

Pressures for change

Change is, of course, endemic in any city that has not become a fossilised museum exhibit, and more than a thousand years of urban development in Europe have depended upon the operation of a cycle of demolition and renewal. The last few generations, however, have witnessed a dramatic acceleration in the pace of change which has replaced a slow piecemeal renewal with the rapid comprehensive development of large tracts of the city. The tendency of the European countries to use their cities as battlefields had resulted by 1945 in the destruction of the central areas of many German, British and Italian cities especially (Hewitt, 1983), but this demolition was quantitatively far less that that wrought upon the city in the subsequent 30 years by planners and developers responding to changes in the demands society was making upon urban land.

A fundamental change is fluctuations in the size, distribution, family structure and wealth of the urban population. Buildings exist to serve the needs of citizens and an increase in population leads to pressures for more or different structures, while a decrease leaves redundant buildings for which new uses must be found or dereliction and demolition ensues.

The well-documented flight of population from the central areas of cities to peripheral suburbs (White, 1984) has left the inner districts of many West European cities, which usually contain a high proportion of the older building stock, with declining residential, retailing and employment functions. The islands in the lagoon that comprise the historic Venice (Figure 7.1) contained 47 per cent of the city's population in 1951 but only 29 per cent by 1970, with almost three-quarters of those aged under 45 moving out during that period. Less drastic, but equally marked, has been the fall in the population of Paris 'intra-muros', Amsterdam within the *grachten*, Stockholm and many other major European cities. This trend removes some of the pressures on the inner city and releases buildings for preservation. It also, however, makes conservation more urgent, as it removes much of the motive for private maintenance of property and invites successive dereliction and comprehensive clearance and redevelopment to serve new uses that realise the economic values of the site. Large parts of even Europe's most acclaimed historic cities have been lost since the war because the time-lag between the flight of the resident population and the appreciation of the possibilities of restoration was long

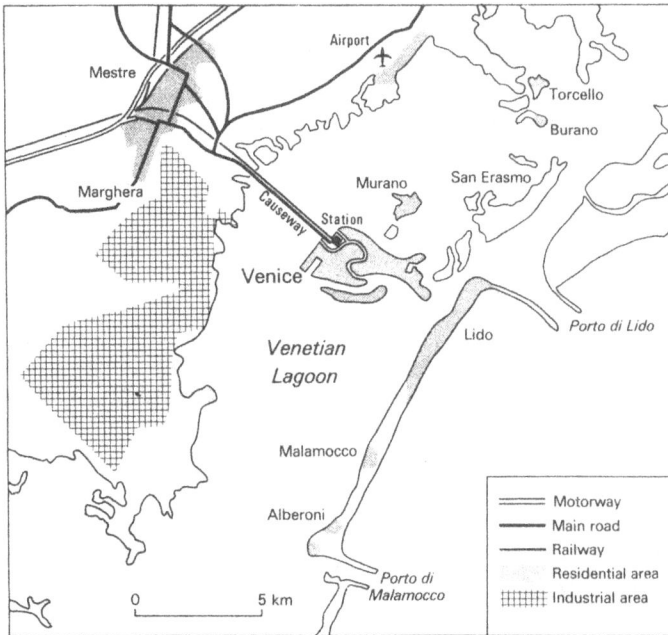

Figure 7.1   Venice urban area (reproduced from *The Geographical Magazine*, London)

enough to make the latter prohibitively expensive. The Southgate and Walcot areas of Bath are typical examples of comprehensive clearance and redevelopment in the 1960s for commercial uses to fill a vacuum created by the outmigration of the resident population over the preceding 20 years (Figure 7.2).

Such vacuums can equally result from the spatial shifts of the centres of commercial or industrial activity which leave zones of discard characterised by structural decay and occupied by low rent uses. Waterfronts in particular provide distinctive and widespread examples of this process (Tunbridge, 1988; Hoyle, Pinder and Husain, 1988).

Although the outmigration of people from the historic city centres often leads directly to pressures for redevelopment, conservation may itself encourage such movement by raising the value of property and enticing existing residential or commercial users to realise the value of their appreciating assets. The relationship between conservation and change in the social functional character of neighbourhoods is imperfectly understood. Is it the expenditure on renovation and rehabilitation that encourages richer residents to acquire city-centre properties in a socially selective, 'back to the city' movement, thus displacing lower income groups? Or is it the movement of the working class, more or less willingly, to newer and more comfortable suburban locations that

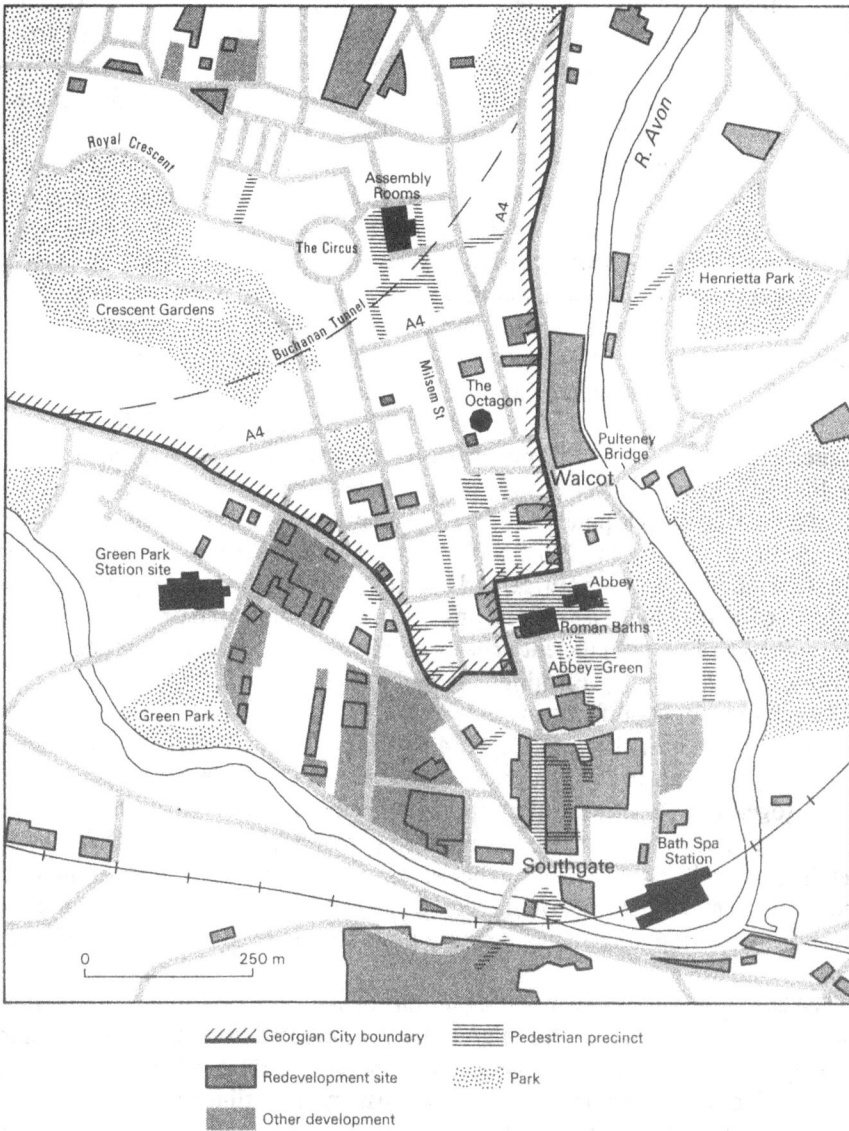

Royal Crescent

Assembly Rooms

R. Avon

The Circus

A4

Crescent Gardens

Henrietta Park

Buchanan Tunnel

A4

Milsom St

A4

The Octagon

Pulteney Bridge

Walcot

Green Park Station site

Abbey

Roman Baths

Abbey Green

Green Park

Bath Spa Station

Southgate

0   250 m

Georgian City boundary   Pedestrian precinct

Redevelopment site   Park

Other development

Figure 7.2 Georgian Bath and proposed developments in the city

leaves a vacuum which in turn is filled by people more willing and able to bear the costs of renovation and maintenance? Case studies from cities in France (Kain, 1975), Italy (O'Riordan, 1975), Britain (Ward, 1968) and Denmark (Skovgaard, 1979) agree that social and functional change and the conservation of the physical fabric of cities are intimately related, but the chronology of cause and effect is generally locally determined.

An important motive for the conservation of cities is the wish to present a contemporary interpretation of selected aspects of a perceived past (Williams *et al.*, 1983). For very similar motives a confident present will wish to leave a record of its existence in the fabric of the city. Thus both preservation and reconstruction can embody ideological motives. The 'Haussmannisation' of Paris, with its wide boulevards and radial street patterns, was to reflect in stone and brick the grandiose policies of the Second Empire as much as to facilitate military control over unruly Parisians (Sutcliffe, 1970). Such wholesale elimination of the physical record of a a thousand years of Parisian building is in a long tradition of politically motivated reconstruction which would include, among the most comprehensive examples, the eighteenth century German Residenzstadt (Mumford, 1961), and the large scale rebuilding of both Rome and Berlin after 1870 as considered appropriate for the capitals of the new Italy and Germany. 'The spirit of Haussmann had also been abroad in London, Manchester and Liverpool, and seems to feel itself just as much at home in Berlin and Vienna' (Engels, 1935). It is not too far fetched to suggest that the large scale office developments and transport systems that transformed the West European city during the 1960s and 1970s were prompted by the need for the visible symbols of a new political order (Esher, 1983) and that the possession of a clutch of central high rise offices, an enclosed shopping mall or inner ring-way was viewed as a sign of efficient urban government, regardless of the cost to the historic morphology (eg Birmingham Bull Ring).

The changes in manufacturing industry, retailing and office employment over the last 20 years described in Chapter 4 have all intensified pressures for change in the physical fabric of the urban environment. The 'deindustrialisation' of cities, together with the outmigration of other commercial functions in search of better access or larger sites, has reduced demand in some areas, threatening dereliction, while paradoxically other areas are threatened with comprehensive redevelopment as a result of increased pressures for space from economic activities requiring single-site, purpose-built structures on a scale previously unknown in the city.

The combination of new building and information processing technologies, the new requirements of an increasingly integrated world economy, and development in the financial system that together triggered the largest and longest office building boom in European history, is outlined in Chapter 4 (see also Bateman, 1985). Its effect upon the physical structure of cities, and especially the inner areas of the handful of cities that developed an international role, is difficult to overestimate. In Brussels, for example, the office boom, initiated by the World Fair of 1958 and subsequently fuelled by the growing international function of the city, resulted in the completion in 1967 of the first of Europe's 'Manhattan' downtown business centres in the Noordwijk district. Other cities followed: Porte Maillot in Paris, the South Bank in London, all involved substantial demolition which could be justified in many cases by the condition of the existing property, but also brought a visible change in the character of these cities, radically altering the scale of building and the resulting skylines. Similarly, new trends in retailing produced the standardised high street facade, and the covered downtown shopping centre and its associated car parking. Hoog Catherijne, built around the railway station at Utrecht to serve as a regional market, is one of the largest of its kind and has resulted in a clear physical and functional division of the city between the new purpose-built large-scale and transport accessible commercial centre and the small-scale, speciality retailing in the conserved pedestrian old city. The changes initiated in the world cities filtered down through the urban hierarchy, eventually having a considerable impact upon the character even of cities such as Bath, cited above, which developed the Southgate shopping centre and high rise office and parking developments.

The most insistent pressure for change, however, can be ascribed to the increasing consumer prosperity of Western Europeans over the last 25 years. Higher standards of space provision for housing, amenity and, above all, circulation and parking for the private motor car (see Chapter 5) have all been demanded and are within financial reach. Policies of restraint for the sake of preserving an historic morphology, which was created to serve quite different demands, deny citizens the accessibility which they have come to regard as a right and which is the main *raison d'être* for urban living; thereby the flight of people, jobs and amenities from the inner city to the suburbs is hastened. The desires of a newly prosperous citizenry for more private consumption of urban space has been paralleled by similar demands for more public provision. In many inner cities currently acceptable standards of public circulation, emergency service access and recreation space cannot be met within the crowded courtyards and narrow streets, and the seeming paradox of selective demolition is an inherent consequence of preservation and rehabilitation.

The reaction to change

Although sporadic examples of the protection of monuments and antiquities can be found world-wide over the last 2,000 years, the emergence of a coherent conservation ethic as an alternative planning strategy is a phenomenon produced by a peculiar set of circumstances in North-West Europe over the last 200 years. It can be attributed to a mix of psychological needs, social and intellectual fashions and consumer prosperity; these led to a concern with the quality of urban life and an association of that quality in Western Europe with a reappraisal of the value of the form of the city, together with the availability of disposable incomes to purchase this.

Local reactions

In Western, as opposed to Eastern, Europe most of the impetus of the conservation movement was local rather than national, and unofficial rather than governmental. The eighteenth century witnessed the birth of an interest among a leisured elite in their physical and cultural environment, including an interest in the architectural relics of the past. This resulted in the establishment of learned societies in capitals such as London, Paris and Berlin, dedicated to their discovery, field-study and, inevitably, preservation of monuments. Such enthusiasm filtered down through the urban and social hierarchies as groups of citizens, predominantly drawn from the articulate middle-classes, formed pressure groups in response to immediate threats to their local urban environment. The Cockburn Society of Edinburgh for example was founded in 1875 to protect the Georgian 'New Town' from the threat of 'improvement'. A popular movement to 'Save Montmartre' – 'the last hamlet in Paris' – was a reaction to demolition proposals in 1911 and led to the more broadly based Association of the Friends of Paris (Fawcett, 1976).

Such pressure groups could for many years do little more than react, usually without much success in defending prominent landmarks, but eventually they shaped a consensus of informed opinion that could be highly effective in particular instances and, more important, put pressure on governments to initiate official national intervention. A well documented and dramatic but not untypical example is provided by the Bath Preservation Trust (Aldous, 1972). The growing demands of traffic led to a proposal from the influential Buchanan consultants for a tunnel to carry the main A4 road underneath the Georgian architectural masterpieces of the Circus and the Royal Crescent (Figure 7.2). This proposal was the catalyst that united popular feeling in defence of the town's heritage and focused a vaguely felt unease on a single issue. A classic confrontation of conservationists with developers was skilfully and successfully managed by the professional talents that rallied to the cry of 'Save Bath'.

Similar local spontaneous action occurred throughout the cities of Europe wherever development was seen as threatening the historical continuity of the built environment. In Brussels the building of the ITT office block in 1972, in itself no more intrusive than many of its predecessors, provoked popular indignation and a last ditch defence to save the little that was left of the eighteenth-century city. Here conservationists were able to find a common cause with local politicians of the Front Democratique Francophone concerned with the destruction of traditional working class communities such as the 'Marolles' and 'Sablons' (Culot, 1974). The responsiveness of governments to such orchestrated popular pressures varied along a spectrum from Scandinavian consensus to French bureaucratic rigidity. The former is illustrated by the sensitive redevelopment of 'Gamla Stavanger' in Norway, which set a pattern for the working relationship of local governments and amenity groups in many other Scandinavian cities. The latter can be appreciated in the largely successful attempts of such groups as 'SOS Paris' since 1973 to introduce some local responsiveness into Parisian redevelopment in the face of the rapid, visually dramatic and uncompromisingly modern developments, such as the Beauborg Arts Centre, Montparnasse Tower, Les Halles redevelopment and left bank motorway, that transformed the inner city in the 1970s (Fabre-Luce, 1976).

In almost all cases, strongly felt but vaguely formulated local reaction to a particular development led to the establishment of local formal associations and coalitions of various interest groups which in turn led to national organisations. One of the most successful, and something of an archetype for others, was the British Civic Trust, born from a sense of 'outrage' (Nairn, 1955), officially established in 1967 and resting on a base of some 1,250 local societies and building trusts.

The international dimension

Notably the two spatial scales of urban conservation activity that have proved to be the most innovative are the local and the international. Architecture and history have an immediate international appeal, and the 'Grand Tour' to view the ruins of classical civilisation has grown into the mass annual migration of foreign tourism. The concept of the existence of an international architectural heritage has provided a major tourism resource, while travel has provided an awareness of the need for conservation across international boundaries. As early as 1963 the Council of Europe recognised that the protection of historic buildings and townscapes was an important part of its work and established 'Europa Nostra' initially to investigate a sample of 50 projects in as many cities (Council of Europe, 1974). Similarly UNESCO summoned an international forum, significantly in Venice, in 1964. As had occurred on the local scale, the international movement was fired by a reaction to specific threats. The 1966 Italian floods on the Arno resulted in the well publicised destruction of an alarming proportion of what was increasingly regarded as European heritage. The vulnerability of some of the most famous European cities was revealed, as was the inadequacy of national governments to meet major catastrophes and the depth of feeling about this in Europe and North America. The sudden realisation that cities such as Florence or Venice, or at least their historic centres, might not exist for the next generation of Europeans to enjoy inspired a crusade in which voluntary relief agencies (with such titles as 'Save Venice', 'Venice in Peril', or 'Arbeitskreis Venedig') funded from abroad, were orchestrated by UNESCO. A series of international conferences, beginning in 1971 in Split, and meeting in successive years in various European cities, called for a new climate of opinion favouring rehabilitation over demolition and stronger national legislation to effect this. European Architectural Heritage Year in 1975 was an attempt to stimulate public opinion and give a continental dimension to the problem as well as to put pressure on laggard national governments to initiate legislation. The tone of this, and subsequent, campaigns was simultaneously both narrow and technical as well as broad and political. On the one hand technical expertise was exchanged, while on the other the conservation of buildings was placed in the wider context of the rehabilitation and renewal of areas and the general improvement in the quality of the urban environment by national and local government sponsored action.

Britain's contribution to this 'campaign for an urban renaissance' included the case of Manchester as well as the less surprising historic city of Durham, which illustrates how far urban conservation had moved from the preservation of monuments towards the renewal of historic areas in multifunctional cities.

The international movement has been effective in levelling up standards, which has usually meant raising the standards of the legislation in Mediterranean countries to the best practice existing in North-West Europe (Council of Europe, 1979). In addition it has financed key restoration projects, acted as a forum for the exchange of technical information, and stimulated public concern. Equally, however, the limits of international action have become clear, especially as a result of experience in Venice, a city regarded not merely as a prime exhibit in the history of European urbanism, but also seen as a model for the city of the future by planners as diverse as Buchanan, Mumford and Le Corbusier.

Twenty years' experience of international effort in the city has revealed that international money and voluntary labour are only effective if action is taken to remove the causes of undesirable change, which in turn depends upon the Italian authorities. Bureaucratic obfuscation and political controversies between the changing national governments, the Veneto region and the Venice commune delayed the implementation of the long awaited plan for preventing a recurrence of the 1966 floods and solving the long term problems of the rising level and increasing pollution of the lagoon (Fay and Knightley, 1976). The still largely inviolable sovereignty of the nation-state within its frontiers removes much of the meaning from the concept of a 'European Heritage', especially when local and international priorities conflict. If Venice and Florence, or for that matter Nürnberg, Berne, Bath, Chartres and Salzburg, are an international responsibility, there must be some way that this responsibility can be exercised other than through national governments.

National reactions

There is a remarkable degree of conformity among the urban conservation policies formulated by the West European countries, perhaps more so than in most aspects of urban planning. The pressures for change, the nature of the local response, the division of responsibilities between governments, pressure groups and commercial interests are all broadly similar from Portugal to Sweden and from Ireland to Austria, despite national differences in wealth, political philosophy and endowment of historic resources. There is a clear sequence of events through which all proceed, although at different paces.

1. Inventorisation
Pressures from the amateur enthusiasts reacting to local threats resulted in the establishment of official bodies charged with the drawing up of a national inventory of surviving meritorious architectural and historical relics. The

French 'Commission for Art and Monuments' began as early as 1790, but was discontinued only to restart in 1889. More typical in timing was the Dutch 'Rijksdienst voor het Monumentenzorg', a department of the Ministry of Culture, established in 1875. Inventorisation inevitably requires the establishment of national criteria whose application generally results in a grading system. Such criteria were nearly always intrinsic to the structure itself, such as qualities of age, and aesthetic or historic value, resulting in such classifications as the British grades 1, 2, and originally also 3, the French 'monuments classes/monuments inscrits', or the Dutch monuments of national, local and street ensemble importance.

2. Legislative protection

It is such a short and bureaucratically obvious step from inventorisation to action to protect what has been listed that the long delay that occurred in most countries needs explanation. Listing originally conceived as the simple task of recording the self-evident, proved longer and more complex than anticipated. The Dutch national list was not declared to be 'complete' until 1908. Even more of an obstacle was the deeply ingrained respect for ownership rights over private property, and consequently resistance to any state constraint over the exercise of these rights. Premature protective legislation (such as in Belgium in 1809 and 1836, and in the Netherlands in 1814 and as late as 1910) was rendered ineffective by a climate of opinion that was not yet ready to accept that 'property entails obligation. Its use should also serve the general well-being', as was written in the West German constitution only 40 years ago. The opinion forming role of the amateur groups and the immediacy of the local threat were critical in preparing society to accept what was a basic infringement of longstanding individual freedoms. It is revealing that the early legislation in a number of countries, such as Belgium, applied only to publicly or church owned buildings. Nevertheless preservation from arbitrary destruction was legally possible, if not always enforceable, for at least major monuments in most European countries, before the Second World War commenced its widespread demolition of them. This legislation included the French 'Law on Historic Monuments' (1913), Danish 'Protection of Buildings Act' (1918), the Dutch 'Monument Act' (1921), Belgian 'Act for the Protection of Landscapes and Monuments' (1931) and Italian 'Monument Act' (1939).

3. Conservation legislation

The following step was for governments to accept the inevitable consequences of their protective actions. Preservation leads to maintenance and ultimately renovation and rehabilitation. This in turn implies the acceptance of financial responsibilities through various forms of subsidy, the shift in the focus of attention from the individual monument to the wider morphological setting and, ultimately, the inclusion of functional characteristics alongside formal ones. In summary conservation, as opposed to preservation, legislation was needed. This step was taken in the Netherlands in 1961 ('Monument Act'),

France, 1962 (the 'Loi Malraux'), the United Kingdom, 1967 ('Civic Amenities Act'), Italy, 1967 ('Urban Planning Act' and Law 512, 1982) and had even reached Turkey by 1973 ('Monuments and Historic Buildings Act').

The main features of this legislation were remarkably similar (UNESCO, 1975). Most included a mix of national subsidy, tax concessions and private financing, added a designation of areas to those of buildings (the French 'Secteur Sauvegarde', the British 'Conservation Area', the Dutch 'protected urban facade') and made some attempt to link preserved forms with intended functions, so as, in Burke's phrase, to 'preserve purposefully' (1976). The results were however by no means uniform. For example a spectrum can be recognised from the centralised French concentration on a small number of set pieces of national importance (around 35,000 monuments and 500 conservation areas) to the British deconcentration of criteria and initiatives resulting in a more numerous and qualitatively varied pattern (around 300,000 monuments and over 5,000 conservation areas) (Kain, 1981). The Netherlands, West Germany and Scandinavia have tended towards the British 'overlisting' while Italy, Belgium and Turkey have been more 'French' in their 'underlisting' (Dobby, 1978).

4. Urban conservation planning practice

In the same way that the national inventories were mistakenly viewed as a once-off recording of an identifiable finite resource, so also was the conservation legislation seen as a comprehensive and permanent framework for the planning and management of the historic urban form. Over twenty years experience in the cities of Europe has revealed this not to be the case and the challenges encountered have been met by a body of evolved planning practice not envisaged in the various Acts.

These challenges, and the practical solutions to them, can be summarised under the three headings of the spatial scale of operations, the balance between public and private roles, and the context of urban conservation in more broadly based urban planning.

The conservation process has tended over time to spread from a concern with the larger, spectacular and usually publicly owned buildings to those that are smaller, more mundane and privately owned. Simultaneously there is a devolutionary trend away from national criteria, selection and finance. In some European countries, such as West Germany, Austria, Switzerland and Belgium, conservation management was early committed to the provincial or urban level, while in others, such as the Netherlands, Denmark and Britain, and more recently even France, practice has encouraged such a shift to the local level. One result is an unevenness within countries in the intensity of the conservation effort and the selection of areas and buildings (as can be seen in Figure 7.3 for the Netherlands). This metropolitan concentration could equally be traced for most West European countries and reflects differences in awareness as much as the spatial distribution of historic architectural resources (Ashworth, 1984).

Secondly, the increase in the number of conserved buildings and the

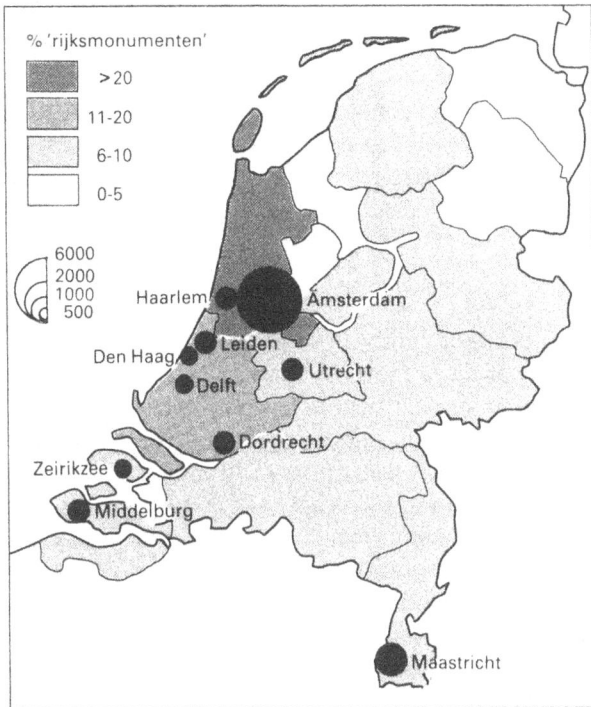

Figure 7.3 The distribution of nationally listed buildings in the Netherlands

increasing proportion of smaller, privately owned and domestic buildings raised questions about the balance between public and private costs and benefits. Similarly, legal protection costs little in itself and even the designation of conservation areas commits minimal public expenditure. However there is a logical if not legal long-term continuous commitment to at least maintenance and repair, if not renovation and rehabilitation. Understandably, therefore, local authorities will wish to encourage financially viable occupation of the ever increasing quantity of premises they have conserved.

Public–private ownership is not a new idea in West European urban conservation. In France, for instance, the restoration work is usually undertaken by a consortium, such as SOREMA in the Marais district of Paris, composed of the central government, local authorities and often private companies (Stungo, 1972). In other countries, such as Britain, the Netherlands and Sweden, investment through the private housing market has in practice assumed much of the accompanying cost once conservation areas have been officially designated.

Thirdly, no longer was conservation concerned only with a few special 'islands' of historic interest within a 'sea' of normality but with most of the central areas of most West European cities. This inevitably brought conservation into contact with urban land-use and development planning, and made it the concern of the urban planner and manager as much as the historian or architect. In a number of countries, such as the Netherlands, the national conservation legislation itself required the subsequent production of a local land-use plan ('Bestemmingsplan'), but even where this was not legally required, in practice it proved necessary unless large tracts of the city were to remain inadequately used as development was frozen.

5. Variety in application

Many examples of both the variety of types of conservation and its integration with urban planning as a whole can be found in West Germany. As a result of war-time destruction, reconstruction is more usual than restoration. A reconstruction that is more a facsimile re-creation of the spirit of the original, rather than a search for detailed authenticity more typical of the cities of Eastern Europe, was the model for Munster's 'prinzipalmarkt', and the central areas of Trier, Bamberg, Freiburg and Nürnberg. Where public support, city initiative and Land money have combined the results have often been spectacular in their scale, quality of detailing, and functional integration with urban activities. The restructuring of the Marienplatz district of Munich between 1967 and 1972 is an example where preservation, renovation and reconstruction have been successfully combined. In contrast, Heidelberg, spared destruction as the 'city of the student prince', demonstrates a systematic and meticulous house by house renovation in which the preservation of form takes precedence over the requirements of function.

Although there was a discernable homogeneity in the national legislation across Western Europe there is equally a clear spectrum of effectiveness between what can be labelled the 'Scandinavian' and the 'Italian' situation, although not exclusive to either of these areas. Scandinavia has the legislation, the money and the will but a shortage of the basic resource, namely the buildings themselves. The use of wood as a building material and the peripheral role of the Scandinavian countries in the development of the European city has left them with less to conserve. The obvious exceptions to this general assertion, such as Copenhagen, Odense, the Jutland and Schlesvig towns, Stockholm, Goteborg and the Scania towns, illustrate the dilemma posed by consumer prosperity and high standards of public amenity provision (Skovgaard, 1979). These make rehabilitation for housing difficult as contemporary standards of living space, vehicle parking and circulation provision, public open space and the like cannot be accommodated without considerable demolition.

The Italian situation is in many ways the reverse of this, with a surfeit of architectural riches resulting from its role as the cradle of European urban civilisation being entrusted to one of the European countries least able economically or in terms of the effectiveness of its government institutions to conserve it. The legislation exists in the Monument Act (1939), Urban Planning Act (1967) and Housing Act (1971) but implementation depends upon the skill and determination of individual city governments, which all too often accommodate rather than oppose entrepreneurial initiative (Fried, 1973) and are incapable or unwilling to implement existing legislation. Comprehensive conservation plans have long existed for such important historical towns as Venice, Genoa, Palermo, Siena, Urbino and Assisi, each embodied in a special law. In each case, however, the obtaining of the necessary enabling legislation has long been delayed. In the case of Rome, Law 512 (1982) enabled a special law to be passed to clean and restore buildings and statues damaged by smog. In the early years scaffolding enshrouded many monuments but now that the programme has been completed, the estimated L9000 billion received by the city from cultural tourism is an adequate return.

The three-tier local government structure has proved particularly inappropriate. The city (commune) has the historic resources and the motive while the province and region have much of the money and the power. Differences of interest, fanned by historic jealousies, between cities, between central cities and suburban municipalities, and between urban and rural areas, can prevent effective action. Small wonder then that Italy provides some of the most tragic as well as the most heartening cases. It was often maintained cynically that natural disasters are needed to get restoration under way. In 1966 Florence was flooded, as was Venice, and Naples received assistance after the 1980 earthquake.

Some of the smaller hill towns such as Urbino and Arezzo in the Marches, and Bergamo in Lombardy, have been skilfully preserved almost *in toto* and are in effect well maintained open-air museums. A different sort of success is exemplified by Bologna, where the communist administration has endeavoured to preserve the inner city through a vigorously enforced conservation area of 430 hectares, without stifling the economic life of this growing industrial and regional service centre of over half a million inhabitants (Figure 7.4). Against such successes must be set the continuing tragedy of cities such as Venice, which still await effective action, and the major cities of Rome and Naples where successive contradictory plans achieved little that was coherent. In Naples, 'Napoli 99', a group of women philanthropists led by Mirella Baracco, have managed to get commercial sponsors for major restoration work since 1982.

Thus more than twenty years' experience in operating the national legislative framework has resulted in the evolution of a body of practice that was not always originally envisaged in it. Decentralised and largely *ad hoc* planning and management at the local scale has replaced the more Olympian visions that motivated the national laws. Significantly few countries have produced new radical legislation and those that have, such as the Netherlands (1987), have generally confirmed rather than arrested these trends. It was implicit in the actions of governments and pressure groups that the urban architectural heritage existed in a fixed quantity and that it was only necessary to identify it, using obvious intrinsic norms, and then protect it by legal classification. Practice has demonstrated that no such immutable stock exists and that the

Figure 7.4   Bologna's *centro storico* restoration proposals

conservation movement, in a sense, creates what it wishes to preserve (Ashworth, 1984). The designation of protected status is the beginning not the end of a planning process. The preservation of form has implications for urban functions, and conservation therefore becomes an instrument and style of urban management.

Consequences of success

It may seem premature to claim success when large intra-continental differences exist in the effectiveness of urban conservation both between and within countries, and when between two and five per cent of the legally protected building stock is demolished annually. However, what began as the enthusiasm of small groups of eccentrics is now encapsulated in legislation in all West European countries and is supported by the major international organisations. The rescue of important urban monuments from imminent destruction is now less significant than extending conservational planning designations over large parts of cities, including Sheffield and Duisburg as well as Chichester and Heidelberg. The Council of Europe has long seen urban environments as part of a broad 'popular culture' (Mennell, 1976) and the historic cities of Europe support a major heritage tourism industry for those who come to enjoy them (see Chapter 9). Even more significant, the conserved city has become a valued part of the life of ordinary citizens. In France (Busson and Everard, 1987) one-third of the population regularly visits monuments and museums and in the Netherlands 61 per cent of the population 'felt involved in urban conservation' and 30 per cent claimed they would personally 'react' to a threat to it in their neighbourhood (Kamerling, 1987).

Difficulties still exist, including serious urban planning problems, but these increasingly stem from the unforeseen consequences of success rather than failure in attaining the aims of the conservationists.

Functional change

A very visible and therefore early recognised consequence of the conservation of the urban form is change in the functioning of conserved areas. Such change may be undesirable because it conflicts with other urban policies, or may be unwelcome merely because it is implicitly arbitrary, unintended and uncontrolled by urban planners. The inner areas of most European cities have traditionally possessed important residential, commercial and industrial functions (see Chapter 4). While many were thankful for the rehabilitation of the former Billingsgate Fish Market in the City of London into offices, it is a matter of regret that the former market will not be accessible to the public. The variety of buildings and the usual inverse relationship between age and cost enabled the older parts of the city to offer relatively cheap and varied premises, especially to those individuals and businesses who required cheap, flexible accommodation accessible to the markets and jobs of the inner city. An unintended result of conservation has been to improve the quality, and thus the land-values, of such areas, thereby altering their economic balance and displacing existing users, who may find increasing difficulty in finding alternative accommodation. The major Dutch 'showpiece' projects for example, such as Maastricht's Stokstraat district (Deben, 1973) and Deventer's Bergwartier (Goudappel, 1978), transformed busy inner city neighbourhoods which had a mix of varied residential, retailing and small office and workshop functions into beautifully renovated precincts serving only the commercial demands of wandering tourists, specialised shoppers and select residents.

Old, poorly maintained and thus low-rent city areas, attractive to conservation, provide housing not only for the traditional inner city working class but for many migrant groups including recently arrived ethnic minorities, young unattached workers and students (see Chapter 6). The physical upgrading of such areas sets in motion a process of rising land values and rents, resulting in practice in the slogan used in Parisian elections in the late 1970s:

'conservation equals deportation' (Castells, 1973). This French reaction stemmed from the experience in areas such as the Marais, which has acquired notoriety as an example of such social change. Here the sequence of restoration, higher rents and social change has been well established with 20,000 residents moving out, and a rather smaller number moving in since the project began. Whether higher land values are a result or a cause of the replacement of working class housing and small workshops by high rent apartments is not clear, but the change in function of the area is.

Table 7.1   Occupation of residents in inner city neighbourhoods in Colmar (after Ashworth and Schuurmans, 1981)

	Neighbourhoods with conservation			
Occupation	*not begun (Vauban)*	*in progress (Cathédrale)*	*'complete' (Tanneurs)*	*Colmar*
Manual Workers	51	56	2	40
Self-employed	20	10	7	20
Artisan	13	22	12	5
Shop Workers	6	–	15	14
Office Workers	2	–	–	7
Professions	4	11	49	10
Others	4	–	15	4

Such changes are not confined to the major cities but can be detected as a result of more modest programmes in provincial towns. Table 7.1, for example, shows the occupational structure of three comparable inner city neighbour-hoods in Colmar. All originally had similar social characteristics to the, as yet, unrestored 'Vauban' quarter, but the results of conservation in process in 'Cathédrale' and completed for some ten years in 'Tanneurs' are dramatic.

A realisation of the seriousness of this problem and a wish to preserve a social balance, or at least to avoid an accusation of expelling the population of the historic centre, have led to various attempts to ameliorate this consequence of conservation. As early as 1954 the master plan for the Christianshavn district of Copenhagen declared that the renovation of the eighteenth-century burgers houses would not lead to a change in use and that the district would remain residential. Despite the selective demolition of around 10 per cent of the building stock to provide garaging and public amenity space, families have tended to move to newer suburban housing and be replaced by students and other child-free groups. Experience in some German cities, most notably Bremen and Hamburg, has indicated that the large houses of nineteenth century merchants can be effectively converted into apartments and a residential function maintained, but that it is far more difficult to find tenants for the smaller one- or two-storey nineteenth-century workers' houses. The work of many housing associations, such as the Dutch 'Stadtherstel', or Ulster's Hearth Housing Association (Shaffrey, 1975), which combine public and private finance to buy, restore and then relet historic buildings on a rolling programme, is effective but small scale.

It is therefore all too easy to view conservation planning in terms of social class. Much of the initiative stems from amenity groups whose membership is largely drawn from the articulate middle classes, who are then seen to be 'gentrifying' the historic quarters of the town. The policies of the communist controlled cities of central Italy are thus especially interesting attempts to conserve both buildings and the communities that inhabit them. In Bologna, for example, the whole of the central city has been effectively conserved since 1969, with both demolition and functional change requiring planning approval. The area has long been important for housing a high proportion of poorer citizens. For ideological as well as practical reasons the decision was taken both to conserve the *centro storico*, which had suffered damage and neglect during and since the war, but also to maintain the function of the inner city as a cheap residential area. The instruments for implementing these policies were large financial grants, strict controls on land-use change and rent levels, and the establishment of neighbourhood consultative councils. The commune can, and will, enforce reluctant owners to restore and maintain their premises and makes the receipt of subsidies dependent upon the offer of the restored accommodation to the original tenants at the unrestored rent (Cervellati and Scannarini, 1973). The whole conservation process is thus linked to the maintenance of pre-existing functional and social patterns as well as to integral transport and land-use plans (Figure 7.4). The ideas developed in Bologna have been applied to some extent in other Italian cities such as Ferrara, Bergamo and Ancona, but of all the fifty Council of Europe 'European Architectural Heritage Year' projects Bologna alone guaranteed continued tenancy to low income residents (Angotti, 1977).

If Bologna shows how conservation and social policies can be combined, it also reveals some of the intrinsic contradictions between them, that renders this combination all but unattainable in most West European cities. The experience of West Germany, Scandinavia and the Netherlands is that social stability can only be attained by controls on the economic consequences which result from higher land values, yet it is precisely this 'speculation' which motivates private investment needed to finance the conservation effort. In other words, conservation areas will tend to become inhabited by those who value historicity sufficiently to pay for it, while others, resident in the inner city for its cheapness and accessibility, who are unable or unwilling to meet the costs of environmental improvement, will suffer voluntary or compulsory 'deportation'.

Costs of conservation

Such social costs are only part of the total cost to the city of conserving its morphology. A fundamental, intractable and growing problem is finding new uses for historic buildings and new economic functions for historic quarters.

The demand that the buildings were constructed to meet usually no longer exists, and most modern functions can be more economically housed in purpose-built structures. The failure to find new uses condemns the city to an existence as an open-air museum, a fate recognised in Bruges 'le Mort' 50 years ago. The extent to which this abandonment leaves the conserved city as a series of historic facades to empty buildings was revealed in the Ministry of Housing and Local Government studies of conservation in York, Chester and Bath, where up to 40 per cent of the floor area of the conserved city, especially the upper storeys, was empty (Esher, 1969; HMSO; 1969, Insall, 1969).

Tourism and recreation are favoured choices for restored buildings which thus become an economic resource. Military ramparts and walls become parks and walkways, as in Copenhagen, York, Berwick; the defensive waterways, the *grachten* of Dutch cities, become decorative amenity areas; castles, palaces, cathedrals and town halls become museums, art galleries, concert halls and exhibition space (Ashworth and Voogd, 1986). The first series of Europa Nostra awards for outstanding conservation projects set a precedent by awarding eight of the nineteen diplomas to new tourist uses in conserved buildings. This ubiquitous, and seemingly harmonious, link between conservation and tourism is not without problems. Although tourists are content to visit the historic city, their demands for accommodation and transport are modern. The Viking Hotel at York or the Beaufort Hotel in Bath illustrate the dilemma of providing twentieth-century accommodation accessible to the historic city without unduly affecting the resource that visitors have come to experience. The physical separation of the hotel district from the historic city (see Chapter 9) only exacerbates the problem of moving tourists to and around the historic city and leads to the sprawl of cars and coaches that surround many a European cathedral and palace. Tourism is in any event highly selective, making use of only a small number of favoured buildings and cities that are becoming proportionately less important as a result of the success of the conservation ethic in planning. Tourism may need the historic city but uses only a fraction of it.

An even more fundamental cost is the constraint conservation imposes upon urban development and the opportunity costs of strategies that are denied. In an extreme form the conflict between alternatives is illustrated in Venice where the survival of the historic inner city is dependent upon restricting the development of the Marghera/Maestre industrial area. The conservation of Venice required controls on ground water extraction, pollution emission, navigation in the lagoon and, even in some schemes, access to the lagoon from the Adriatic (Fay and Knightley, 1976). All these would impose high costs on the main source of employment in the urban region. Even the development of new employment possibilities within the historic city, in an attempt to halt the flight of people and jobs, presents problems. A large number of sites are available, most notably the 'Arsenale', but any prospective entrepreneur must accept not only high maintenance costs of structures that cannot be altered but also the inconvenience of the absence of road and public transport, for which *vaporetti* are a poor substitute. Schemes for improving transport to and within the lagoon city, beginning with the Miozzi Plan of 1956, foundered on the opposition of the conservation lobby (Rogatnick, 1971).

A more positive approach is to regard the conserved city not as a constraint on development but as a marketable economic resource, either directly as a primary tourist attraction (Ashworth, 1987) or indirectly in its contribution to an attractive living and working environment. 'City Marketing' either narrowly conceived as the promotion of the city as product to potential investors, entrepreneurs, tourists and residents, or as a philosophy of market planning (Ashworth and Voogd, 1988), frequently depends heavily on preserved monuments and a general atmosphere of historicity.

Continuity and change

The city needs both continuity and change 'so that the comfort of the past may anchor the excitement of the future' (Lynch, 1972). In the years since the end of the Second World War more change was wrought upon the fabric of the European city than had occurred in the preceding few centuries. This change was rapid, occurring within a single generation, and comprehensive in its clearance and rebuilding of whole districts and creation of quite new skylines. This abrupt break with the past provoked first local reaction, then national conservation legislation in the 1960s and 1970s, and ultimately a body of planning experience through the 1970s and 1980s.

This experience in turn warned that the absence of change can lead to fossilisation, where the preserved form becomes the city's main or only function. Bruges, Stratford-on-Avon and the city-state of San Marino are as unifunctional as any colliery town or fishing port. Preservation is a particularly likely fate for cities of homogeneous age. Georgian Bath and, on a smaller scale, seventeenth-century Willemstad, or the IJsselmeer towns of Elburg and Hardewijk, find evolutionary change difficult and become locked into contemporary visions of their past. Flexibility in response to change is an urban quality that must itself be conserved. If conservation planning is 'the management of change' (Ford, 1978), then there is a need to plan for cities which are capable of evolution and can welcome the future and accommodate the present without severing the thread of continuity with the past.

Small wonder therefore that conservation planning is an increasingly contentious issue in the urban political arena. Confrontation may be between groups concerned with residential amenity, and those concerned with employment opportunities and the availability of low-cost housing; or between 'developers', whether public or private, and 'preservers' for the individual or common good. These sorts of alignments are complicated by a paradox. Successful implementation of conservation is dependent upon controls over land use which are typically associated with the left of European politics, and communist (as in the cities of the Marches) or social democratic (as in

Stockholm) councils have been among the most successful. Although it is the political right that often has the strongest motives for conservation, it is the left that has the most effective means of executing such policies. The result can be a confusing and seemingly inconsistent alignment of political groups in particular cities or on particular issues. In Bologna, for example, communists and socialists tend to support conservation, while christian democrats, which include many small businessmen and shopkeepers, usually oppose it. Similarly, in Venice, republicans and communists, at opposite ends of the spectrum, have been among the strongest supporters of control over the damaging effects of mainland industry, while christian democrats, socialists and liberals have generally combined to oppose such attempts.

The process of monument preservation evolved through its own success into the creation of the 'historic city' (Ashworth and de Haan, 1986). This is more than the preservation and renovation of historic forms, or even planning for a sector of urban life: rather it is a holistic philosophy of planning that recognises the distinctive heritage of the urban morphology and its historic associations as a central characteristic of West European cities. The planning and management of the historic city involves a series of technical and ultimately political decisions whose implications are only now becoming apparent but which are critical to the future of the city and its citizens. The reclassification of urban morphology as 'heritage' implies the existence of a market. The essential dilemma of conservation planning in the European city for the next generation is the determination of whether the legatee for such an inheritance shall be the tourist, inner-city resident, gentrifier, entrepreneur or commercial investor.

References

Aldous, T. (1972) The continuing battle of Bath, *Built Environment*, 1(7), pp. 480–484.
Angotti, T. (1977) *Housing in Italy: Urban development and political change*, Praeger, New York.
Ashworth, G. J, (1984) The Management of Change: Conservation policy in Groningen, the Netherlands, *Cities*, pp. 605–616.
Ashworth, G. J. (1987) Marketing the Historic City: the selling of Norwich, in R. C. Riley (ed.) *Urban Conservation: International Contrasts*, Portsmouth Polytechnic, Department of Geography, Occasional Papers, 7, pp. 51–67.
Ashworth, G. J. and de Haan, T. Z. (1986) Uses and Users of the Tourist-Historic City, *Serie Veldstudies* 10, GIRUG, Groningen.
Ashworth, G. J. and Schuurmans, F. (1981) Colmar: Aspects of form and function in a conserved city, *Serie Veldstudies* 3, GIRUG, Groningen.
Ashworth, G. J. and Voogd, H. (1986) Marketing van het Europese Erfgoed: een ekonomisch hulpbron, *Plan*, 9, pp. 28–34.
Ashworth, G. J. and Voogd, H. (1988) Marketing the city: concepts, processes and Dutch applications, *Town Planning Review*.
Bateman, M. (1985) *Office Development: a geographical analysis*, Croom Helm, Beckenham.

Burke, G. (1976) *Townscapes*, Penguin, Harmondsworth.
Busson, A. and Everard, Y. (1987) *Portraits économiques de la culture*, Notes et études documentaires, **4846**, Paris.
Castells, M. (1973) *Luttes Urbaines et Pouvoir Politique*, Maspero, Paris.
Cervellati, P. and Scannarini, R. (1973) *Bologna: Politica e Metodologia del Restauro nei Centri Storichi*, Bologna.
Council of Europe (1974) The future of our past, *Ekistics*, **39**, pp. 139–142.
Council of Europe (1979) *Monument Protection in Europe*, Kluwer, Deventer.
Culot, M. (1974) The rearguard battle for Brussels, *Ekistics*, 37, pp. 101–104.
Deben, L. (1973) Waar zijn de Stokstaaters gebleven?, *Nonen. TA/BK*, **8**, pp. 21–22.
Dobby, A. (1978) *Conservation and Planning*, Hutchinson, London.
Engels, F. (1935) *The Housing Question*, Martin Lawrence, London.
Esher, L. (1969) *York: a study in conservation*, HMSO, London.
Esher, L. (1983) *A Broken Wave: the rebuilding of England 1940–1980*, Penguin, Harmondsworth.
Fabre-Luce, H. (1976) SOS Paris, *Urbanisme*, **153–4**, p. 104.
Fawcett, J. (1976) *The Future of the Past: attitudes to conservation*, Thames and Hudson, London.
Fay, S. and Knightley, P. (1976) *The Death of Venice*, Deutsch, London.
Ford, L. (1978) Continuity and change in historic cities, *Geographical Review*, **68**(3), pp. 253–273.
Fried, R. (1973) *Planning the Eternal City: Roman politics and planning since World War 2*, Yale University Press, London.
Goudappel, H. M. (ed.) (1978) *Tien Jaar Stadsherstel: Bergkwartier, Deventer. 1967–77*, Hermes, Deventer.
Hewitt, K. (1983) Place annihilation: area bombing and the fate of urban places, *Annals of Association of American Geographers*, **73**(2), pp. 257–284.
HMSO (1969) *Bath: a study in conservation*. A report to the Ministry of Housing and Local Government and Bath City Council, HMSO, London.
Hoyle, B. S., Pinder, D. A. and Husain, M. S. (eds) (1988) *Revitalising the Waterfront: International dimensions of dockland redevelopment* Heinemann, London.
Insall, D. and Associates (1969) *Chester: a study in conservation*, HMSO, London.
Kain, R. (1975) Urban conservation in France, *Town and Country Planning*, **43**, pp. 428–433.
Kain, R. (1981) *Planning for Conservation: an international perspective*, Mansell, London.
Kamerling, J. (ed.) (1987) *Burger en overheid in de Monumentenzorg*, Heemschut, Amsterdam.
Knox, P. L. (1985) *The Geography of Western Europe: a socio-economic survey*, Croom Helm, London.
Lynch, K. (1972) *What time is this place?* MIT, Cambridge, Mass.
Mennell, S. (1976) *Cultural Policy in Towns*, Council of Europe, Strasbourg.
Mumford, L. (1961) *The City in History*, Penguin, Harmondsworth.
Nairn, I. (1955) Outrage, *Architectural Review*, Special Edition.
Oaks, R. (1974) Conservation in Italy, *Town and Country Planning*, 4, pp. 270–275.
O'Riordan, N. (1975) The Venetial ideal, *Geographical Magazine*, 47(7), pp. 416–426.
Rogatnick, A. (1971) Venice, problems and possibilities, *Architectural Review*, **149**(891), pp. 261–273.
Shaffrey, P. (1975) *The Irish Town: an approach to survival*, O'Brien Press, Dublin.
Skovgaard, J. (1979) Conservation planning in Denmark, *Urban Studies*, pp. 519–539.
Stungo, A. (1972) *The Malraux Act 1962–1972*, Royal Town Planning Institute, London.
Sutcliffe, A. (1970) *The Autumn of Central Paris: The defeat of town planning*, Edward Arnold, London.

Tunbridge, J. E. (1984) Whose heritage to conserve? Cross cultural reflections on political dominance and urban heritage conservation, *Canadian Geographer*, **26**(2), pp. 171–180.

Tunbridge, J. E. (1988) Revitalisation of the Waterfront, in Hoyle, B. S., Pinder, D. A. and Husain, M. S. *op cit.*

UNESCO (1975) *The Conservation of Cities: studies commissioned by UNESCO*, Croom Helm, Beckenham.

Ward, P. (1968) *Conservation and Development in Historic Towns and Cities*, Oriel Press, Newcastle.

White, P. (1984) *The West European City: a social geography*, Longman, Harlow.

Williams, N., Kellogg, E. H. and Gilbert, F. B. (1983) *Readings in Historic Preservation: why, what and how?* Rutgers University Center for Urban Policy Research, New Brunswick.

CHAPTER 8

Urban Recreation Planning

Planning for the leisure city

The interest of public bodies in the planning and management of leisure in its various forms was largely a response to the growth in demand for it. The limited attentions of national governments to this topic until some 20 years ago is reflected in a handful of scattered pieces of legislation which were essentially attempts to channel and control the effects of activities that had developed spontaneously. Meanwhile local governments slowly acquired a large collection of diverse leisure interests through the accretion of functions, undirected by any overall concept of the role of city government in this field.

In fact the city had at least three distinct roles to play in planning for leisure. The city could be seen in its traditional role as a service centre providing for the recreational needs of its citizens, as a generator of recreational demands upon both the city and a wider region, and as a recreational resource in itself, serving an economically important set of leisure industries. These quite different functions not only required the formulation of different policies but represented quite different approaches to planning for urban leisure.

The period from the middle of the 1960s to the early 1970s posed a special challenge to city authorities throughout Europe. A combination of popular growth, increases in disposable income and leisure time, and the emergence of a society in which leisure activities were particularly valued, led to a very rapid increase in almost all types of leisure demand. This 'fourth wave' (Dower, 1965) in social development was in essence an urban phenomenon in Western Europe for the simple reason that the strongest leisure demands were for activities in, around and within day trip reach of the home, and cities were the home of most Europeans. In addition, this surge in demand occurred at a time when it was popularly felt that it was the responsibility of public authorities to satisfy it. The provision of public libraries, swimming pools, playing fields and public gardens had long been provided by city governments throughout Europe for diverse educational, health and social reasons. To these were now added a whole range of new facilities including sports centres, boat marinas, caravan parks, picnic sites and even theatres and concert halls. The pressure of citizens' expectations both spurred local authorities to respond to their electorates and also put pressure on central governments to encourage and subsidise this response. In

Britain, for example, the Acts of 1968 (Access to the Countryside) and 1969 (Development of Tourism) were both national responses to existing leisure demands from and on cities that were largely dependent upon local authorities for their implementation.

Thus the urban planner was faced with a new function. The traditional trilogy of fields of interest – work, housing, transport – had been joined by recreation. This new set of tasks posed a number of basic questions to those assuming the responsibility of planning for them. These can be summarised as:

(1) *What is to be provided and, as a necessary collorary, for whom?*
Within the range of possible leisure activities, it was necessary to define the objectives of public provision and thereby delimit what was seen as the proper spheres of commercial and municipal enterprise respectively. The origins of local authority leisure planning and the resulting lack of a coherent philosophy resulted in wide differences existing between and within counties. In West German, Dutch and French cities, for example, direct municipal enterprise or sponsorship in the Arts was accepted more easily than in Belgium or Britain, while British seaside resorts, often with Conservative councils, had freely engaged in direct municipal trading, on a scale unknown in the Mediterranean resorts.

(2) *How much should be provided?*
The rapidly rising levels of demand for most leisure activities, together with an open-ended public commitment to provide facilities for them, and a demand that was itself often stimulated by the supply of these facilities, made this an urgent question. To discharge its assumed responsibilities the city therefore needed to know if it was under or over-provided, so that policy to achieve 'enough' could be implemented, but unfortunately no certain means of doing that existed.

(3) *Where should it be provided?*
Once quantities have been determined, the location of recreation facilities within the city in relation both to the spatial patterns of latent demand, and of other urban functions, becomes an important local planning problem. The solution to such problems involved in part the creation of spatial models of various sorts and in part their application within the particular individual urban situation.

(4) *How should it be provided?*
This question raises problems of organisation of provision. The incremental growth in local authority responsibilities frequently necessitated restructuring. This generally involved the establishment of new administrative structures, such as the amalgamated 'leisure directorates' created in many British cities after 1974, and a 'professionalisation' of local authority management in fields as diverse as sports, arts and tourism. However, the creation of new public recreation apparatus did not eliminate the many problems and possibilities of integration both between public departments and between public and private enterprises in recreation provision.

(5) *Why should it be provided?*

Although such a fundamental question would logically precede provision, in practice much local authority planning for recreation has been a reaction to external pressures, and justifications have frequently only been sought for existing provision. Among the very varied answers to this question found in the European city are justifications based on welfare and equity, on stimulation and development of local economies, or even policies designed to defensively contain and channel rising demands and mitigate their perceived deleterious effects.

Any survey of planning for this aspect of urban life faces a fundamental dilemma. On the one side it is necessary to adopt an integrative approach. Planning for recreation cannot be divorced from planning for all the other aspects of the city described in this book, but should be considered alongside housing, transport, commercial activities and the like. On the other hand, most leisure research has isolated this particular aspect of behaviour for systematic study, and leisure planning is the responsibility of specific departments and agencies. In addition neither the demand for recreation nor the supply of

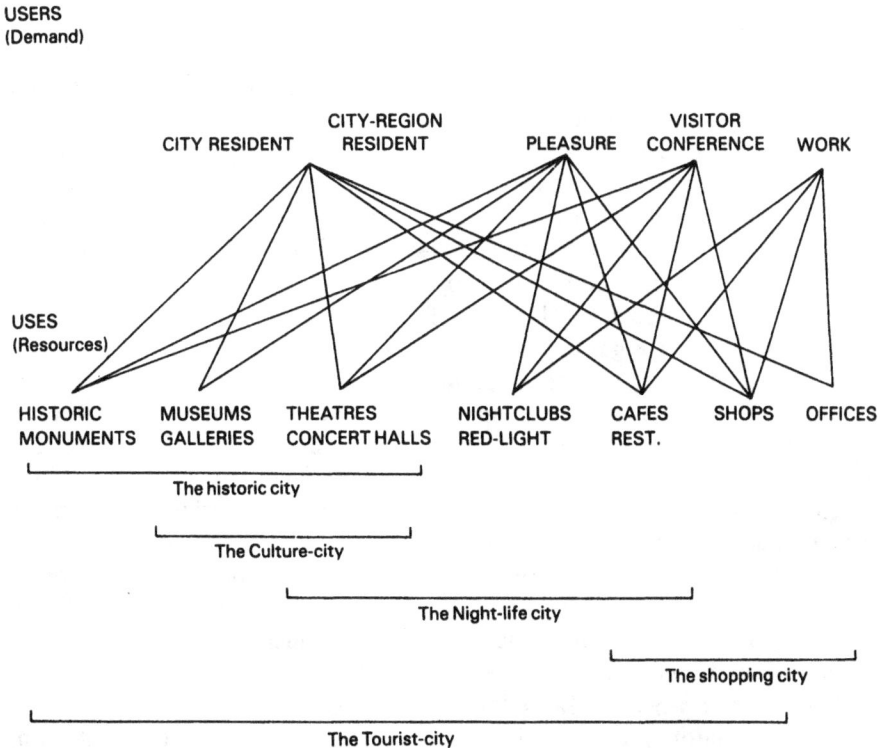

Figure 8.1 Some interrelations of recreational users and uses in the tourist-historic city

potential recreation resources in the city are homogeneous but are usually specific to a particular set of activities. Figure 8.1 shows something of the complex relationships between groups of recreational users and categories of uses in the city. Therefore the planning problems of the leisure city and the policies derived for their solution are equally specific and urban planners have had necessarily to focus upon a particular set of demand and resource conditions existing in identified recreation situations. It is the most important of these situations which provide the structure for this chapter, and the next, although their wider context in the functioning of the city must never be forgotten.

Planning for outdoor recreation and sport in the city

The legacy

The contemporary West European city has inherited a varied collection of outdoor recreational and sporting spaces and facilities. It is this legacy, created for whatever past motives and conditions, that provides the basis for the recreation planning of the modern city.

The park is probably the oldest feature of the city created specifically to serve the recreational needs of citizens. Its ancestry can be traced to both the common and the private garden: the one providing the idea of free public access and the other that of relaxation among the cultivated works of nature. The medieval city had its private gardens for leisure as well as food. The sixteenth and seventeenth centuries added extensive tracts of landscaped park to the city for the enjoyment of monarchs. The royal parks of London (Hyde Park, St James's Park, Kensington Gardens), Paris (Tuileries, Luxembourg), Berlin (Charlottenburg, Tiergarten) and those of lesser dignitaries, ecclesiastical authorities and universities have become a major element in the land-use of the central areas of cities of all sizes, and once opened to the public, these relics of past privilege have made a substantial contribution to the general welfare. The slow process of the opening up of the private parks to the citizenry spanned the seventeenth to the nineteenth centuries. New parks in imitation of the private estates, and provided on a commercial basis to satisfy the demands of the middle-classes, led to the pleasure gardens such as London's Vauxhall or Ranelagh and Copenhagen's Tivoli. A fusion of the private gardens of the rich and the public fair produced the pleasure park, such as Battersea Gardens (opened in 1951) and ultimately the theme park perfected by Disney in the USA and now to be opened at Marne-la-Vallée in 1993.

The vast majority of the inhabitants of the 'Coketowns' of nineteenth-century Europe had access to neither private gardens nor public commercial pleasure parks. A report of the health commissioners in Lancashire in 1842 stated that 'there is only one public park in the county and that is at Preston'

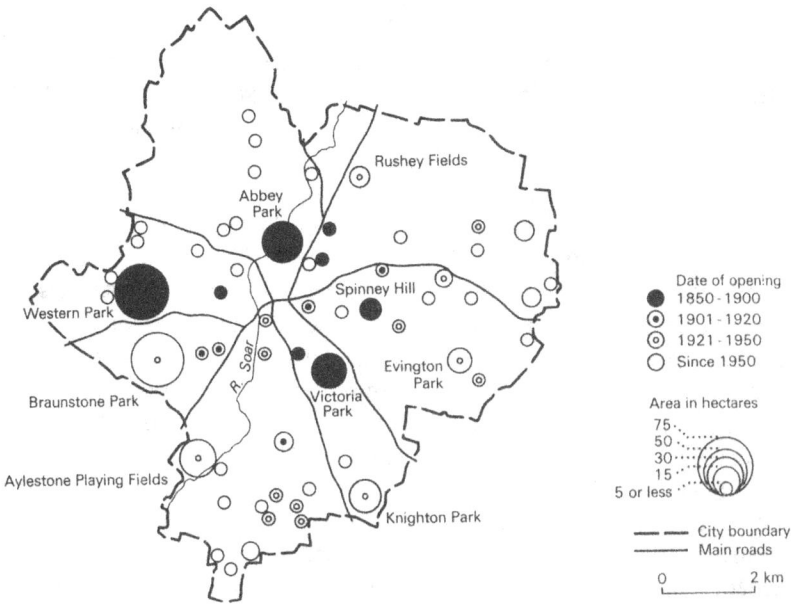

Figure 8.2 The age of parks in Leicester

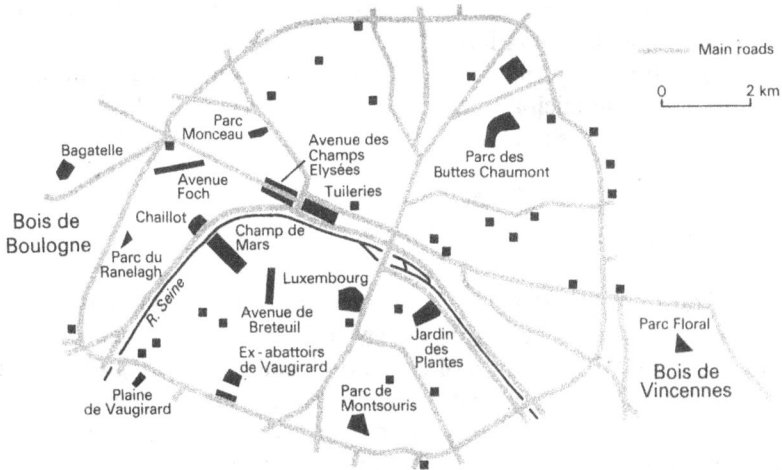

Figure 8.3 The distribution of green space in Paris

(quoted in Ward, 1976). The Victorian municipal park was created to serve these people for a mixed set of motives which included a utilitarian belief in the value of recreation to labour productivity, a romantic desire to bring nature back into the cities from which it had been excluded for the first time in European history, and civic pride. In London, Victoria Park, situated between Hackney and Bethnal Green, was established in 1845 in a deliberate attempt to provide in the East End some parks to balance the royal parks in the West. This was a forerunner of many other 'Victoria Parks' established in the provincial cities of Britain over the next 50 years, so that by the end of the century most British cities had acquired, through private philanthropy or municipal enterprise, a set of public parks freely open to all citizens. The City of Leicester for example (Figure 8.2) created a network of six parks between 1882 and the end of the century, encouraged by no fewer than four Acts of Parliament (Bowler and Strachan, 1977). Similarly on the continent Haussmann was building the Parc Monceau (1861) and Parc des Buttes Chaumont (1864) for the citizens of Paris (Figure 8.3) and the Volksgarten movement was inserting green spaces into the expanding industrial cities of Germany. Such parks were laid out as small scale imitations of the private estates of the gentry for a public expected to promenade and quietly appreciate nature rather than engage in more active pursuits, and were regulated by a battery of byelaws, enforced by railings, notices and uniformed officials. The legacy of space for active sports was generally less, and more fortuitous. Available spaces were used for sports wherever they occurred, such as the open areas around the defence works of continental fortified cities, and the notion of providing specialised spaces or facilities at public expense for public land use was late to develop. Private associations and educational institutions had in the course of the nineteenth century acquired playing fields for their members, which eventually prompted a public response for municipal authorities concerned about the health, fitness or even contentment of their citizens.

The assumption of new responsibilities

Thus the West European city inherited a variety of public open spaces, created by private or municipal design or merely by a tradition of public usage of commons, disused fortifications, or canal, river and lake frontages. It is hardly surprising that the quantity, type and distribution of such spaces did not correspond to the patterns of demand for it in the modern city. The changing volume of use in response to increases in leisure time during the day and at weekends, the demand for facilities for more active recreational pursuits, and the redistribution of the urban population as well as changes in its life styles, all posed new challenges. A few enlightened European cities, such as Leicester (Figure 8.2) or Barcelona, and those experimenting with the garden village movement for new suburban developments (see Chapter 2), such as Amsterdam, created new peripheral parks to cater for rehoused populations in the inter-war period. But in most other cities inaction was supported by a body of informed planning opinion that dismissed much of the nineteenth-century inheritance as irrelevant to the twentieth-century city. The revitalisation of the urban park can be credited to, first, the survival of romantic ideas about the value of 'nature' – now expressed in such terms as 'lungs for the city' and politicised through the 'green' lobby – and, second, the emphasis on the importance of space, especially green space, to children's development, stressed by psychologists with increasing insistence in the post-war world (Mercer, 1976).

The responsibility for the systematic provision of publicly accessible sports facilities has only recently been assumed by city authorities, and then more usually as a result of pressure from the local electorate than in response to specific government legislation. As with the planning of parks, the first problem is to determine need and priorities, and the second to devise policies to satisfy those selected needs that can be accommodated among the existing land-uses. The demand for sports facilities is varied, specific to particular activities, often prompted by the existence of a supply, and frequently volatile; all of which complicates the problem of what to provide. In particular the questions of what should be provided by public endeavour as opposed to private associations, individuals or commercial enterprises, and whether investment should be in sporting excellence for the gifted few or 'sport for all', were answered in different ways by different cities in Europe.

In addition, spending on sport has generally been low in the priorities of local authorities; many facilities are too demanding of land to utilise small infill sites, and large multipurpose sports complexes have proved very expensive to build and operate. Consequently there is a great variety in provision between and within cities, with the presence of facilities being often a result of political will or the adroit opportunism of planners in responding to the fortuitous availability of sites (see Ashworth, 1979 for an account of just such developments in recreation facilities in the British provincial city of Portsmouth).

The question of sufficiency

Urban authorities attempting to review the nature of their recreational provision in the light of their newly assumed responsibilities were immediately confronted with the question of adequacy. How much public recreational supply should exist, of what sort, and for what purposes? One set of answers is provided by the creation and acceptance of conventional standards of provision, which enable deficiencies or surfeits to be identified. As early as 1925 the British National Playing Fields Association suggested a need for 2.8 hectares of public open space, including 0.4 hectares of ornamental gardens, for each 1,000 of the population. Similar figures for hectares of playing fields, kilometres of recreation paths or square metres of swimming pool have been produced at various times. Such standards were seized upon over the next 50

Table 8.1 Hierarchy of public open space in the Netherlands

Type	Responsibility	Requirement	Size	Range	Transport Facilities	Other features
Neighbourhood	Gemeente	4 square metres per inhabitant	1–4 hectares	up to 0.5 kilometres	Paths	—
Local	Gemeente	8 square metres per inhabitant	6–10 hectares	0.5–1 kilometres	+ Cycle tracks	Toilets/kiosk
District	Gemeente	16 square metres per inhabitant	30–60 hectares	1–3 kilometres		Café
City	Gemeente	32 square metres per inhabitant	200–400 hectares	3–5 kilometres	+ Car parking	Restaurant
City-region	Gemeente/Regional authority	65 square metres per inhabitant	1000–3000 hectares	5–20 kilometres	+ Public transport access	
'Country'	Province/state	125 square metres per inhabitant	10000–30000 hectares	50–100 kilometres		Hotel/campsite
National	State	250 square metres per inhabitant	50000–10000 hectares	100+ kilometres		

Source: Rijksplanologische Dienst, Netherlands, 1965

years in numerous urban plans (Hampshire County Council, 1978), or were reconstituted as a planning aim. For example, the London County Council adopted a standard of 1.7 hectares per 1,000 population supplemented by 1.3 hectares outside the LCC boundary. This was modified in the 1951 development plan which proposed a 2.9 hectares standard within the county boundary. The 1967 Greater London Development Plan increased this figure overall but varied it in different areas, reducing it to 1.6 in the densely populated inner boroughs (GLC, 1967a). None of these targets were realised.

Most West European countries derived similar standards which could be extremely detailed, as for example in the Netherlands where amounts of public open space, sports fields and even 'leisure gardens' were evolved as planning for cities of varying size (Table 8.1). Such standards, however, were more often arbitrarily conceived than based on any detailed understanding of the nature of the latent demand for recreation space. They have, however, been used effectively to point out gross inequalities of provision between cities. In Paris for example (Figure 8.3) the target of 100 square metres for each citizen has been used to highlight an overall deficiency, which revealed that this city devoted only 7 per cent of its area to 'green space', half as much as London or New York (Conseil de Paris, 1971). Similarly a study of the nine largest West German cities showed wide variation in provision from the relatively lavish 32 square metres per citizen in Bremen, 24 in Cologne and 22 in Munich, to the more parsimonious provision of 15 in Dusseldorf, and 14 in Essen and Stuttgart: there is no general explanation for these differences which relate to different city development (Stadt Koln, 1978). In the same way a British study comparing all towns with more than 100,000 inhabitants showed that not only did very few approach the 60 year old NPFA standard, but also that large discrepancies existed at the extremes, which were not explainable by reference to employment structures or age of development (Lever, 1973).

Discrepancies within cities could similarly be revealed. For example in Cologne (Stadt Koln, 1978) a detailed inventory revealed discrepancies which ranged from 14 square metres of public open space for each inhabitant of the Altstadt to 177 in the spacious suburban Pesch district. In Hannover (Ulfert-Herlyn, 1977) a similar disparity in provision was compared with other variables, and a paucity of parks was strongly correlated with, and compounded by, an absence of private gardens, and with a whole range of other neighbourhood amenities. Park planning was thus seen as only one facet of planning for 'positive discrimination' in favour of the traditional working class inner suburbs such as Linden-Mitte and Kleefeld.

Consideration of the quantity of public open space was rarely matched by an examination of its use.

> In orthodox city planning, neighbourhood open spaces are venerated in an amazingly uncritical fashion, much as savages venerate magical fetishes. (Jacobs, 1962)

Studies in many different European cities during the late 1960s and early 1970s were broadly agreed that distinctly different markets existed. Parks, for example, were open-air dining rooms for the central business district office workers, open-air play rooms for small children, and open-air sitting rooms for the old in search of company during the day and for the teenager in search of privacy in the evening. Frequency of habitual use varied widely between cities, suggesting an element of national custom. In Paris (Leusse, 1976), for example, only a minority of citizens made a regular visit to a park, while in London (GLC, 1976b) there were 1,004 visits for each 1,000 of the population; although the average distance travelled to neighbourhood parks was broadly similar.

The question of accessibility and location

Such standards are of limited use in planning the location of recreational open space within the city. Table 8.2, for example, appears to show that the city of Portsmouth is reasonably well provided in comparison with other similar cities. However it cannot be assumed that both supply and demand are homogeneous, without variations in the nature of the facility and of the needs resulting from the age and family structure of the population. Similarly the relationship between supply and demand will depend not only on the varying propensity of different population groups to engage in particular recreational activities but also on their willingness and ability to travel. Portsmouth's inherited stock of

Table 8.2 Comparison of standards of public open space in Portsmouth and other South Coast towns

Towns	Total land area (1000ha)	resident population (1000)	Total area of open space (ha)	Open space as % of land area	Standard of open space per 1000 population
Portsmouth	3.7	200	385	10	1.9
Southampton	5.3	214	445.2	9	2.1
Plymouth	8.1	250	429	5	1.7
Brighton	5.9	164	1104.8	19	6.7
Bournemouth	4.7	149	469.4	10	3.2

Requirements in Portsmouth

	Size of playing units (Sports Council) ha	Units required per 200 000 population	Area ha	Existing units in Portsmouth
Soccer	0.90	81	74	56
Rugby	1.10	7	8	4
Cricket	1.80	36	64	16
Tennis	0.06	100	6	106
Hockey	0.60	7	4	12
Bowls	0.20	33	8	17
Netball	0.06	3	0.2	17

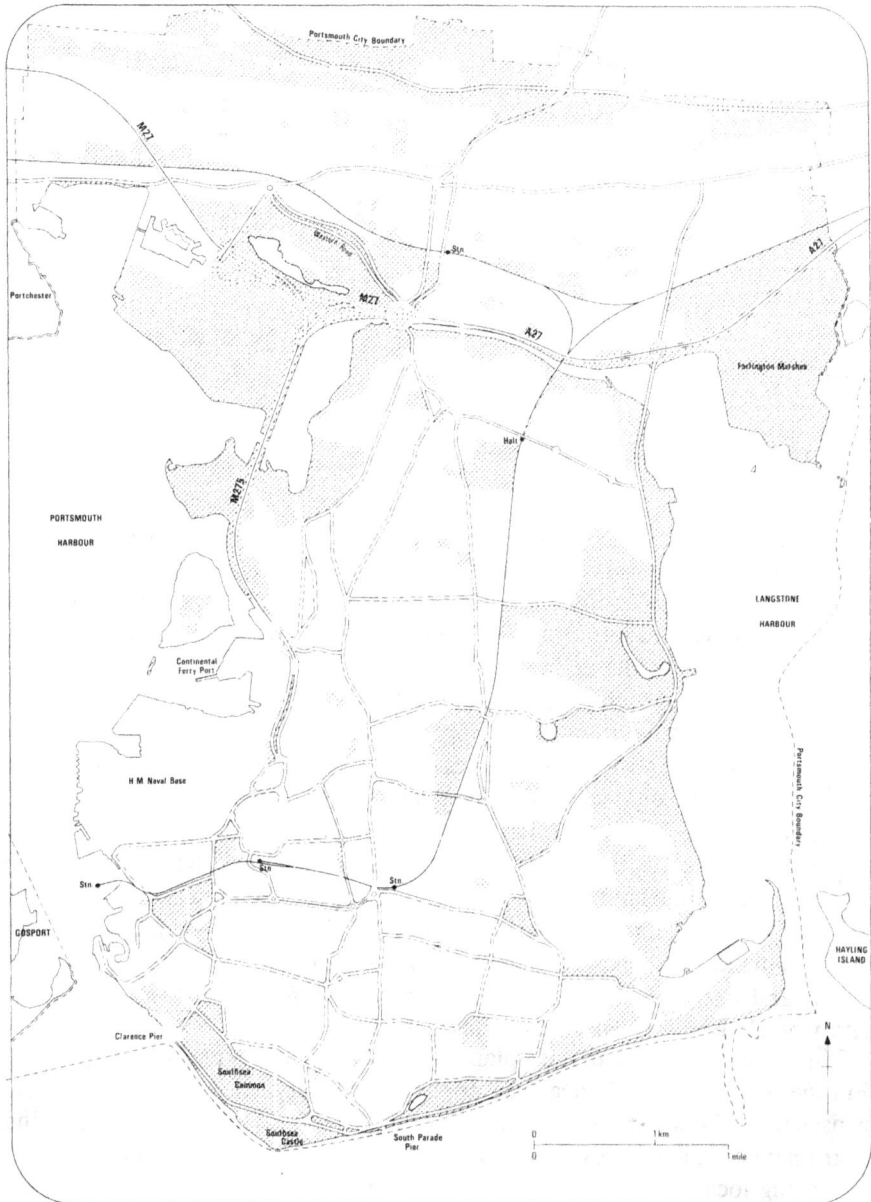

Figure 8.4 Portsmouth's open space 1990

Figure 8.5   A hierarchical park system

public open space (Figure 8.4) is clearly both extremely varied in character (including a nature reserve, reclaimed coastal marshes, some open downlands, a seafront common, some old fortifications and various small parks and playgrounds inserted into the build-up area over the last 100 years) as well as largely peripheral to the main areas of demand.

Consideration of such problems led planners in a number of countries away from a search for overall standards of provision towards the creation of an ideal hierarchical park system that combines the size and nature of facilities provided with the range and character of the demands for them, which results in the neat spatial pattern in Figure 8.5. Dutch urban recreation planners had produced and calibrated such a theoretical park hierarchy by the middle of the 1950s. It was seriously applied in the West Netherlands conurbation during the 1960s and early 1970s, most comprehensively in the province of Zuid-Holland, which includes the Rotterdam agglomeration (Provinciale Bestuur van Zuid-Holland, 1976), and examples of similar planning can be found well into the 1980s (Provinciale Bestuur van Zuid-Holland, 1983).

This approach to urban park planning was attractive in many other parts of Europe and its central idea was applied with varying degrees of comprehensiveness. New towns in particular used such hierarchies as planning instruments; Tapiola in Finland, for example, consciously reproduced a series of nesting locational patterns, based on Christaller, for its public recreation provision (Pigram, 1983). National and even local differences in activity preferences, and distances travelled, clearly affect the nature of the facilities provided at different levels in the system, so that the plan for Bologna (Figure 8.6) must of necessity vary greatly from that for Rotterdam. The largest city to make a conscious attempt to plan for a complete hierarchy was London. The Greater London Development Plan (GLC, 1967a) based its recreational open space proposals squarely upon a hierarchy of provision that ranged from small

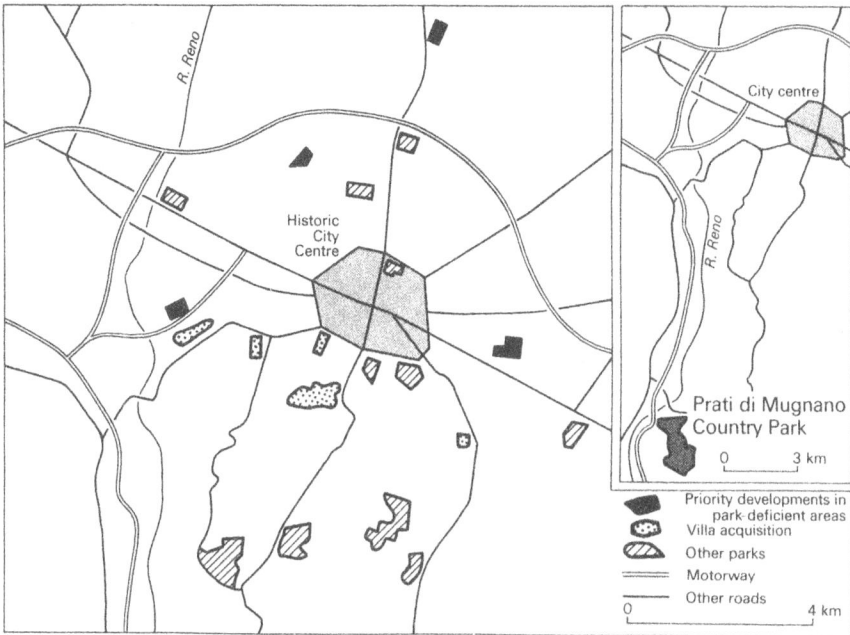

Figure 8.6 Public open space plan for Bologna

neighbourhood parks drawing visitors to simple facilities from less than 2 km, to major, well provided, city-region parks such as the Lee Valley Regional Park. Recreation planning mirrored in this respect very similar approaches to other urban facilities such as shopping centres.

The problem of the spatial scale of administrative control makes the application of such a theoretical schema difficult in practice. While neighbourhood parks and playgrounds are usually provided by the city authorities, city-region facilities are usually the responsibility of counties or provinces. In addition the very variety of recreational activities determined that responsibility for provision was dispersed through different local and national government departments.

However, it was not these administrative difficulties alone that cast doubt on the universal utility of such ideas as a central instrument of urban recreational planning. The public enquiry into the GLC Development Plan concluded that

> both old (ie global norms) and new (ie hierarchical provision systems) have defects in that they largely ignore the density of population and accessibility to public transport. (GLC, 1967b)

This argument against over-generalising the nature of the demand can be taken further, and in detailed practice it has long been realised that demand and supply are not independent variables. If, in this as in many other aspects of recreation, there is no finite fixed quantity of demand that can be measured and then satisfied, then almost any additional provision, in almost any location, will meet, or call forth, a need. It is not surprising therefore that most European cities have proceeded in an opportunist manner, utilising relict space, wherever it occurs, as well as using more effectively existing provision, and seeking only subsequently post-hoc theoretical justifications. In the 1970s and 1980s efficient management of existing urban recreational resources, whether in the public or private sectors, was more important than identifying and fulfilling perceived needs.

In practice, therefore, outdoor recreation planning has been more successful in identifying areas of gross deficiency within the city and ameliorating these, rather than creating idealised patterns. The lack of subtlety matters less when the anomalies between districts within the city are so large. In Paris, for example, it is not so much the overall deficiency of public open space compared with other cities that has prompted action as the demonstrable discrepancies in provision between parts of the city. The inner city (especially the 9th to 12th arrondissements) has less than 0.3 square metres of public open space for each inhabitant, compared with the better served western districts (1st, 5th and 6th arrondissements) with more than 3.5 square metres (Figure 8.3). The task of the planners here is to insert green space into a densely developed environment when the opportunity occurs. This policy in Paris can be traced back to Haussmann during the Second Empire. He created not only a number of large parks, in the poorer districts, but also attempted to recreate in Paris the small parks of London's West End squares, many of which had begun as 'subscription gardens'. According to Chadwick (1966) his efforts earn him the title of 'the creator of the first real urban park system'. More recently, similar initiatives in Paris have used a large part of the old Citroen works, the military training area at Ile St Germain, the Sèvres quarries and even the roofing of the underground development of the Les Halles site in the heart of the city. The largest park development is that at La Villette and the smallest the Parc de Belleville, both in eastern Paris.

It has often proved necessary to demolish existing buildings and lower the existing density of housing in order to insert new public open space. Thus part of a district is cleared so that the remainder can be raised to modern standards. Frequently this has involved retaining the buildings flanking the residential blocks while clearing the back streets for parks and other amenities, as in Vienna's Lichtental district (Figure 8.7). Similarly in the 'bridge quarter' of Copenhagen, the policy of retaining both the historical facades and the residential function necessitated the clearance from the inner courtyards of the more modern accretions and the insertion of a series of 'mini-parks' (Christansen, 1978). Thus recreation planning becomes an integral part of the more comprehensive inner-city renewal programmes described in Chapter 6.

Studies to determine the propensity of populations to engage in active sport, and to establish the distances participants were prepared to travel, proliferated after the middle of the 1960s, and in turn led to the creation of theoretical

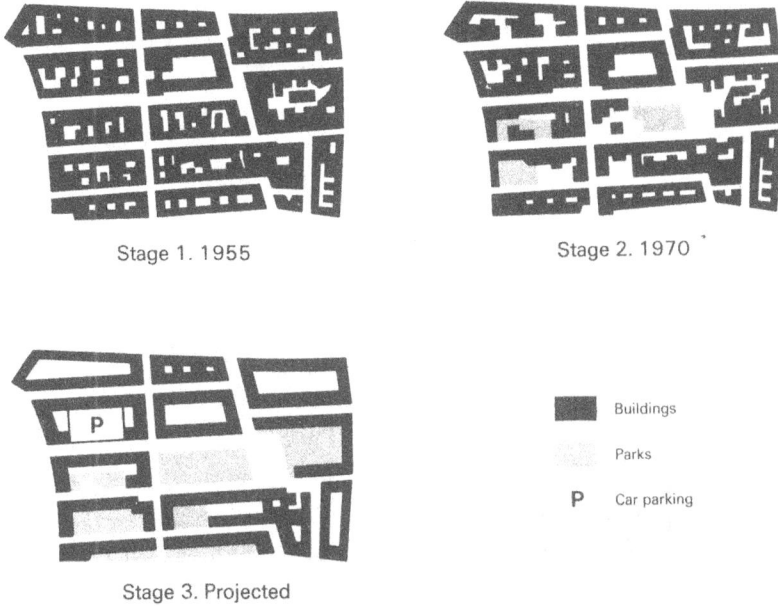

Stage 1. 1955

Stage 2. 1970

Stage 3. Projected

Buildings

Parks

P   Car parking

Figure 8.7   Park insertions in Lichtental district, Vienna

hierarchies of provision based on various optimum catchment areas. A study in Edinburgh (Cargill and Hodgart, 1978), produced a series of locations for a variety of sports facilities that minimised travel distance for potential participants. Robertson (1978) demonstrated that a variety of different location patterns could be produced in Glasgow if different variables were optimised (Figure 8.8) so that planners could choose between minimising travel distance or equalising the size of catchment area or population, and between concentrating facilities in large multi-purpose complexes or dispersing them into smaller, less well equipped local centres. Similarly, in many medium-sized West German cities, a comprehensive range of standards of provision has been devised in terms of space required for various sports for each inhabitant to be provided within a defined distance of home (Stadt Koln, 1978). A simple but completed example of the execution of such policies can be found in Zurich, where a comprehensive cover of the city using hinterlands of 15 to 20 minutes travel time has been attained.

Such analyses of demand are, however, prone to the difficulties in application that have already been mentioned in the discussion of public open space planning, in particular that the demand for and supply of sports facilities are not independent variables. In any event it becomes clear in planning the location of sports facilities that accessibility was dependent upon more than spatial factors alone. To the barrier of distance can be added other equally significant barriers of familiarity, knowledge and money.

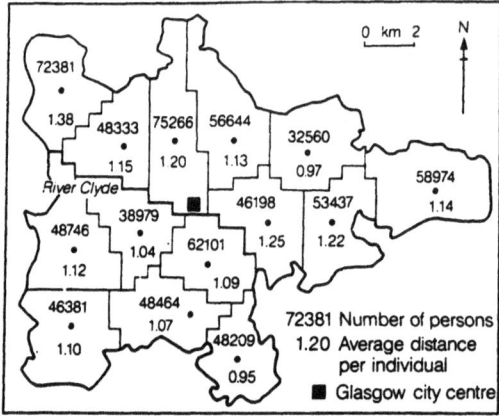

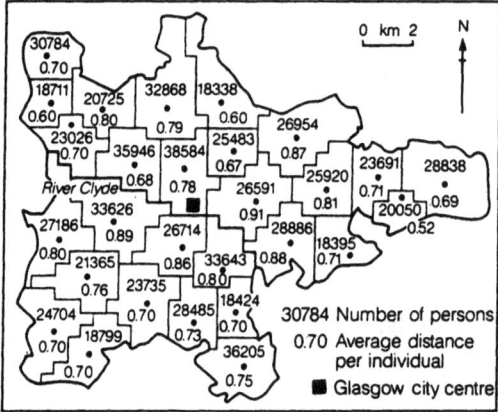

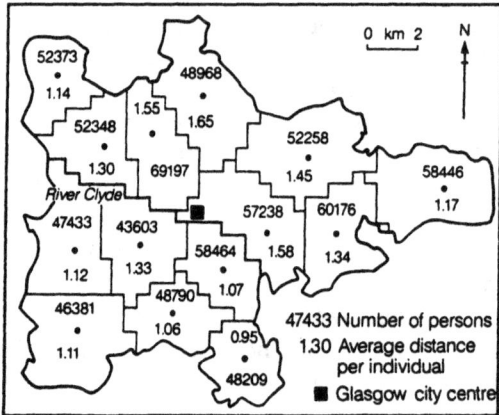

Figure 8.8 Location patterns in Glasgow

In Britain, the Wolfenden (1960) and Albermarle (1981) reports gave central government encouragement to extending sports provision and the especially favoured instrument for this was the building of large sports centres, but the initiative still remained with the local authority. The first of the projected 815 of these was opened in Harlow in 1963 (Molyneux, 1972). A growing feeling that the concentration on building large multipurpose centres with a city-wide hinterland was bound to produce an uneven pattern of supply, together with the greater financial stringency of the late 1970s, prompted a major shift in policy in Britain away from the expensive sports centre with its specialised equipment and car serviced hinterland towards the lower end of the hierarchy. The Sports Council's 'Sport for All' campaign coincided with government encouragement for relatively simple neighbourhood facilities usually in converted premises, located close to the demand and serving neighbourhoods of not more than 25,000 people. The fact that this figure also happens to be the size of a secondary school catchment points the way to a more effective dual use of facilities at the community level. Although Britain, and especially the urban authorities of central Scotland, were early to appreciate these possibilities, the cities of Scandinavia, the Netherlands and West Germany provide similar examples where 'access' to sports facilities was increasingly resolved by making better use of existing supply by removing barriers to its use as much as seeking new locations for new facilities. In contrast, many cities elsewhere in Europe are still vigorously pursuing policies of large centrally located multipurpose sports centres.

Planning for culture and entertainment

Planning for the provision of entertainment and cultural facilities is again an aspect of urban life that was not regarded as a responsibility of local authorities until quite recently. Although the entertainment of residents and visitors has always been an important function of cities, and cities have always been centres of cultural productivity, most facilities for these activities have been provided by private enterprises. Even when public subsidy has been involved, often little attention has been given to planning considerations despite the obvious importance of these facilities to the form and functioning of cities.

Planning for culture

In the course of the 1970s the Council of Europe took an interest in national variations in cultural provision and instituted a long term study of a sample of fourteen European towns (Mennell, 1976). The gist of their argument was that they 'expect local authorities increasingly to shoulder the burden of public patronage of culture and leisure – highbrow, lowbrow and middlebrow ... so how is a town to distribute its limited resources to best advantage?' Attempts over the study period to estimate desirable levels of provision, isolate sections and areas of underprovision, and establish the legitimate areas for public policy intervention, encountered the same problems as with other aspects of recreation. There exist large variations in the popularity of entertainment media, and in the responsibility assumed by urban authorities. For example, regular theatre-goers formed one per cent of the inhabitants of Stavanger but 10 per cent in Stockholm and only one half of one per cent in Bologna. Yet local government per capita expenditure on public entertainment was four times higher in Bologna than, for example, in Exeter. After 10 years study it was concluded that it was not possible 'to develop one ideal cultural policy as a model to be copied by other towns. Obviously that is impossible – towns and people vary too much for that.' But policy in this relatively unfamiliar area of urban planning may still be assisted by the development of a 'set of methods by which the effectiveness of cultural policies can be judged and evaluated'. (Mennell, 1978)

Attempts to allocate cultural facilities, especially those with a national or regional catchment area, vary from country to country. In the Netherlands in the course of the 1960s, for example, the state accepted a responsibility, together with its financial consequences, for improving accessibility to the performing arts. In spatial planning terms this resulted in a system of regional orchestras and theatrical and operatic companies based on the major provincial centres with an outreach touring programme. Diminishing levels of public subsidy during the 1980s have weakened this system in a number of respects, but the principle that national and local governments have a responsibility to make culture, however defined, available and accessible remains intact. In France, the central government inspired and substantially financed a network of *maisons de culture* from the mid-1960s to serve as regional culture centres. That at Grenoble, for example, served a region of more than a quarter of a million, while more recent foundations at Caen, Bourges and St Etienne have been rather less successful, perhaps owing to the lower level of enthusiasm for the project. If French policy suggests a missionary endeavour by the central government in provoking 'animation' in the provinces, the British government, in contrast, relies heavily on supporting local initiatives, usually from the local authorities. Although the Arts Council have developed some guidelines, especially in their sponsorship of 'regional' theatres in provincial towns, and through the International Festivals Society of regional arts festivals, there is little to compare with the French, Dutch or Scandinavian policies of seeking spatial equity in arts provision.

These sorts of national and local policies were usually implicitly justified on the grounds of social justice or educational value, but more recently quite explicit economic motives have become more apparent. These are related not only to the tourism industry in which major cultural facilities play an important catalytic role, but more generally as a central city-marketing attribute in which the possession of cultural facilities is a major factor in shaping a civic image attractive to potential investors, residents and entrepreneurs (Whitt, 1987).

Art and music festivals have proliferated. The well established events at such recognised cities as Edinburgh, Venice and Bayreuth have been joined by less obvious competitors. Düsseldorf/Duisburg has its 'Oper am Rhein', Liverpool its Garden Festival, and Glasgow was declared cultural capital of Europe in 1990. Publicly financed campaigns to replace existing urban images with those which are seen as more favourable for attracting new economic activities are now commonplace, and culture and recreation broadly defined are major components of such images.

At the intra-urban scale the cultural function has a distinct tendency towards spatial clustering, if only because the various facilities are serving the same 'night-out' market. This tendency has been reinforced by local planning, especially in the cities with large enough catchment to support a range of specialised cultural facilities. The result in the large international arts centres can be major concentrations such as London's South Bank and Barbican, and Les Halles in Paris, in which facilities for the performing arts serve as a focus for specialised catering, shopping and other related activities in marked 'culture quarters'. Smaller cities rarely have the markets to support high order cultural facilities on this scale. Some cities merge facilities, eg Düsseldorf and Duisburg operas as 'Deutsche Oper am Rhein'. Niche marketing can give culture a place, such as the Moers theatre which was established in 1974 to stage rarely performed works. Short festivals as in Arundel (population 4,000) can provide a range of events over a short time span to local and visitor populations. However major regional capitals, such as Munich, Salzburg, Milan, Edinburgh and the like, which make a conscious use of cultural events and facilities in the shaping of their civic identity, while not having planned cultural 'quarters' at least use such facilities as elements in their land use planning.

Planning for popular entertainment

If public authorities in most cities have been only recently and sporadically interested in the allocation of cultural facilities within and between cities, they have rarely given any specific attention to popular entertainment. Yet many of these facilities, such as cafés, restaurants, bars, night-clubs, cinemas, have pronounced spatial characteristics and effects. They tend to cluster together in response partly to the behaviour of customers who seek out a district where comparisons between establishments can be made and where the appropriate atmosphere can be sampled. Examples can be found in Bonnaine-Moerdyk's (1975) 'gastronomic geography of Paris' or Smith's (1983) demonstration of the different clustering patterns of various types of fast-food outlet. Each type of entertainment facility provides only a part of the total night-out package assembled by the customer; therefore, different sorts of entertainment will tend to cluster near one another, so that together they form the entertainment quarter or night-life district that the customer seeks. Once established this concentration in a particular district will be reinforced both by popular sentiment that advertises the area to those in search of such entertainment and by planning decisions that will endeavour to contain 'bad-neighbour' land-uses in a defined area.

The existence of the nightlife district has long been an important function of the European city, and for many visitors is a major motive for their visit. While many of the city's residential and employment functions have fled to the suburbs, the city's function as meeting place – for personal interaction, for business, cultural enlightenment, sexual encounter, political protest or just pleasure – has remained so firmly rooted in the centre of the city as to have become synonymous with the term 'downtown'. A closer investigation of the component elements and local planning reactions to them in a variety of cities is needed.

Various sorts of night-life districts have existed in West European cities, frequently in the same areas for many centuries. Concentrations of bars, restaurants, night-clubs, together with cinemas and theatres, form what can be termed 'West End' entertainment districts occupying relatively expensive sites in high prestige areas. In the larger cities there may be a number of such areas, as in Paris, north of the Champs Elysées, Montmartre and on the left bank around Boulevard St Michel; in Amsterdam, Rembrandts and Leidse Plein; in Copenhagen the streets immediately west of the Tivoli in Vestgaarde.

A different sort of night-life district may exist near to, or spatially quite distinct from, the 'West End'. This is the 'red light' district offering cheaper and more explicitly sexual entertainment including commercial prostitution. Such districts are a traditional, if largely unacknowledged part of the European city, and although citizens may prefer to ignore their existence, planners must increasingly include them in their plans. In Amsterdam, for example, the 'Walen' district near to, but one street away from, the main commercial thoroughfare has entertained visiting seamen since the seventeenth century, and today continues that tradition for a varied international clientele. Figure 8.9 traces both some of the internal functional specialisation of such areas, and also some locational associations in relation to 'legitimate' entertainment and other commercial functions of the city.

The response of city planners and politicians to such areas has been generally ambiguous. On the one hand they are frequently related to a range of social problems and associated with illegal drug use, crime and personal insecurity. They are the areas of social malaise, recognised by Burgess's famous designation 'vice district' on his functional map of Chicago (Burgess and Park, 1925). On the other hand they are extensions of the legitimate commercial entertainment function of the city and in major cities often also an important part of the tourist industry. 'Clean-up' policies generally result in a shifting of such locations around the inner city, while containment policies intensify their concentration.

The existence of red-light districts in large cities such as London (Soho) and Hamburg (St Pauli) is well known but their presence in smaller towns is less well documented despite the major dilemmas they pose to urban planners. In

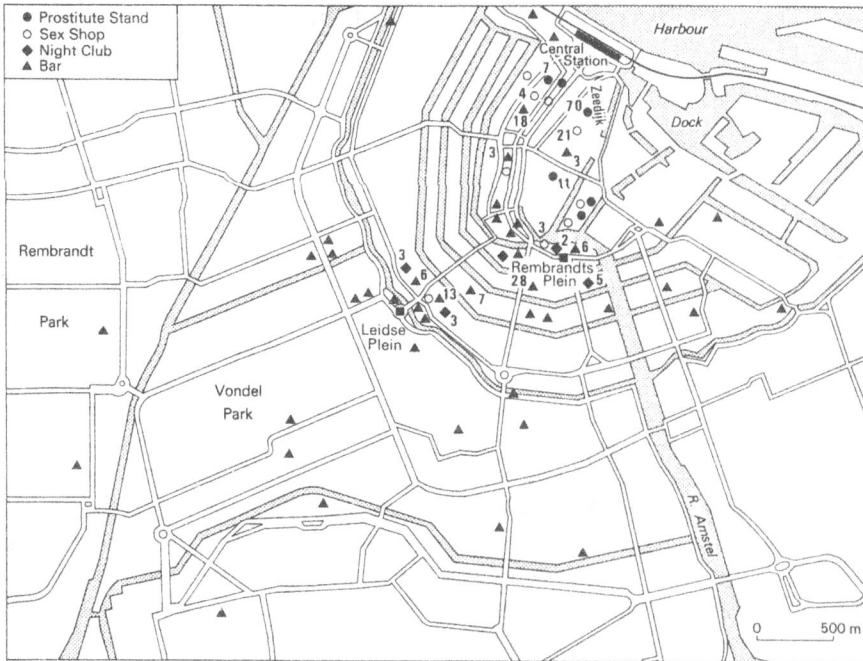

Figure 8.9 Nightlife in Amsterdam. Numbers refer to the total of each function at that location

Arnhem, for example, a city of only 150,000 inhabitants, the 'Spijkerkwartier', a district of the inner city just to the north-east of the central business district, demonstrates the same sorts of planning dilemmas as in the larger cities. Originally developed in the late nineteenth century for middle class housing, the buildings were largely subdivided for low rent apartments for small households composed principally of young people and more recently ethnic minority immigrant groups. Prostitution is concentrated into a clearly defined area, occupying the ground floors which are used for display purposes. The Spijkerkwartier, Arnhem, has the problems of declining population, ageing building stock and the lack of a stable community typical of many such inner city wards, but its rehabilitation is made more difficult by its red-light function which tends to repel other commercial uses, hinders the renovation and gentrification of the residential function (Gemeente Arnhem, 1975) and locks the area more firmly into its role of accommodating the poor, immigrants and the socially deprived. Rehabilitating the area by renovating the buildings, and removing undesirable functions as part of urban renewal schemes, would change the social and commercial character of such areas, but would equally merely chase the social and moral problems to new areas of the city.

There is little reason to assume Arnhem is unique in this respect. The Guillemins district of Liege, the Waalkade of Nijmegen, the streets south of the

Kurfurstendam in West Berlin, and the Derby Road area in Southampton, all house similar functional mixes and pose similar planning problems, the solutions to which depend more on national legislation, policing practice and cultural attitudes than on a local planning response (Ashworth *et al.*, 1988).

Recreation planning for the city region

A tradition of recreation use

There is a long established tradition of citizens using the rural environs of the city for recreation. Even before the dawning of the age of mass leisure the European city was typically surrounded by the hunting lodges and rural retreats of richer citizens. The royal palaces of Nonsuch, Versailles or Potsdam, and further down the social scale the country seats of seventeenth-century Amsterdam merchants on the sandy ridge of Het Gooi, were all attempts at combining access to the urban centres of political and economic power with the recreational opportunities of the countryside. The modern day-trip in the country and the ownership of a holiday home, caravan or even 'leisure garden' outside the city are the reactions of a newly prosperous urban population, motivated by the same desire for space, landscape beauty or a nostalgic temporary return to a lost rural past, as their social superiors in a previous age.

The preservation of recreational space

Understandably, therefore, the relationship between urban and rural areas, and the nature of the interface between the two, have been a preoccupation of urban planning since its inception. The experience of utopians from Plato to Thomas Moore was of cities small in physical extent, and they understandably assumed that the built-up area, contained within its defensive walls, would be surrounded by a countryside that was easily accessible on foot to all citizens. Many later visionaries, including those as diverse in other ways as Howard and Le Corbusier, implied that placing the city in a rural setting was in itself sufficient to provide for the demands of citizens upon the rural hinterland.

The designation of a 'green belt' around London, proposed in 1935, legalised by the 1938 Green Belt Act and demarcated as part of Abercrombie's 1944 plan, gave a name to the city's rural hinterland and initiated a continuing discussion on its purposes. As early as 1580 a ban on building on new sites within three miles of the gates of London had been instituted, although ineffectively enforced, principally in order to create a *cordon sanitaire* that would check the passage of disease. The 1938 legislation was somewhat similar in form, defining a green belt as 'an area of land near to, and sometimes surrounding, a town, which is kept open by permanent and severe restriction on building' (Ministry of Housing and Local Government, 1962), but it had wider purposes. Its two main functions were first to prevent the outward spread of the city by containing it, and second to provide a peripheral zone with space for the recreational needs of citizens. Although both purposes were mentioned with almost equal emphasis in the 1935 proposals, the later Act and the Abercrombie plan in practice stressed the former.

After the 1947 Town and Country Planning Act other British cities were empowered to establish green belts and by the early 1960s they were included in the plans of seven city-regions. In all, however, there was a tendency to regard them negatively, as means of containing growth and conserving rural landscapes, rather than as opportunities for a conscious development of the urban periphery for recreation. In practice many sites in the green belt, such as Box Hill or Epping Forest, had long been favoured by London trippers and many London educational institutions, local authorities and private clubs located their sports and other outdoor recreation facilities in this area. Although around 10 per cent of the total area of green belt land is actually occupied by such uses, there was in fact no automatic right of access for the public on land which remained dominantly in private ownership (Munton, 1981). The implementation of the conservation objective could even frustrate the construction of recreation facilities, such as camp and caravan sites and commercial entertainment centres, on the grounds of these being developments incompatible with the character of the green belt, and the preservation of an attractive residential environment in short supply would in any event tend to price land beyond the reach of many land extensive outdoor recreational uses.

The London green belt had many imitators outside the United Kingdom, although the scale and the emphasis upon the twin purposes of physical containment and recreation supply were frequently quite different. The Copenhagen Land-Use Zoning Plan of 1962, for example, designated a belt, about 50 km wide, around the rapidly suburbanising conurbation from Helsingor in the north to Koge in the south, which stretched completely across Sjaeland, including about half the island. Within this area new residential development was restricted and specific provision was made for public open spaces, day recreation centres and even second home complexes.

In Paris, a modified form of green belt planning was included in the 1965 Schéma Directeur for the city. Green spaces, known as 'zones de discontinuité', were established between the main axes of centrifugal urban expansion. After a decade's experience of the operation of this plan, a reassessment (Préfecture de la Région Parisienne, 1975) concluded that such barriers to urban expansion had proved to be too vulnerable to development pressures. In addition it was clear that the mere preservation of green space without a consideration of its use did not in itself create a recreation zone. A further problem, not unique to Parisian planning, but more explicit here than elsewhere, was that the rural area was being treated as empty space to be manipulated in accordance with the needs of the city, whereas such zones had substantial populations, planning problems and political objectives of their own. Inner urban demand for, and urban fringe supply of, recreation space

Figure 8.10   Green space planning in the Paris region

could only be brought together through city-region planning when the political and administrative structures were also effective at this scale.

The 1975 proposals reasserted the role of rural-urban fringe as a barrier to the undesirable outward expansion of the city's built-up area. A series of 'fronts ruraux' marking a sharp but narrow green buffer were established at points around the edge of the city where outward expansion would be resisted. The previous 'zones de discontinuité' were renamed, and redrawn in detail, as 'zones naturelles d'équilibre', and given a more clearly conceived set of functions and a positive strategy for development. The six zones (Figure 8.10) together accounted for 250,000 hectares or 30 per cent of the area of the Ile de France planning region. The new zones were to include positive provision for outdoor recreation, as well as attempts to accommodate the needs of the existing settlements and employment structure, especially argiculture. The new recreational functions were grafted onto the traditional land-uses by a process described as 'controlled evolution', but the results were little different in essentials from the more uncontrolled evolution already described for the London region.

It is clear, however, in the large conurbations that a peripheral green belt bestows its benefits of recreational and residential amenity unequally upon citizens. Accessibility to it is easiest from the outer residential areas, which are usually relatively well endowed with public and private open space provision, and most difficult from the central districts of the conurbation where the need

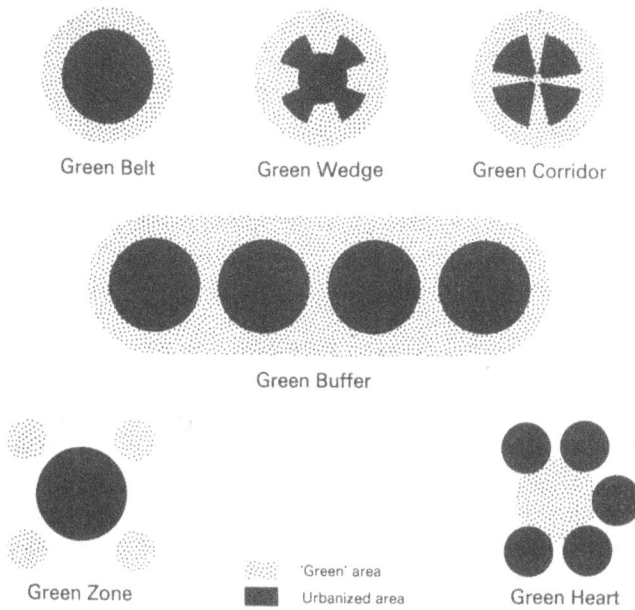

Green Belt	Green Wedge	Green Corridor

Green Buffer

Green Zone	Green Heart

'Green' area
Urbanized area

Figure 8.11 A typology of 'green' planning

for recreation space is typically high and levels of car ownership are typically low. The problem is how to bring the green belt closer to all parts of the city. One solution was the 'green wedge' driven deep into the built-up area (Figure 8.11). Cities which had developed outward radially along transport routes found it relatively easy to designate the less built-up interfluves between the developed 'fingers' as green wedges. In Bologna, for example, the site with its steep range of hills to the south and west threaded by the settled valley of the Reno river allowed the conservation of the upland spurs as recreation space quite close to the city centre and well served by public transport.

If such wedges can be continued right across the city to join the green belt on the other side, then clearly even better access to green space can be achieved especially for the central districts. Such an arrangement, known variously as 'green corridors', 'green binding zones' or 'green axes', is particularly attractive when an additional objective is to prevent the morphological coalescence of a number of distinct urban nodes within a conurbation. It is not surprising therefore that such ideas have been most extensively pursued in the two largest multicentric conurbations of Europe, the West German Ruhr and the Dutch 'Randstad'.

In the Ruhr the problem of poor access to open space caused by the rapid population growth in the years before the First World War was particularly acute, and the threat of the coalescence of Dortmund into a continuous built-up conurbation was very real. Fortunately a regional planning authority, the

Siedlungsverband Ruhrkohlenbesirk (SVR), had existed since 1920, and the idea of forcing green corridors between the urban nucleii was implemented in the course of the 1960s (SVR, 1966). The overall plan designated general conservation measures in a wide zone surrounding the conurbation, which in practice has been more effective in the north and south than in the east and west. The alignment of the Ruhr towns broadly along an east-west axis has determined that the corridors separating them shall run from north to south. Six of these were developed, most of which exist as through routes, although the corridors narrow dramatically to a width of only a few hundred metres in their central sections.

In the West Netherlands there was a similar need to provide simultaneous easy access to open country for the leisure pursuits of city dwellers and preserve the spatial integrity of the individual towns. Again both objectives were seriously threatened in the decade after the Second World War by rapid population growth and the concomitant increasing land demands for housing, industry and transport. The arrangement of the major urban centres of the West Netherlands in the approximate shape of a horsehoe suggested the concept of 'randstad', which was propounded in detail in the first national structure plan of 1962 (Riley and Ashworth, 1975). This plan had three main objectives with regard to the maintenance of green open space. The character of the relatively less developed areas beyond the urban rim (or 'rand') would be maintained as a sort of green belt, which would include the Rhine Delta islands of South Holland and Zeeland and the eastern and southern sandy heathlands of Utrecht, Gelderland and Noord-Brabant. Thus the West Metropolitan Netherlands would be kept spatially apart from the expanding conurbations of the Rhine-Ruhr and Brussels-Antwerp. Secondly, the cities of the rim would be separated from each other by 'green buffer zones'. In some cases this function was performed by the rivers, as south of Utrecht, while in others planners hoped to prevent development by the creation of narrow zones of a few kilometres width, such as between Delft and Rotterdam, or Rotterdam and the Hague. Thirdly, the area within the rim, optimistically designated as an 'open-heart', was to be maintained for agriculture and for recreation. Thus an amalgam of relatively small separate towns in what has been called 'green-heart metropolis' (Burke, 1966) promised a novel and imaginative juxtaposition of town and country.

This promise unfortunately was seriously threatened. Growth in the demand for land in the 1960s, not least within the 'open-heart' itself, led to a radical rethinking of the Randstad concept in the second and third National Structure Plans (1966, 1968) and its replacement by an arrangement of built-up 'wings' (or 'vleugels') and open corridors (Figure 8.12). The northern wing recognised the existence of a 'Greater Amsterdam', stretching from the North Sea at Ijmuiden through the Gooi ridge to Utrecht. Similarly the southern wing extended from the sea at The Hague/Scheveningen, through Delft and Rotterdam to the 'Drechtsteden' on the rivers. Urban expansion was allowed for eastwards to Arnhem-Nijmegen, south-eastwards through the Brabant

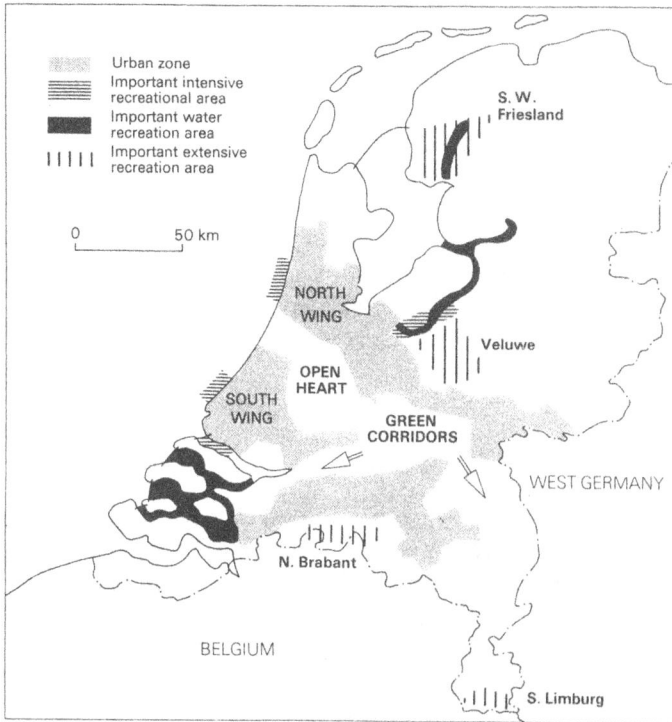

Figure 8.12 The West Netherlands open space plan

towns to Den Bosch and Eindhoven, and northwards in Kennermerland to Alkmaar. The 'open-heart' had thus become a series of corridors running through the metropolitan region and linking it to the more extensive recreation areas on the national periphery, such as the watersports areas of the Delta and Friesland, and the heaths of Brabant and the Veluwe.

The comprehensiveness of such schemes for the successful fusion of peripheral green belts and green corridors in open space provision for the urbanised Western Netherlands, as outlined in the National Plans and implemented at the provincial and districts levels (Steigenga, 1968), is unequalled in Western Europe. A sign of its success is that the fourth structure plan (1988) contained few departures from the basic physical pattern laid down more than twenty years earlier. The stress is less upon channeling expansion than upon consolidation and intensification within the existing urbanised areas, including the provision of outdoor recreation facilities within or close to the urban areas.

Many other urban areas, including cities much smaller than the metropolitan Netherlands, have attempted to plan for what they term 'green systems'. Land both on the periphery and within the city, whether designated formally for recreation or for other uses which tolerate an informal recreation use, is linked both functionally and spatially. A large number of the medium-sized West German cities are active proponents of the idea, including Bremen, Munich, Nürnberg and Mannheim. The last named city has applied the idea most rigorously in the city-region plan (Stadt Mannheim, 1977); it included not only a carefully graduated hierarchy of open space provision from the regional to the neighbourhood scale, but also introduced 'green binding zones' that link the parks not only with each other but also with the pedestrianised areas of the central city, thus providing continuous walkways through both the natural and built-up environments. The idea has been implemented in towns as diverse in their size and history as Vienna and new towns such as Harlow and Evry.

Planned recreation provision around cities

Until quite recently planners in many cities considered that they had discharged their obligations to the recreational needs of citizens in the rural periphery once they had conserved areas designated for this purpose, and that the recreational facilities themselves would be provided either by private enterprise, or that recreation could be accommodated on an informal basis on land officially designated for some other purpose, such as agriculture or forestry. In the Ruhr, for example, until the mid-1960s the KVR considered its function to be to secure land on the periphery upon which casual and unorganised recreation might take place, and only since 1967 has it attempted actively to encourage and organise recreational activities, in a hierarchy of recreation centres. Similarly in the London green belt, the conservation of land in itself created neither community ownership nor public access, and facilities for recreation were provided by private clubs and institutions. Golf courses, both private and municipally owned, are a particularly favoured use, the extreme case being Edinburgh with its 22 courses in the green belt within the city boundary (Edinburgh District Council, 1977).

By the middle of the 1960s it was becoming clear to many that this largely unplanned provision of facilities in green belts and zones was inadequate on two counts. First, it was feared that a rapid increase in demand by urban populations for rural recreation would overwhelm a more slowly growing supply. In particular the growing demand for the day trip to the countryside was strongly associated with the increase in motor car ownership in the decades since the war. Wippler (1966) produced the classic study of day-visitor demands of urban populations, using the case study of Groningen in the Netherlands, and formulated a series of predictions, later replicated for Rotterdam (Provinciale Bestuur Zuid-Holland, 1969), that caused considerable alarm. In summary it was predicted that between 20 and 30 per cent of households in Dutch cities of over 100,000 inhabitants would seek recreation in a zone from 15 to 30 kilometres from the city centre during a typical summer weekend. Although pioneered in the Netherlands this type of study was rapidly imitated in many European cities. It was not surprising therefore that planners responded to this increasingly clamorous chorus of predictions, warning that urban recreation demands were about to overwhelm the countryside surrounding the cities.

Secondly, the tradition of accommodating recreation on land designated for other uses was breaking down as a result of both the increasing numbers of recreationists, and also of changes, especially in agriculture, that were rendering these informal multi-use schemes less acceptable to both visitors and landowners. By the early 1970s, the National Farmers' Union in Britain was complaining that informal recreation was effectively sterilising agricultural land on the immediate periphery of urban areas. Increasingly recreation was to be included as a land-use in its own right needing specific planning policies.

Around many European cities this planning took the form of establishing recreation 'honeypots' at defined distances from the city to 'soak-up' the visitor exodus. In the Ruhr, for example, a hierarchy of recreation centres was planned. At the apex of the hierarchy are five regional parks each of 25–35 hectares, set in a further 25–60 hectares of open space, in which a range of sports and cultural facilities are located. The first was established at Herne Gysenberg, which was followed by others at Duisburg, Dortmund, Essen and Bottrop: they attract 7.5 million visitors annually. A second type of centre are the 300-hectare water parks at Kemnade and Xanten, administered jointly by the KVR (Kommunalverband Ruhrgebiet) and private companies. The six 'recreation' focal points are smaller and contain more limited facilities. Finally there are sixteen 'recreation sites' of around 10 hectares. Thus by 1990 the Ruhr region could boast of no less than 3,900 hectares of park, 4,700 hectares of leisure gardens (allotments), 12,400 hectares of recreational water, 3,065 public playgrounds and 75,200 hectares of forest, as well as 92 'fitness centres' and 44 zoos, bird sanctuaries, and game reserves (Kommunalverband Ruhrgebiet, 1982). Similarly in the West Netherlands high capacity day recreation centres have been specifically designated in the countryside at the critical distances from the city revealed in the research on visitor demand. In the Netherlands, these distances tend to be both shorter and better served by public transport than in the other major European conurbations. Sites such as Brielsemeer for Rotterdam, Kennermerduinen for Amsterdam and the Loosdrecht Lakes for Utrecht were sited in the 1960s in the peripheral green belt and interior 'green heart'. Later plans in the 1970s continued this policy of providing intensive 'honeypots' but the outward growth of the urbanised area and increasing car ownership led to the second generation being developed further from the western cities along, for example, the Ijsselmeer 'randmeren', the Gelderland rivers district and in the Veluwe.

Both London and Paris were some years behind the Ruhr and the West Netherlands in day-recreation planning. In Britain it was not until the 1968 Access to the Countryside Act that local authorities could be reimbursed for the creation of what were termed 'country parks' and the smaller 'picnic sites'. These were publicly owned and managed recreation centres offering a range of facilities to car-borne day visitors within around an hour's driving time of the cities. Although individual local authorities were already active in developing such demand-orientated facilities – most notably the Greater London Council with projects such as the Lee Valley Regional Park – and the English aristocracy had turned their country seats at Beaulieu, Longleat, Woburn and elsewhere into broad based day-recreation centres, the 1968 Act was a spur to the development of a rational network of such centres around British cities.

Such planning has however been by no means confined to the large metropolitan areas, and similar ideas on more modest scales are now found in the plans of most medium sized cities. Many have *de facto* green belt policies and have created facilities in them for the recreation of their citizens. Hannover has its Steinhuder Meer, Edinburgh its Pentland Park and Turin its Villa Real.

The coincidence of a variety of a new concerns in the course of the 1970s, including the temporary increasing costs of petrol, an awareness of the needs of the carless inner city populations and the environmental costs of transport, together with increasingly severe limitations on local government expenditures in a number of countries, led to a reassessment. This resulted in a re-emphasis upon the immediate rural-urban fringe zone. In Britain for example the 1974 White Paper on sport and recreation, reflecting the views of the Countryside Commission, called for more attention to the possibilities of this zone rather than the creation of major honeypots distant from the city (Davidson, 1978). The provision of sport and other recreation facilities was viewed increasingly not so much as a social service responding to need but also as a factor in urban and regional regeneration capable of providing if not direct, economic returns then at least indirect, economic benefits. The distinctions between public and private provision, and economic and social objectives, became increasingly blurred in many countries, as recreational provision in the city region was seen in wider economic, planning and even promotional contexts.

In the Paris region, the 1975 plan (Préfecture de la Région Parisienne, 1975) made specific suggestions for the development of day recreation centres usually within the six 'zones naturelles d'équilibre'. Twelve of these large, multifunctional 'bases de loisirs' were established at distances from 20–40 kilometres from central Paris (Figure 8.10). They were intended to serve both a regional day recreation market and also the local needs of the new towns, thus playing a role in the regional restructuring of the urban system as well as satisfying purely recreational demands.

The need to link, physically and functionally, the three 'green' elements (inner urban parks, urban fringe facilities and rural attractions) resulted in the schema for the city-region shown in Figure 8.11. Although this is drawn from a Dutch source it is equally applicable elsewhere.

Thus, the major North-West European conurbations have successively recognised the significance and then conserved their respective rural peripheries, and made a positive contribution for sport and day recreation in them. However it should be noted that the differences between these cities and those of Southern Europe are possibly more pronounced in this aspect of planning than in most others discussed in this book. In particular the chronology of the developing relationship between city and countryside related here stems largely from 'northern' experience, and cities in Iberia, much of Italy, and Greece are in general still in the earlier phases.

Planning the wider pleasure periphery

Beyond the urban fringe the city's inhabitants make demands on the surrounding region for a wide variety of sports, and for the more specialised activities often at a considerable distance from the city. Numerous examples of the link between particular cities and resource orientated sports facilities can be found and whole sets of spatial models have been developed (among others by Miossec 1976; Lozato 1985) relating central cities to their 'pleasure peripheries' which can be many hundreds of kilometres in extent. The sporting potential of mountains links the cities of South Lancashire with the climbing potential of North Wales, the Lancashire and Yorkshire conurbations with the caves of the Pennines, and the cities of Piedmont and Lombardy with the ski slopes of the Val D'Aosta and Lombardy Alps respectively. The marinas and associated boat facilities of the Solent coast are strongly patronised by boat owners from Greater London, as are those of the Zeeland lake from Rotterdam and IJsselmeer randmeren from Amsterdam. The car-borne day visitor may travel some 30 to 60 kilometres for general day recreation but twice that for more specialised pursuits or for weekend activities, thus extending the city's recreational hinterland beyond the sports facilities of the rural-urban fringe and the picnic sites around the conurbations into a 'weekending zone' of second homes, yachtsmen's cottages, camp and caravan sites as shorter and more flexible working hours become more widespread. The use of the Veluwe, Kempen and Frisian Lakes by the citizens of the Western Netherlands, New Forest by Londoners, Bavarian Alps by citizens of Munich, and the Ardennes by inhabitants of both the Sambre-Meuse and Rhineland towns, creates what are in effect national, and increasingly international, rather than strictly urban recreation systems.

Planning for recreation for city regions whose effective demand is so extensive poses major administrative problems and often there are distinct differences in approach by the government of the region and the country; while the city may regard the region as its legitimate playground, the region may view the influx of retreating citizens in terms of local costs and benefits and conclude that it is an unwelcome and unprofitable invasion. Urban recreation needs, and the facilities and infrastructural demands they generate, may conflict with national conservational and land-use plans and investment priorities. Thus in few areas of planning has the need for coordination between levels in the spatial planning hierarchy become more necessary as an item for the planning agenda of the next decade.

The planning problems at the end of the century

Within a period of less than a generation leisure activities in European cities have ceased to be a marginal concern of urban planners occupying attention, finances and space left over from other functions and have become a central expectation of citizens and preoccupation of their governments. The recognition that recreation in its many forms required public authority planning was accompanied by a rapid and interrelated growth in both demand for, and supply of, facilities in both public and private sectors. Few planners would now question the conventional wisdom that cities should provide a wide range of recreational facilities accessible to their inhabitants in the performance of what is recognised to be a central function of the city of the twenty-first century. Discussion is focused upon the questions of what should be provided, by whom, and where, rather than why.

Throughout the wide field of leisure provision, planners in cities all over Europe found the problem of defining their role and the essentially spontaneous rather than regulated nature of the activity. Planning for housing, transport or employment has a longer history and therefore a tested body of practical experience. Consequently local authorities have acquired reponsibilities for the very diverse activities outlined in this chapter without any clear idea about their function. American and even Eastern European experience is based upon assumptions too different to be of comparative value, thus the West European city has had to devise its own justifications, methods and policies.

Although large differences can be detected between cities within Western Europe, attributable to differences in economic priorities, political traditions and social preferences, urban planners have responded in recognisably similar ways. Administrative fragmentation of responsibilities, which has occurred as a result of the *ad hoc* expansion of planning interests, was ameliorated in Britain and Italy through integration at city level, whereas in Germany the *Land* has tended to assume this role, while in both France and the Netherlands strategic planning has been largely assumed by the national and regional authorities leaving the city with the task of implementation. The attempt to establish standards of provision, although based on little more than intuition, has highlighted comparative deficiencies or surpluses in facilities as diverse as parks, sports fields and theatres. Attempts to create international standards have really only been tried in the arts and here with little success. The monitoring of the adequacy of provision led to the study of the effective range of demand, the estimation of catchment areas, and ultimately the creation of scale hierarchies of provision. Finally, studies of adequacy led in turn to policies of territorial justice, although such concepts of spatial equity are found most commonly for public open space provision at the intra-urban scale, where deficiencies are usually most obvious and demonstrably socially damaging.

Serious problems remain to be confronted in the last decades of the century but these can largely be attributed to this very success. The naive assumption that demand was a more or less fixed quantity of needs that could be satisfied by the provision of a calculable quantity of sports halls, swimming pools and green spaces is giving way to the realisation that demand for leisure provision in general, as opposed to the demand for any particular recreational activity, is, like the other higher order services of education and health, in practice insatiable. At the same time the clear cut distinction between 'social' recreation facilities provided by public bodies for welfare goals and 'economic' recreation facilities, for which private provision should be made, is breaking down. This is in part a result of the admission by local authorities that their commitment was limitless while their resources were not, and partly the realisation that many leisure activities, whether sports, the arts or entertainment were major economic functions of the city, with impacts upon the urban economy at least equal to those of the traditional manufacturing and commercial service sectors.

The justification of the choice of activities undertaken by public planning authorities is little more than, 'historical accident plus the concept of "worthy" leisure . . . private enterprise has offered us the frisbee, grouse shooting and Summerland, while Epping Forest, the Royal Festival Hall and our local tennis courts are by courtesy of the public purse'. (Roberts, 1974)

The respective roles of public and private sector organisations, already unclear, are becoming even more blurred, and it is increasingly difficult to distinguish collective social welfare from commercial market objectives. Regulatory land-use planning designed to resolve spatial conflict gives way to development planning aimed at stimulating rather than controlling change, and public investment to meet perceived needs is replaced by the search for public-private investment and risk sharing partnerships, as demonstrated for instance in the Ruhr parks. The basic questions posed earlier, of who should provide what, have returned but in a different guise with the answers being sought in political compromise, what Veal called the 'pragmatic approach' (1981) to planning for leisure, rather than in mathematical norms or spatial optima.

References

Albemarle Report of the Committee on Youth Service in England and Wales (1961), Cmnd 929. HMSO, London.

Ashworth, G. J. (1979) Leisure, in R. Windle (ed) *Records of the Corporation 1966–1974*, Portsmouth City Council, Portsmouth.

Ashworth, G. J., White, P. E. and Winchester, H. P. M. (1988) The red-light district in the West European city: a neglected aspect of the urban landscape, *Geoforum*, 19(2), pp. 201–212.

Bonnaine-Moerdyk, R. (1975) L'espace gastronomique, *L'espace Geog.*, 2, pp. 113–126.

Bowler, I. and Strachan, A. (1977) *Parks and Gardens in Leicester: a survey*, Recreation and Cultural Service Department, Leicester City Council.

Burgess, E. W. and Park, R. (eds) (1925) *The City*, University of Chicago Press, Chicago.

Burke, G. (1966) *Greenheart Metropolis, Planning the Western Netherlands*, Macmillan, London.

Cargill, S. and Hodgart, R. (1978) *A Strategy for the Provision of Squash Courts in the Lothian Region*, Department of Geography, University of Edinburgh.
Chadwick, G. (1966) *The Park and the Town*, Architectural Press, London.
Christiansen, M. (1978) *Park Planning Handbook*, Wiley, London.
Conseil de Paris (1971) *Les Espaces Verts de Paris*, Paris.
Davidson, J. (1978) Growing importance of urban fringe, *Surveyor*, 151(4479), p. 19.
Dower, M. (1965) Fourth Wave, *Archit, J.*, 141(3), pp. 122-190.
Edinburgh District Council (1977) *Planning Department Report*, Research Section, Edinburgh.
Gemeente Arnhem (1975) *Stadsvernieuwing Arnhem*, Sociografische Dienst Gewest Arnhem, Arnhem.
Glasgow County Council (1975) *Open Space and Recreation*, Planning Policy Report, Glasgow.
Greater London Council (1967a) *Greater London Development Plan*, GLC, London.
Greater London Council (1967b) *Greater London Development Plan, Enquiry*, GLC, London.
Greater London Council (1976) *Greater London Development Plan*, GLC, London.
Hampshire County Council (1978) *Policy Review: Recreation*, Hampshire County Council, Winchester.
HMSO (1975) *Sport and Recreation*, Cmnd 6200, HMSO, London.
Jacobs, J. (1962) *The Death and Life of Great American Cities*, Jonathan Cape, London.
Kommunalverband Ruhrgebiet (1982) *Changing for the Future*, Essen.
Leusse, M. (1976) *Le Quotidien*, 20 April.
Lever, W. (1973) Recreation Space in Cities, *J. Town Plan. Inst.*, 59, pp. 138-40.
Lozato, J. P. (1985) *Géographie du Tourisme*, Masson, Paris.
Mennell, S. (1976) *Cultural Policy in Towns*, Council of Europe, Strasbourg.
Mennell, S. (1978) *Strategy for the New Project*, Council of Europe, Strasbourg.
Mercer, C. (1976) *Living in Cities: Psychology and the Urban Environment*, Penguin, Harmondsworth.
Ministry of Housing and Local Government (1962) *The Green Belts*, HMSO, London.
Miossec, J. M. (1976) Elements pour une théorie de l'éspace touristique, *Cahiers du Tourisme*, C36, Aix-en-Provence.
Molyneaux, P. (1972) The context for planning indoor sports centres, *Built Environ.*, 1(8), pp. 523-526.
Munton, R. (1981) Agricultural Land-Use in the London Green Belt, *Town and Country Planning*, 50(1).
Pigram, J. (1983) *Outdoor Recreation and Resource Management*, Croom Helm, Beckenham.
Préfecture de la Région Parisienne (1975) *Schéma Directeur d'aménagement et d'urbanisme de la Région Parisienne*, Paris.
Provinciale Bestuur Zuid-Holland (1967) *Vrije Uuren in de Vrije Natuur*, Rotterdam.
Provinciale Bestuur Zuid-Holland (1969) *Rotterdammers op Zondag*, Provinciale Planologische Dienst, Den Haag.
Provinciale Bestuur van Zuid-Holland (1983) *Openlucht Recreatie*, The Hague.
Riley, R. C. and Ashworth, G. J. (1975) *Benelux: an economic geography of Belgium, the Netherlands and Luxembourg*, Chatto and Windus, London.
Roberts, R. (1974) Planning for leisure, *Building*, 15(3), pp. 98-102.
Robertson, I. M. L. (1978) Planning for the location of recreation centres in an urban area: a case study of Glasgow, *Reg. Stud.*, 12(4), pp. 419-428.
Siedlungsverband Ruhrkohlenbezirk (1966) *The Ruhr Plans, Progress and Prospects*, SVR, Essen.
Smith, S. L. H. (1983) Restaurants and Dining out: geography or tourism business, *Annals of Tourism Research*, 10, pp. 515-49.

Stadt Koln (1978) *Stadtentwickelungsplan, Freizeit, Sport, Freiraumplanung*, Cologne.

Stadt Mannheim (1977) *Grunordnung und Stadtplanung in Mannheim*, Mannheim.

Steigenga, W. (1968) Recent planning problems of the Netherlands, *Reg. Stud.*, **2**, pp. 105–118.

Ulfert-Herlyn, V. (1977) Soziale Ungleicheiten in der Stadischen Freiraumversorgung, *Landschaft und Stadt*, **9**(2), pp. 49–57.

Veal, A. J. (1981) *Planning for Leisure: three approaches*, Papers in Leisure Studies, Polytechnic of North London.

Ward, B. (1976) *The Home of Man*, Penguin, Harmondsworth.

Wippler, R. (1966) *Vrije Tijd Buiten*, Economische-Technologische Instituut, Groningen.

Whitt, J. A. (1987) Mozart in the metropolis, *Urban Affairs Quarterly*, **23**(1), pp. 15–36.

Wolfenden Committee of Sport (1960) *Sport and the Community*, Central Council of Physical Recreation, London.

CHAPTER 9

Planning The Tourist City

Tourism as a concern of urban planning

The reception of visitors is a traditional function of cities. Cities both generate most tourism demand and supply important resources to satisfy it. It would seem self-evident therefore that this important, long standing and ubiquitous aspect of the urban scene would have received significant attention from urban governments and planners, especially in Western Europe. The consistent neglect of tourism as a legitimate preoccupation of planning in most towns until recently (Ashworth, 1988) therefore needs explanation.

It has been convenient to divide cities into two categories: resorts where tourism is a dominant feature of the urban economy, and the rest, in which the very continuity, dispersion and ubiquity of the tourist function tended to render it invisible and therefore ignored by urban governments. Although towns had always accommodated visitors for business or pleasure, this was viewed as either an inevitable temporary inconvenience to be borne by citizens or conversely a windfall economic gain to some existing facilities and a free bonus to the prosperity of the urban economy as a whole. In practice, however, all resorts are to an extent multifunctional, and all towns have some tourist functions and are thus, to an extent, resorts. The implications of this are that all towns must plan for tourism, but also that tourism must be considered as part of urban planning as a whole.

Neglect of tourism was a result in part of the conviction that it is an external influence over which the city had little control. The size, timing and nature of the tourism 'invasion' was determined by national or international factors and managed by marketing agencies whose success was measured by their ability to attract ever larger numbers of visitors regardless of the impact, welcome or not, upon the cities they frequent. Most cities therefore did not consider it to be within their competence or authority to influence the size, direction or composition of the tourist flow and, in any event, were organisationally and politically incapable of doing so.

So far as inadequate planning for tourism is concerned, the structure of local government is largely to blame; the management of the cities is run on behalf of the residents by elected city councils or their equivalents. They undoubtedly equate the needs of the city with the needs of the city's residents and would suffer electoral embarrassment were they to do otherwise. (Young, 1973)

In addition tourism is difficult to isolate as a separate industry or even activity but is deeply embedded in the various aspects of the leisure city outlined earlier in Figure 8.1. Tourists make some use of many urban accommodation, catering, shopping, entertainment and transport services, but a dominant use of very few. It is not surprising therefore that in most cities tourism was regarded as merely a marginal additional profit or cost, rather than a separate set of demands or facilities that required public sector intervention.

It was not until well into the 1970s that planning authorities, outside a few specialised seaside, spa and winter sports resorts, began to regard tourism as an important urban function to be planned for alongside others, rather than a phenomenon over which they had minimal control. This change in attitude can in turn be related to fundamental changes in both the nature of tourism and the reaction to it within the European city.

The series of imperatives that underlie the significance of planning for tourism in the West European city can be simply stated. The demand for tourism, whether domestic or foreign, and whatever the motive for the visit, has expanded consistently on a global scale over the past 30 years faster than most other economic or social activities. Within this overall expansion Western Europe is the largest generator and receiver of visitors: of the around 300 million foreign trips made globally, between two-thirds and three-quarters originated from Western Europe and were to Western European destinations. Within Western Europe the cities are the most important destinations for foreign visitors and one of the most important for domestic visitors. Such bland assertions conceal enormous variations between countries and particular cities (which Pearce (1988) has described in more detail) but the recent, dramatic rise in tourism as an 'industry', a land-use and an urban function makes it difficult to overstress the importance of cities, and the cities of Europe in particular, to world tourism. The large world cities, such as London, Paris, Rome, Copenhagen, Athens and Brussels, attract as many foreign tourists as whole Alpine or Mediterranean resort regions.

The very recentness of mass tourism renders it necessary to outline the nature of the phenomenon, and its impact upon the Western European city, before evaluating the planning reactions to it. Therefore this chapter will consider first the nature of the demands made upon the West European city by tourism, which requires a brief outline of some relevant characteristics of the tourist, and secondly a consideration of the city as tourism 'resource' and tourism 'product'. These considerations of tourism demand and supply, customer and product, are a necessary preliminary to the development of urban planning strategies specifically designed for managing the urban tourism function.

The urban tourist

The reasons why cities in general and West European cities in particular exercise this powerful attraction for visitors are of less immediate importance in this context than those generalisations about the nature of this visitor invasion that will allow urban authorities to plan to accommodate it, satisfy it and profit from it.

The diversity of visitor motives, characteristics and behaviour, resulting from a similarly wide diversity of urban resources and opportunities, is central to both such an explanation of the phenomenon and the beginnings of a description of its character and thus of the planning response to it.

Four characteristics of visitors to West European cities form the basis for understanding and planning.

Motivation

People visit cities for a wide variety of reasons, and the separation of these motives, whether business or pleasure, is difficult and generally unnecessary for planning purposes except in so far as identifiable groups of visitors make specific demands on particular sets of resources. It is this very diversity of motives that is a distinctive characteristic of urban tourism and which provides the opportunity for management.

Unlike other tourist areas a high proportion of visitors to the cities are there for business instead of, or as well as, pleasure. The amount of business travel is strongly related to the economic importance of cities and their role in the international economy. The major financial centres (such as London, Frankfurt and Zurich), the international governmental and administrative centres (such as Brussels, Strasbourg, Geneva and Luxembourg City), the period trade fairs (such as Utrecht, Frankfurt and Hannover), arts festivals (Bayreuth, Salzburg and Edinburgh) and international conferences and exhibitions, all both generate large amounts of travel and concentrate it in a limited number of favoured cities. These centres with an international appeal are paralleled on a lesser scale by the many smaller multifunctional European cities whose familiar shopping, historical, cultural as well as business facilities attract an inflow of visitors proportionately as important. A recent study of visitors to a selection of medium and small sized Dutch towns, for example, has stressed the importance of such prosaic urban features as periodic markets, cafés and shops in tourist activities (Jansen-Verbeke, 1988).

Timing

Visitors to cities do not generally stay long. The typical business trip is short and holidaymakers can grasp the attractions of a large city in a few days (the average length of stay of foreign visitors in Amsterdam is less than two days, Paris around three days), and a small city, such as Trier, Canterbury, Chartres or Bergamo, however historically renowned, in a matter of 2-4 hours. The important implication of this brevity is that for the visitor all but the largest individual cities are only one element in a wider package. Even world renowned centres such as Leiden, Versailles, Bruges or Stratford-on-Avon will provide only brief entertainment for tourists on circuits of similar attractions. Effective planning therefore for the tourist city needs to operate at the regional, national and even international scales as well as locally.

The profitability of tourism facilities depends in part on seasonality. Urban tourism is less dependent than beach or winter resorts on weather, more likely to consist of short breaks throughout the year than a single long summer holiday, and business visitors have a weekly rather than yearly cycle. Thus this less pronounced seasonal peaking provides a regular base-load for hotels and other facilities.

Distance

Visitors attracted to the tourist resources of cities travel further than visitors to other sorts of holiday destination (*interalia*, Pearce, 1987). The significance of this point is the relationship between distance travelled and such characteristics as length of stay, choice of accommodation and attractions, which in turn can be related to visitor expenditure.

A higher proportion of the total visitors to the West European city are foreign compared with visitors to inland rural and beach resorts; 86 per cent of visitors to Brussels are foreign, for example, compared with only 9 per cent of visitors to the Belgian coast. In West Germany 68 per cent of all tourism overnight stays are in towns of between 20-100,000 people but only 18 per cent of these are foreign. However in cities larger than this, which account for 32 per cent of total visits, 23 per cent are foreign while in Munich and Frankfurt foreign visitors make up 40 per cent and 52 per cent of the total respectively. Similarly in Britain, although domestic holidaymakers outnumber foreign visitors by more than three to one, just under two-thirds of all visitors to London are from overseas. The importance of the foreign element in the tourism industry increases sharply with the size of the city, with the large West European capitals forming an exclusive league of world tourism centres. Capitals are endowed with a disproportionate share of accommodation, museums, theatres and national art collections, and serve as national showcases, so that a visit to Britain but not London, or Denmark but not Copenhagen, becomes less likely. Small wonder that more than half of all foreign visitors to the Netherlands stay in Amsterdam, and two-thirds of all foreign visitors to Britain visit London, with 40 per cent visiting nowhere else (ETB, 1981; CBS, 1987). Only Germany and Switzerland with their lack of a clear cultural capital, Austria with its capital so distant physically and culturally from its Alpine tourism resources, and Spain with its dominant littoral tourism developments, are exceptions to this pattern.

Not only does the city appeal to foreign as well as domestic tourists, it also attracts a high proportion of intercontinental as opposed to near neighbours. In Paris, for example, between a quarter and a third of all tourists are North American, about the same as all Germans, Dutch and Belgians together. While in the coastal resorts European visitors outnumber North Americans by more than ten to one. Much the same pattern is evident in Vienna, London, Brussels and Rome where the tourist industry is dominated by the long-haul American, Australasian and East Asian visitors.

Expenditure

The importance of tourism in urban economies is underlined by visitors to cities spending around twice as much per day as visitors to other destinations such as beach resorts. In part this is a reflection of the relatively high age and socio-economic status of visitors attracted to the urban heritage, shopping and entertainment facilities, but also in part a result of the pattern of accommodation choice (hotels account for more than half the visitor-nights in London, Paris and Amsterdam, in each case more than twice the national average) and importance of high spending business and conference visitors.

The overall picture that emerges is of short staying but year-around, high spending, hotel based visitors, a high proportion of which will be not only foreign but intercontinental, with a wide variety of mixed business and pleasure motives. Such generalisations apply essentially throughout West Europe with the distinctions being not so much between countries as between the major multifunctional world cities and the smaller provincial towns, and the specialised resource centres.

Urban tourism resources

The answer to the seemingly simple question of what have cities to offer visitors is complicated by the variety of urban facilities used by visitors and by their exclusive use of so few. The identification of tourism resources as a preliminary to their planning is thus very difficult. There have been many attempts to draw up general inventories (such as those for West German (Maier, 1972) or French cities (Mirloup, 1984)), but although having some comparative value in signalling relative deficiencies in facilities these are rarely either comprehensive or particularly useful as a basis for planning. More specialised descriptions of urban cultural, historic (Ashworth, 1986) or shopping resources (Jensen-Verbeke, 1989) may serve to emphasise not only the variety but also the relationship between visitor and resident facilities and services.

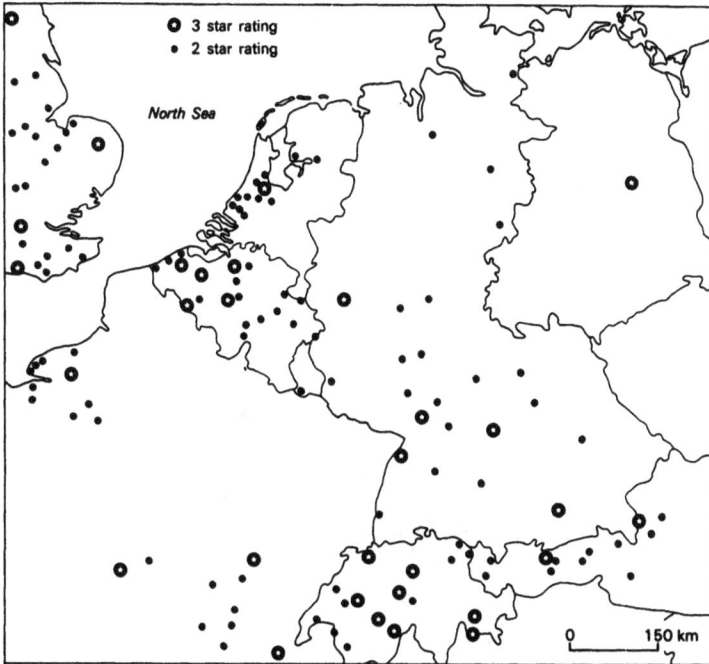

(a) After Michelin

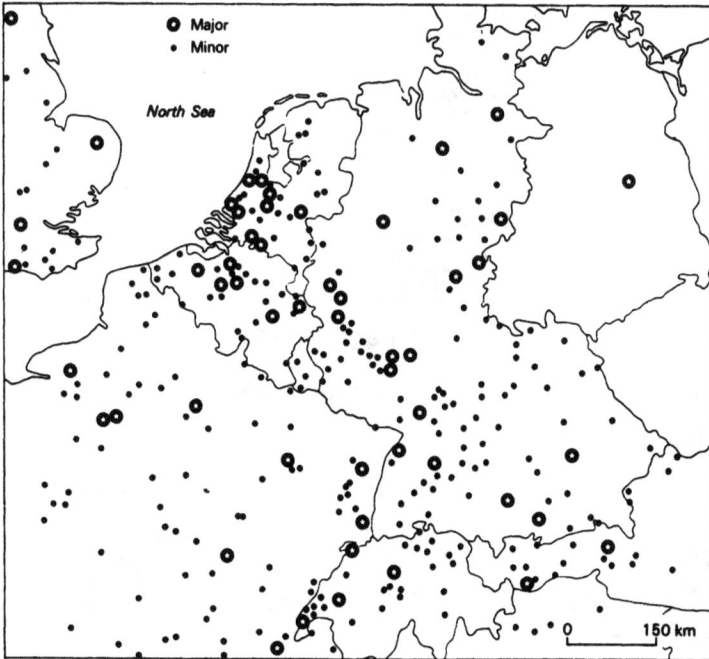

(b) After Ritter

Figure 9.1 West European tourist resources based on alternative sources

The two inventories of the tourism resources of North-Western Europe (Figure 9.1) show not only the importance of cities, and especially the tendency towards agglomeration of tourist cities in recognised tourist regions and circuits (such as the 'Art Cities of Flanders', or the 'Romantische Strasse' in Bavaria), but also the possibility of the same resources being perceived differently by different groups of visitors, in this case a French and a German view.

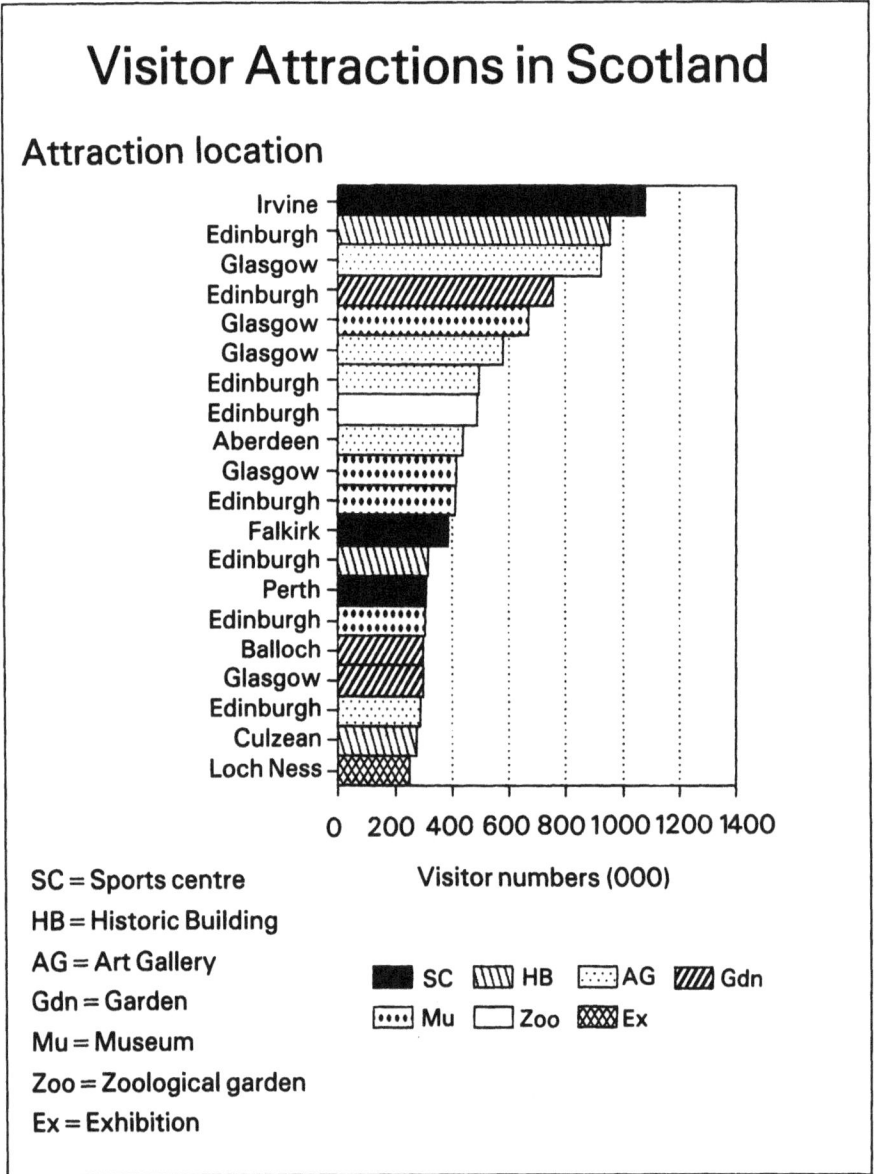

Visitor Attractions in Scotland

Attraction location

SC = Sports centre
HB = Historic Building
AG = Art Gallery
Gdn = Garden
Mu = Museum
Zoo = Zoological garden
Ex = Exhibition

Visitor numbers (000)

SC   HB   AG   Gdn   Mu   Zoo   Ex

Source: Scot. Geog. News 1990

Figure 9.2   Visitor attractions in Scotland

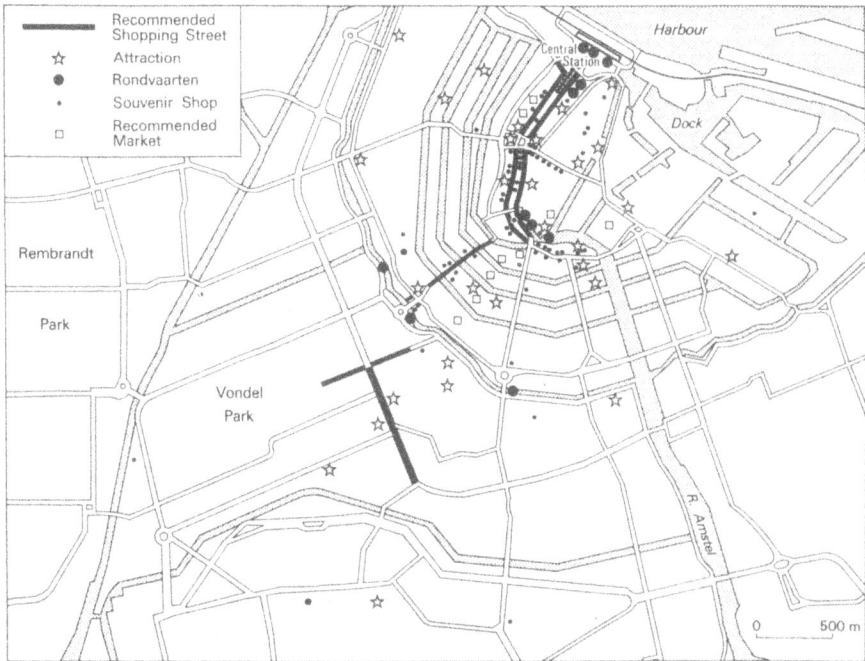

Figure 9.3 Major tourist attractions in Amsterdam

Figure 9.2 shows the most visited tourism sites in Scotland. All but three are urban, but the coincidence of locational concentration with variety in content could be found in most European countries.

Figure 9.3 illustrates the nature of these difficulties at the level of the individual city, and the resulting 'tourist's city' for the case of Amsterdam which, with its annual 2.5 million foreign tourist-nights, 0.5 million domestic visitors and 2.5 million day visitors (VVV Amsterdam, 1987), can be regarded as typical of West European big city tourism.

The intensity and nature of the impact of tourism upon the city is determined largely by its spatial and temporal concentration and by its functional associations with other urban attributes. It is these two conditions rather than the absolute numbers of tourists or the locations of individual tourism facilities that present most of the opportunities and problems that challenge local planning. Thus an understanding of the locational patterns of tourists and tourism facilities within the city in relation to the form and function of the city as a whole is critical. The spatial concentration of tourism activities can be related to three distributions in the city: the location of the tourism attractions that have drawn the visitors to the city ('the primary attractions'), the location of the services that support, accommodate and entertain them while they are there ('secondary attractions'), and the spatial patterns of actual usage of these facilities by visitors in the city.

Primary resources

The tourism resources shown in Figure 9.3 include the selected museums, galleries and historic monuments that dominate 'heritage tourism'. The concentric growth of most cities has left most of the historic attractions concentrated into a small area of the historic inner city core, often clearly morphologically delineated by old defensive walls or waterways. These resources could be regarded as fixed, but a degree of selection has been exercised by both visitors and urban managers, resulting in attraction clusters reinforced by the location of museums, craft and souvenir centres, and cultural and tourism information services that locate near the historic attractions they illustrate and document. Distinctive 'tourist-historic cities' (Ashworth and de Haan, 1986; Ashworth and Tunbridge, 1990) may have emerged, whether by the design of planners or not. In the Amsterdam case, the 'historic-city' can be identified in the streets around the 'Dam'. Larger cities may have a number of such tourist-historic districts whose physical attractions are thematically promoted (eg London's City/Westminster/Covent Garden/Bloomsbury and many more).

In the smaller towns the pattern is often simpler with the contemporary commercial centre and the concentration of tourist-historic resources coinciding, as in Delft, Heidelberg or Florence. The migration of the central business district can leave the redundant historic centre to tourism functions alone. This is particularly obvious in hilltop Mediterranean towns such as Bergamo and Arezzo or Carcassonne, where the modern residential service functions have migrated to more convenient low level sites leaving the 'upper town' to tourism, or in waterfronts where commercial activities have similarly migrated (Tunbridge, 1988).

In addition, in Amsterdam a distinct 'culture quarter' was developed about 2 km to the south-west of the city centre, formed of the Rijksmuseum complex of three major art galleries, the Concertgebouw concert hall, associated parks, and a linking shopping street specialising in artware and antiques. The Les Halles–Beaubourg quarter in Paris provides an illustration of the mix of functions attracted by the presence of major cultural facilities. In this case the close-packed streets between the two main cultural complexes of Les Halles and Centre Pompidou contain a range of cultural and entertainment services (three theatres, three cinemas, 28 private art galleries), 'appropriate' shops (including 59 boutiques and 14 book shops), catering establishments (67 restaurants, 27 cafés), as well as hotels with 240 beds in a mix of restored and purpose-built premises. Similar examples of such agglomerations of cultural and entertainment resources can be found in the EUR in Rome, Cologne's museum and cathedral complex, Brussels' Kunstberg, London's South Kensington museums complex or South Bank and Barbican cultural complexes, and similar features at various scales in almost all the European capitals.

Apart from historic and cultural clusters and areas, tourists make use of a large number of linear facilities such as specialised shopping streets like

Kalverstraat (Amsterdam), the Champs Elysées (Paris) and Köningsallee (Düsseldorf), canal excursions, river embankments, trails and routes through selected parts of the city.

Thus the pattern of tourism resources, as can be seen in the case of Amsterdam, is concentrated into a small fraction of the total urban area, with clusters of generally inner city historic resources but including selective use of shopping, cultural and entertainment districts that principally serve local citizens.

Secondary resources

The distinction between the primary resources, which attract visitors to the city, and the secondary resources, which support them, varies between individual visitors, and the range of services used by tourists includes those shopping,

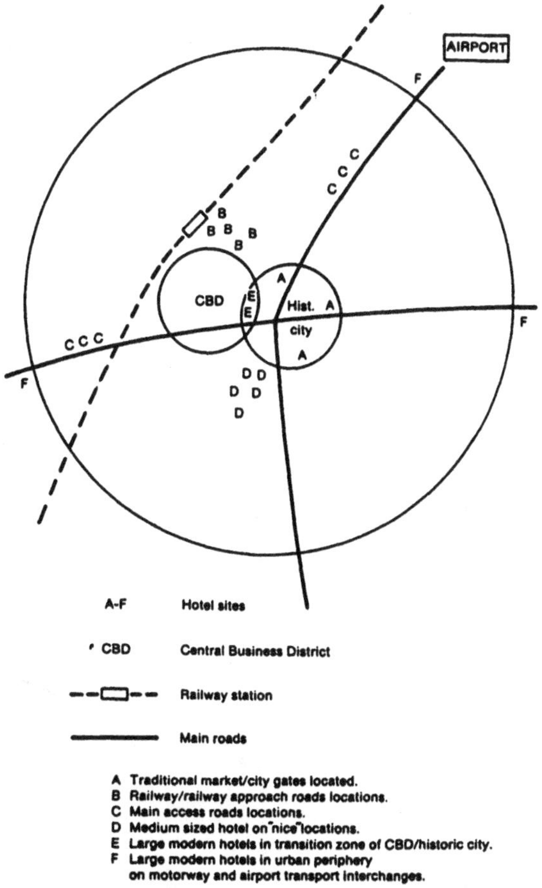

A-F	Hotel sites
' CBD	Central Business District
--☐--	Railway station
———	Main roads

A Traditional market/city gates located.
B Railway/railway approach roads locations.
C Main access roads locations.
D Medium sized hotel on nice locations.
E Large modern hotels in transition zone of CBD/historic city.
F Large modern hotels in urban periphery
   on motorway and airport transport interchanges.

Figure 9.4 Hotel location model

transport, recreation and entertainment facilities discussed in previous chapters, which also and principally serve the demands of residents. Hotels serve a wide variety of visitors as well as providing many ancillary services to residents and this ubiquitous feature of the urban scene can thus serve as an illustration of secondary resources.

The traditional location for hotel accommodation is as an integral part of the central business district in close proximity to the historic and commercial attractions ('A' sites in Figure 9.4), and this pattern remains typical of many small and medium sized towns (as described in Nürnberg for example by Vetter, 1975). Even in a city the size of London, 45 per cent of all hotels are located in the single borough of Westminster and a further 20 per cent in Kensington and Chelsea. The tendency for hotels to become larger, in response to changes in the economics of the industry, and the high costs of expansion on city centre sites, together with changes in the mode of travel of visitors, have encouraged migration to new sites where land is cheaper and accessibility better (Ashworth, 1989). This is by no means a recent process, as Gutirrez (1977) has traced for Madrid and Ritter (1985) for Cologne. The peripherally located railway station often attracted substantial hotel development in the late nineteenth century ('B'

Figure 9.5   Hotel accommodation in Edinburgh

sites) as did, rather later, the main access roads into the city ('C' sites), as can be recognised in Edinburgh in Figure 9.5. The distribution of hotels in pre-war Berlin for example (Figure 9.6) shows a distinct clustering around the three main railway stations, which themselves were located just outside the historic Altstadt. Some diffusion of the hotel quarter westwards along the Tauentzien-strasse and Kurfurstendamm can be seen before 1945, but by 1976 this trend, strongly reinforced by the political division of the city, had resulted in the virtual abandonment of the railway station sites, especially in Kreuzberg in favour of Tiergarten in particular.

More recently the importance of access to motorways and airports has encouraged the establishment of out of town hotels ('E/F' sites) sometimes in association with congress facilities, such as Europahalle in Trier and GMEX in Manchester. These trends were reinforced in some instances by planning policies for the inner city. In Amsterdam, for example (Figure 9.7), a large number of hotels remain in the city centre but severe restraints on new hotel development and new access requirements of visitors led to new hotel development on the urban periphery, especially around the southern motorways; and, like Schiphol airport, every airport with international flights whether it be Hannover, Glasgow or Amsterdam attracts some hotel development. Similarly in Paris the hotels of the central *arrondissement*, such as those off the Boulevard St Michel on the left bank, tend to be small, located in converted rather than purpose built buildings and family owned and managed. By contrast the larger purpose built hotels constructed since the war have largely been located at such sites as Porte Maillot and La Défense, with access to the national motorway system and the Boulevard Périphérique. In London the inner boroughs imposed many restrictions on hotel building or conversion during the 1970s while many of the peripheral boroughs and districts encouraged new clusters, such as around Heathrow and Gatwick airports. In the course of the 1980s however the simple pattern of inner restraint/outer en-couragement became less sharply defined as the value of including hotel accommodation in multipurpose redevelopment schemes in inner city locations (frequently 'C' sites on Figure 9.4) was realised. In Paris the Fronts du Seine, in Amsterdam on Zeedijk, in London's Docklands, or even in a provincial city such as Manchester's Salford Quays, hotel development is viewed as an essential part of area regeneration projects.

Visitor accommodation is clearly homogeneous in neither quality nor function, and these differences lead to distinct variations in location within the city. In particular the small hotel and pension sector usually forms a distinct quarter near, but outside, the historic core, and often associated with areas of large older housing suitable for conversion. Planning policies restricting changes of use have tended to further concentrate accommodation in clearly defined quarters. This feature can be seen in Nürnberg (Stadt Nürnberg, 1976) while in Amsterdam it is located especially around the Vondel park, about 2 km from the city centre (Figure 9.7).

In many cases therefore the accommodation resources of the city are located some distance from the primary attractions, and this distance between the two

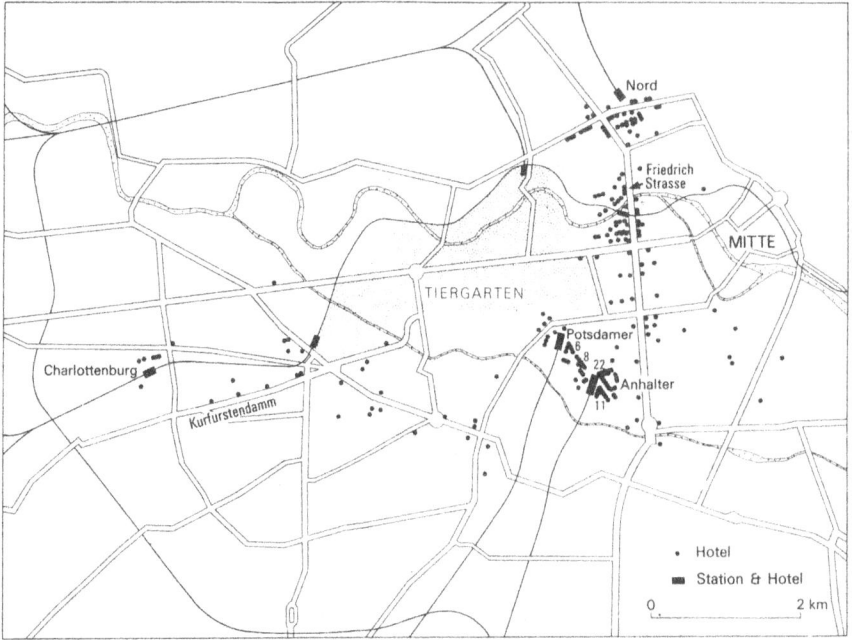

(a)

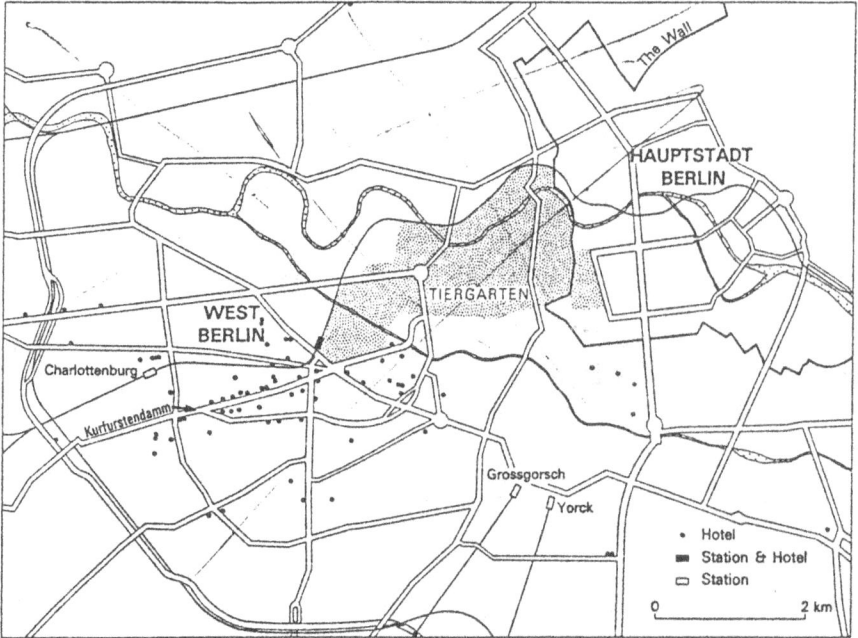

(b)

Figure 9.6   Changing hotel locations in Berlin 1943–76: (a) hotels (1943) in
Greater Berlin; (b) hotels (1976) in West Berlin

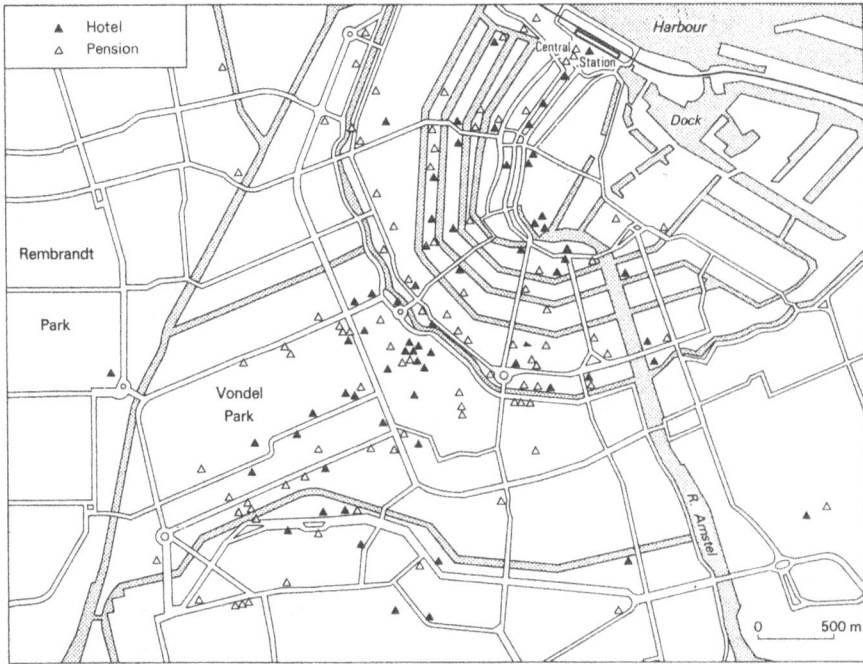

Figure 9.7 Hotel accommodation in Amsterdam

has tended to widen as a result of planning policy. This in turn poses problems of traffic management for the movement of visitors frequently by coach, from hotels to attractions and within the city, and congestion problems around key sites, such as London's Parliament Square or Amsterdam's Rijksmuseum complex.

Usage

The third variable affecting the agglomeration of tourism within the city is the behaviour of visitors themselves. The visitor's image of the city is likely to be restricted to a few world famous sights that are fixed in the popular imagination by guidebooks and tourist publicity (they are 'marked', to use MacCannell's (1976) terminology). The longer the distance visitors have travelled, the more their expectations are likely to be reduced to a handful of essential experiences. Therefore the Louvre, Strøget, Dam Square, the Vatican, the Prado and similar 'highlights' will attract a very high proportion of total visitors, and conversely few will extend their search much further afield. As many North American visitors to Amsterdam take the traditional *rondvaart* as are registered in the city's hotels. Similarly in Paris intercontinental visitors make up less than one-third of the city's foreign tourists, but over two-thirds of foreign visitors to the Louvre. As well as a fragmentary mental map, the visitor has little time and limited mobility. The absence of personal transport and unfamiliarity with public transport further confines the visitor to the areas accessible on foot or included in coach tours.

Consequently the 'tourist city' covers only a small fraction of the area of the 'residents' city'. In smaller towns, with clearly defined attractions, where the length of stay is measured in hours, visitors will confine themselves to an area of a few hundred metres (as reported for Delft (VVV Delft, 1972) and Lourdes (Chaudfaud, 1981)). In the large cities this 'tourist city' may consist of a collection of such concentrations, linked in some cases by tourist 'corridors', such as, for example, the Latin Quarter, Notre Dame, the Louvre and Etoile in Paris, linked partly by the Champs Elysées, and The Mall and Whitehall linking Buckingham Palace, Trafalgar Square and Parliament Square in London; or on a more modest scale Edinburgh's Royal Mile, Amsterdam's Damrak or Barcelona's Ramblas linking Plaza de Cataluna with the waterfront.

It is possible therefore to conceive of a 'Central Tourist District', conceptually similar to and spatially overlapping with Stansfield and Rickert's (1970) commercial 'Recreational Business District', which occupies only a small proportion of the city's area but contains most of the facilities used by tourists. In all but the most important tourist centres such a district is likely to share the city centre with other central business district functions.

Planning for urban tourism

The word 'variety' has been used above to describe an essential feature of both tourism demands on the city and the urban tourism product. Planning will obviously be complicated by such diversity and in particular in the European city the devising of strategies for tourism occurs against the background of organisational fragmentation. In institutional terms there is the fundamental, if ill-defined, distinction between the fields of operation, working methods and objectives of public agencies and of private commercial firms. Frequently this schism is reflected in tourism through a distinction between those organisations concerned with the maintenance and management of many primary urban tourism attractions and those bodies concerned with selecting and promoting the tourism product. Not only does public planning for tourism involve a large number of local authority departments, at various spatial scales in the government hierarchy, each responsible for providing different sets of tourism facilities or coping with aspects of tourism demand, but in addition tourism as a commercial activity is dominated by a large number of relatively small catering, accommodation, transport and retailing enterprises.

The possibility of achieving harmonious coordination, or at least the absence of contradictory policies, is therefore organisationally more difficult than for most urban planning functions described in this book.

Defensive policies

The planning problems posed by tourists were viewed until recently by many city authorities as a matter of accommodating, with as little possible expense or political disturbance, a flow of unknown size over which it had no control. Even in ostensible resort towns this defensive cost-reducing approach was often adopted by the planning authority, while the possible benefits were assumed to accrue to the private sector. Whatever the national benefits to be gained, tourism was all too often viewed at the city level as a threat to be reduced if not diverted, and a set of problems, revealed by a string of commentators (Hall, 1970; Young, 1973; Eversley, 1977; Chenery, 1979), to be solved. Many costs are imposed by tourists on the local area, but those that city planners were expected to mitigate concern the allocation of urban resources, especially land, and the pressures imposed by visitors on an urban infrastructure that was designed for residents alone.

A common problem in the first category was the 'homes or hotels' controversy that has featured in the local politics of many cities, most virulently perhaps in London and Amsterdam. The rapid increase in the demand for hotel beds in the major European cities in the 1970s, coupled with the existence of rent controls on residential rented property in some form or other in most West European countries (see Chapter 6) made the conversion of property to serve visitors rather than residents economically attractive. In some countries the existence of government subsidies for the creation of new hotel beds increased the incentive. Districts composed of large, old, multiple-occupancy housing close to the city centre proved especially vulnerable to what was termed 'creeping conversion' (GLC, 1978). A situation where in London, for example, some 1,000 'residential bed spaces' were being 'lost' to hotels annually, had most of the ingredients of a popular political issue. Relatively rich tourists could be seen to be usurping an increasing share of scarce resources from relatively poor residents.

Reality however was more complex and tourism was only one part of the more general trends in population, employment and housing noted earlier. The districts of the city ripe for conversion in this way were usually those located in the transition zone that was already losing population, what Youngson (1966) had termed in Edinburgh the 'tattered fringe' of the historic conserved area (Figure 9.5). Although the number of hotel beds available in the West European cities more than doubled between 1975 and 1985, most of these were provided in large, purpose-built developments as part of more extensive urban revitalisation projects.

Defensive planning measures were hampered by the absence of basic information on the size and characteristics of the tourist flow and the capacity of the city to accommodate it. Planners' attempts to equate supply and demand were frustrated by an inadequate knowledge of the dimensions of either, and thus tended towards reactions to specific perceived crises rather than strategic forward planning. In addition tourism demand is essentially volatile, with

visitor numbers, as well as seasonal, accommodation and locational preferences, fluctuating dramatically, while supply facilities tend to be both immobile
in location and inflexible in quantity. For example the rapid increase in demand
for high quality hotel space from transatlantic visitors to London in the late
1960s prompted the Hotel Incentive Scheme, which subsidised the creation of
40,000 new hotel beds in the city, but the benefit of this investment was not felt
until the middle 1970s, corresponding with a period of slackening of intercontinental demands, which was replaced by demand from West European
visitors, many of whom favoured cheaper or self catering accommodation
forms, outside as well as inside London (Lavery, 1975). Individual cities have
little ability to either predict or control the nature of the visitor demands for
them.

A rather more sophisticated approach is to attempt to influence the size,
timing and composition of the tourist flows in order to achieve a desired
cost-benefit ratio. In practice one of the few possibilities open to urban
managers was to attempt to spread tourism demands on the facilities of the city
in time and space, thus both relieving pressure at congested periods and raising
the levels of annual occupancy, and thus profitability, of tourist facilities. The
timing of holidays is influenced as much by factors in the place of origin of
visitors as by destinations, but the very varied, all-weather range of urban
attractions has allowed cities to be more successful than most holiday
destinations in identifying and capturing specialised and off-season markets,
such as conferences, and short shopping, cultural and entertainment breaks.
Encouraging a wider spatial spread of visitors through the city, or between
cities, is usually more difficult. Visitors have a clear preconceived idea of what
they have come to see and do, and the heights of Belleville are no substitute for
the Eiffel Tower, nor Southwark Cathedral for Westminster Abbey, regardless
of the intrinsic merits of the attractions.

New attractions are likely to be most successful if they reinforce existing
patterns rather than endeavour to shape new ones. The St Katherine's dock
tourism development in London, for example, adjoins the existing 'Tower'
complex of foreign visitor attractions. Similarly the competition for foreign
visitors is between the major international centres rather than between cities in
any one country: tourists dissuaded from visiting Paris are more likely to divert
to Rome or Amsterdam than Lyons or Marseilles, and centres outside the
confines of existing 'milk-runs' (whether Windsor–Oxford–Stratford for
London, or Enkhuizen–Hoorn–Haarlem for Amsterdam) will have difficulty in
being accepted. Nevertheless the restricted fields of information of visitors, and
their relative immobility, mentioned earlier, within the city, present local
tourism planners with a potent instrument of management. By influencing the
flow of information through official guidebooks, information centres, adver-
tising and signposting, tourists can be encouraged, or discouraged, from
visiting sites and channelled along trails and corridors. Promotion can be thus
more than a crude method of attracting visitors; it can also be an instrument for
their management.

Stimulational policies

It is technically a short step from policies for restraint or control, through those aimed at minimising costs and maximising benefits, to the deliberate stimulation of selected aspects of tourism in pursuit of national or local economic goals. In terms of political will and organisational structures, however, such a step proved to be more difficult to take, in part because it required a redefinition of the role of public agencies in general and city governments in particular.

Defensive strategies were well suited to statutory, regulatory planning bodies. Stimulation required public agencies to become, if not entrepreneurs themselves, at least partners in a development enterprise. At the national level this has involved the distribution of selective subsidies (such as has been possible in the UK through 'Tourism Development Grants' since 1969), or direct state investment (such as in the so-called 'Areas of Special Attention for Recreation and Tourism' (TRAGs) in the Netherlands). In both cases national state intervention was intended to identify potential development opportunities, some of which were in urban areas, to be exploited by the private sector. At the local level the coordination and partnership role of the city authorities may be reflected in the establishment of development departments, joint public–private enterprises and promotional ventures, such as Garden Festivals at such ostensibly unlikely locations as Gateshead in 1990, or new planning initiatives. Typical of the last are the 'Tourist Development Action Programmes' that have been produced by a number of British cities, most notably Norwich (1988), for the development of heritage and shopping tourism, and Portsmouth (1988), for the stimulation of a mix of maritime heritage, conference and water-associated tourism. These differ markedly from traditional structure plans in that they identify sets of objectives, inventorise existing resources in the context of a market analysis and thus highlight public investment deficiencies, which are to be rectified by committed public or predicted private investment.

Tourism is a source of employment and both individual and collective incomes, as well as a land-use, but in many cities even this economic addition was seen until recently as more a danger than an opportunity. The large number of seasonal, part-time and often low paid jobs was seen as a 'threat to the local employment structure' (Young, 1973). 'The danger is that tourism will call for precisely the sort of job opportunities we do not want' (GLC, 1971). In addition the prosperous cities of Western Europe imported much of this labour from low-wage Mediterranean countries, and tourism could thus be charged with encouraging the creation of a pool of cheap inner city labour and even contributing to social and ethnic conflict.

In reality cause and effect were often reversed, with tourism functions mopping up existing surplus labour that had been marooned in the inner city by the flight of jobs described in Chapter 4. It also became clear to many in the course of the last decade (ETB, 1981) that tourism jobs, despite their shortcomings, provided one of the few labour intensive, growing, low skilled possibilities in the inner areas of cities. 'Candy floss jobs' (Williams and Shaw, 1988) were better than none and less directly tourism facilities, tourist spending and even the physical presence of tourists could provide a stimulus for the more general revitalisation of the inner areas of cities with few other viable economic functions.

This line of thinking has led to the use of tourism services (generally luxury hotel and congress facilities) as an integral part of comprehensive downtown redevelopment projects, that also include a selection of shopping, catering, office or residential facilities. The well documented experience of such projects, especially in the downtown areas of the cities of the east coast of North America, is one of the few instances in this book where public planning practices have migrated eastwards rather than westwards over the Atlantic. It should be stressed that these are rarely tourism projects as such nor are they the result of traditional land-use planning, although such traditional planning as pedestrianisation, building conservation and area rehabilitation is often supportive of them. They are the use of tourism functions within a package of financially mutually supporting facilities with private investment being coordinated by public development agencies. The 'Festival Markets' of Boston (Quincy), Baltimore (Harbor Place) or Halifax (Privateers Wharf) are increasingly finding an echo in similar developments, although often on a smaller scale, in the 'fun-shopping' developments as described by Jensen-Verbeke (1986) for many West European cities. Such a physical integration of tourism facilities, reflecting their functional interrelationships within the multifunctional inner city, both capitalises on the benefits of tourism demand while avoiding many of the costs that result from monofunctional tourism areas mentioned earlier.

If tourism is seen as an economic activity to be encouraged as an instrument in urban economic development strategies then the 'product' created by such an 'industry' must be identified. Such an urban tourism product is of course composed of elements drawn from the facility supply described above but is more than such an aggregate. The success of stimulation strategies will depend largely upon the accurate identification of a particular product, whose sale on a specified market will achieve defined objectives. Such a 'commodification' of tourism has led to the similar commodification of tourism destinations as 'place products' (Ashworth and Voogd, 1990) which in turn has allowed the development of market planning as a public sector urban management technique, most especially, though not exclusively, for tourism (Goodall and Ashworth, 1989).

Such market planning is not only a matter of the selection of a range of tourism products, whether heritage, culture, shopping, or entertainment, but also the shaping and promotion of an appropriate image of the city as a tourism destination for targeted markets. Relevant aspects of the images of different West European cities held by visitors (Figure 9.8) can be influenced by tourism agencies. Such 'city-promotion' for tourism cannot be divorced from other

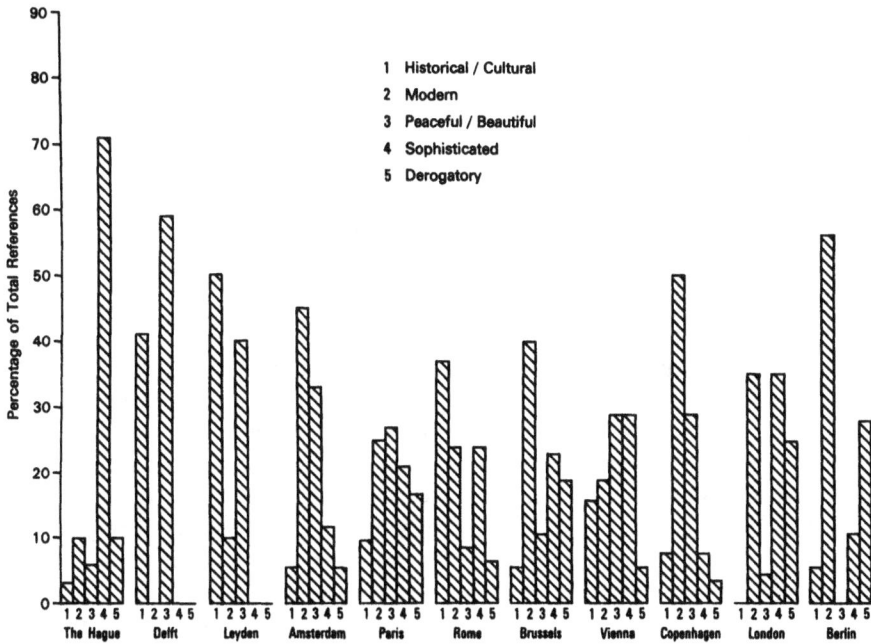

Figure 9.8 Images of West European cities

supporting or contradictory city images. Similarly the pursuit of a policy of stimulation using city-marketing instruments raises the questions of what sort of policy, in pursuit of which marketing aims. Table 9.1, for example, shows some of the many possibilities for devising marketing strategies for tourism development appropriate to both the nature of the existing demand and the capacity and objectives of the city.

Table 9.1 Types of tourism market planning strategy (Adapted from Kotler, 1975)

Nature of Demand	Marketing Strategy	Example
Full demand	Maintenance marketing	Heidelberg/Bruges
Latent demand	Development marketing	Aarhus/Groningen
Negative demand	Conversional marketing	Lille/Duisberg
No demand	Stimulation marketing	Lyons/Frankfurt
Faltering demand	Re-marketing	Baden-Baden/Scheveningen
Excess demand	(Selective) De-marketing	London (Westminster) Venice (Lagoon) Munich (Oktoberfest)

Changes in marketing strategy have been forced, for example, upon Lugano where rising property prices, fuelled by the city's success as a banking centre and increasing numbers of second homes, have put pressure upon the supply of hotel accommodation and raised its cost beyond the expectations of traditional lake-side tourism. The average length of stay dropped from 4.2 days in 1970 to 2.8 in 1987. This, together with a loss of tourist appeal owing to increasing traffic congestion, has prompted a series of new promotional initiatives based upon a redefinition of the tourism place product in the context of changing demands for it: short stay but high cost cultural and festival holidays are now marketed.

Reconciliation

The uni-dimensional nature of both defensive planning strategies and many short term stimulational policies has become increasingly apparent, and unease has surfaced in the last few years in a number of guises. This has led to the development of policies for urban tourism which attempt to reconcile some of the confrontational dualisms intrinsic to both the above types of policy. In particular reconciliation has been sought between the economic justifications for the provision of tourism facilities and the social welfare objectives determining much recreation provision; between the idea of the city as a saleable tourism product and as an environment for living and working; and between tourism project development and more broadly based urban development policies. Although logically none of these pairs of objectives need be exclusive alternatives, much tourism policy through the 1970s, and currently in many cities, assumed that such choices were both necessary and largely self-evident which in turn both created planning conflicts and led to tourism development strategies that failed to realise much of their potential.

It is evident from the varied pattern of behaviour of visitors to cities outlined above that tourists will make use of many cultural, sporting and public utility facilities whose provision, and thus financing, location and management operational philosophy was devised for serving residents. Similarly many public services can receive substantial support from tourism demands. Such mutuality may be manifest on the scale of the sport and leisure complexes built in the 1970s in a number of seaside resorts from Scheveningen to Great Yarmouth, or the major culture, leisure, entertainment and shopping complexes such as the London Docklands' 'Gunpowder Dock scheme' or Rotterdam's 'Waterstad' quayside developments.

The Dutch Tourism and Recreation Development Plans (Ashworth and Bergsma, 1987) drawn up on a provincial or city level are novel in their attempt to regard the total recreational provision of the planning area as a potential tourism resource and to consider their development in the service of the demands of both visitors and residents. At best, for example the 'TROP' for the city of Maastricht, these plans can lead to the more effective utilisation of existing facilities, the identification of largely untapped urban tourism resources and the highlighting of deficiencies in investment. They can, however, be little more than extended inventories of existing and desirable facilities with minimal market analysis and an over-optimistic dependence upon private investment for their implementation.

One of the many problems in constructing a cost–benefit balance sheet for tourism is that each side of the equation tends to accrue to quite different groups within the city so that the 'tourism interests' and the 'residents' interests' become irreconcilable opponents. The phrase 'community tourism planning' has become popular (Murphy, 1985) to describe development planning which attempts to combine provision for tourism with the needs of the local community. This has often proved in practice to be a question of scale and sensitivity to various local capacities that are easier to appreciate and are more critical in rural than urban communities.

Finally it has become clear from much of the experience with large scale tourism development projects in inner cities that the chance of success for both the narrow profitability of the project itself, and for the wider revitalisation aims of the city, depends upon the degree of integration between tourism and other urban functions. Tourism services may play a catalytic role within cultural, entertainment, catering and shopping services, provide a justification for heritage planning, environmental improvements and urban design embellishments, and lead to the shaping of new urban images and a renewed self-confident civic culture; all of which contribute to what has been called 'the new urban renaissance' in urban contexts as ostensibly unpromising as Manchester (Law and Tuppen, 1986) and Glasgow. Urban development should not be tourism project led, but tourism projects can play a vital role in urban development.

Thus in a remarkably short period of time, urban authorities in Western Europe have passed from an attitude of indifference towards tourism, through a series of defensive reactions, to a realisation of its potential for aiding the achievement of a number of planning goals, and even showed signs of moving towards a coordination of organisations and sectors in order to reconcile conflicting interests. It would be surprising if such a progression had occurred at the same tempo simultaneously in all regions of Western Europe and in the many different types of city. The planning response has varied according to the perceived needs, strength and nature of tourism impacts, national supportive policies and opportunist management skills of the individual cities. Long established tourism resorts, whether health resorts (such as the archetypal spa) or seaside resorts (such as most of the French, British, Belgian and Dutch Channel and North Sea resorts) have often reacted with a complacency bordering on indifference, world famous heritage centres (such as Bath or Heidelberg) have been more concerned with containing demand than stimulating it, the large multifunctional capitals (such as Copenhagen, Rome or Madrid) have, with only a few lapses, generally had the in-built flexibility and diversity to accommodate tourism growth as part of a continuous process of organic change, while it has frequently been the ailing monofunctional economies of nineteenth-century industrial towns, such as Bradford, Liège or Mulhouse, which have sensed the opportunities and engaged in vigorous policies of attraction. In West Germany, in addition to the longstanding importance of tourism in towns such as Bayreuth, Augsburg, Bamberg, Munich or Nürnberg, the *land* of Rheinland-Halz has the promotion of tourism as the fourth highest priority in its regional plan.

Whatever the nature of the emerging post-industrial city that European society will require in the twenty-first century, it is clear that tourism will play a major role in it as a land use, an economic function, a provider of demand for urban services, and a participant in the shaping of new urban images and civic self-esteem. The basis for that role is being forged by West European cities of almost all types and location during the last two decades of the twentieth century, largely by local authority planning agencies, to whose regulatory functions have been added the new responsibilities of development and promotion. None of these changes involves tourism alone, but in all of them tourism has played, and will continue to play, an essential part.

References

Ashworth, G. J. (1988) Urban Tourism: an imbalance in attention, in Cooper, C. (ed.) *Progress in Tourism*, Belhaven, London.

Ashworth, G. J. (1986) Accommodation in the historic city, *Built Environment*, **15**, pp. 92–100.

Ashworth, G. J. and de Haan, T. Z. (1986) Uses and users of the tourist-historic city, *Field Studies*, **10**, GIRUG, Groningen.

Ashworth, G. J. and Bergsma, J. (1987) Policy for tourism: recent changes in the Netherlands, *Tijdschrift voor Economische en Sociale Geografie*, **78**(2), pp. 151–5.

Ashworth, G. J. and Tunbridge, J. E. (1990) *The tourist-historic city*, Belhaven, London.

Ashworth, G. J. and Voogd, H. (1990) Can places be sold for tourism?, in Ashworth G. J. and Goodall, B. (eds) *Marketing tourism places*, Routledge, London.

Chaudfaud, M. (1981) *Lourdes: une pélérinage, une ville*, Edisud, Aix-en-Provence.

Chenery, R. (1979) *A comparative study of planning considerations and constraints affecting tourism projects in the principal European capitals*, British Travel Educational Trust, London.

English Tourist Board (1981) *Planning for tourism in England*, London.

Eversley, D. (1977) The ganglion of tourism: an unresolvable problem for London, *London Journal*, 3(2), pp. 186–211.

Greater London Council (1971) *Tourism and hotels in London*, GLC, London.

Greater London Council (1978) *Tourism, a paper for discussion*, GLC, London.

Goodall, B. and Ashworth, G. J. (1988) (eds) *Marketing in the Tourism Industry*, Croom Helm, London.

Gutirrez, R. S. (1977) Localización actual de la hosteleria Madrilena, *Bol. de la Real Soc. Geografica*, **2**, pp. 347–57.

Hall, P. (1970) A horizon of hotels, *New Society*, 12 March, 445.

Jensen-Verbeke, M. C. (1986) Leisure, recreation and tourism in inner cities, *Netherlands Geographical Studies*, **58**, Utrecht/Nijmegen.

Kotler, P. (1975) *Marketing for non-profit organisations*, Prentice-Hall, Englewood Cliffs.

Lavery, P. (1975) Is the supply of accommodation outstripping the growth of tourism?, *Area*, **7**(6), pp. 289-96.

Law, C. M. and Tuppen, J. N. (1986) *Urban Tourism Project: final report*, Dept. of Geography, University of Salford.

MacCannell, D. (1976) *The tourist: a new theory of the leisure class*, Schoken Books, New York.

Maier, J. (1972) Munchen als fremdenverkehrsstadt, *Mitt. der Geog. Ges. Munchen*, **57**, pp. 51-59.

Mirloup, J. (1984) Tourisme et loisirs au milieu urbain et periurbain en France, *Annales de Géoegraphie*, **520**, pp. 704-18.

Murphy, P. E. (1985) *Tourism: a community approach*, Methuen, London.

Pearce, D. G. (1987) *Tourism Today*, Longman, London.

Ritter, W. (1985) Hotel locations in big cities, in Vetter, F. *op. cit.*

Stadt Nurnberg (1976) *Nürnberg: Plan teil 1: bericht zur Entwickelung der Altstadt*, Nürnberg.

Stansfield, C. A. and Rickert, E. J. (1970) The recreational business district, *Journal of Leisure Research*, **2**(4), pp. 213-25.

Tunbridge, J. E. (1988) Policy convergence on the waterfront? A comparative assessment of North American revitalisation strategies, in Hoyle, B. S., Pinder, D. A. and Husain, M. S. (eds) *Revitalising the waterfront: international dimensions of dockland redevelopment*, Belhaven, London.

Vetter, F. (1985) *Big city tourism*, Reimer Verlag, Berlin.

V.V.V. Delft (1972) *Vreemdelingenverkeer in Delft*, Delft.

Williams, A. and Shaw, G. (1988) *Tourism and economic development: Western European experiences*, Belhaven, London.

Young, G. (1973) *Tourism: blessing or blight?*, Penguin, Harmondworth.

Youngson, A. (1966) *The making of classical Edinburgh*, Edinburgh University Press, Edinburgh.

CHAPTER 10

Urban Planning: National and International Influences

City planning in West Europe has been part of a growing trend towards intervention apparent in almost all European countries in the period since 1945. Government intervention in most areas of the economic and social life of the nation on behalf of citizens has had both direct and indirect effects on city planning. While direct involvement by central government in city planning is rare, for almost all the West European nations it is true to say that various forms of economic policy, such as those towards regional development or national transport, have been incorporated in city policies. Similarly, social policies towards housing, education and welfare are generally implemented by local authorities with the consequence that the city structure plans cannot avoid these impositions of national ground and planning law.

A clear and often dramatic example of this is provided by the university towns of Europe, where national educational institutions, centrally financed, have a massive impact on the economy, society and physical appearance of their host cities, and become a major item in the attention of local planners. When a large national institution is located in a relatively small city the result is often spectacular. In Groningen, for example, the state university has 15,000 students in a town of 160,000 inhabitants. The employment of about 4,000 staff makes it the largest employer not only in the city but in the entire north of the country, while the spending of £400 million by the university and another £100 million by students makes it the most important channel of government money in the region (de Jong, 1969). The impact on the demographic structure, with three times as many people in the 20–24 years age group as in the 40–44 or 0–4 years age groups, is obvious on the streets, but equally pervasive is the effect on the housing and land markets, retailing and social attitudes. The traditionally conservative calvinist city now has a strongly entrenched socialist city council, and the highest proportion of declared 'non-religious' of any Dutch city (51 per cent). Although this is perhaps an extreme case, it is not unique. National educational institutions play a similar role in cities as varied as Münster, Edinburgh, Florence or Louvain. Similarly other sorts of national institutions, such as defence installations and major hospitals, as well as government departments and agencies, will exercise local influence but be largely outside the competence of the local planners.

Thus the ideology and actions of the country's governing party can influence urban developments even in cities governed by parties subscribing to alternative ideologies and committed to alternative actions. Nevertheless there are certain national planning policies and goals which have evolved in a particular fashion to provide a relatively distinctive and continuous impact on urban development within each country. It is also true that there is an even broader scale of physical planning that has emerged at an international level with the developing role of the enlarged European Economic Community.

International influences

The movement towards European integration has focused attention on the urban system of the European Economic Community especially in the highly urbanised core of the original community of six. Until after the Second World War attempts to view the national urban systems were beset by problems of the intervening national boundaries and very little attempt was made to recognise the impact of extra-territorial urban developments on the cities and towns of the country. Therefore the few city planning statements that existed were usually focused on the city itself rather than its relationship either to the broader region or to other cities. For instance, the interdependence of cities in the Maastricht–Liège–Aachen triangle, which is readily recognised today, was affected by the nationalistic views of territory in the early twentieth century. In addition, the problems of cities located on the national periphery, a long way in both physical and time distance from the core of the state, has slowed the development of many cities in the past. This has been the claim of Nancy, Metz and Saarbrücken (Burtenshaw, 1976a and 1976b).

With integration in the European Economic Community, many authorities have begun to consider the cities of West Europe as the parts of a potentially integrated system. Given the acceptance of this concept, it is possible to see each individual urban plan within the context of a much broader vision of urban development. Kormoss (1976) recognised this in the area covered by the North-West European Physical Planning Conference (Belgium, Netherlands, Luxembourg and parts of West Germany and France). He noted that there were 43 conurbations of increasing dominance in the region, from the Ruhr at one end to Leiden at the other, which housed 40 per cent of the region's population.

The cluster of clearly interdependent towns and interrelated planning problems that exist on either side of the arbitrary border between Belgium and the Netherlands was an obvious testbed for early experiments in international co-operation. Indeed the Benelux treaties of 1949 made particular reference to joint planning policies and established a governmental committee at a senior level to encourage their creation and implementation. The need was emphasised by a frontier in Limburg that separated the city of Maastricht from its hinterland to the west and south, and separated home from workplace for many in the settlements along the Maas. To emphasise that, the two regions of Dutch Limburg and Belgian Limburg, although peripheral to their nation-states, could be viewed as central when taken together. The joint region was called the Benelux–Middengebied.

The experience of co-operation has, however, not fulfilled the expectations. Study groups have reported (Verbeck, 1973) and plans have been proposed, but in the last analysis the future of the two regions has until now always been determined by the policies of their respective central governments. The quite different strategies for the development of the divided Limburg coalfield settlements on each side of the frontier is a major example. Such co-operation that is apparent on the ground, such as the Belgian–West German motorway link across Dutch Limburg or the joint recreation proposals for the Maas valley, could, one is tempted to believe, have occurred without the panoply of joint committees. If planning co-operation between two friendly countries linked by treaty has proved so difficult in practice, there are grounds for pessimism about the results of similar joint projects elsewhere in Europe.

The concept of interdependent national urban systems was taken a stage further by Michels (1976) who drew together the various national and regional urban systems to produce an international system of development axes and growth points for north-west Europe. He examined the Dutch planning structure with its proposals for grouped deconcentration (*geburdelde deconcentratie*) and hierarchies of centres (Buursink, 1971). Similar policies for an urban hierarchy and sectoral growth poles for particular developments in Belgium together with a linking transport net, were considered alongside German and French studies. While differences between individual national policies, which we will examine later, were apparent in terms of the number of levels in a hierarchy, and the French use of the concept of discontinuously urbanised axes, it was possible for Michels to propose a basic axis development point system (Achsen–Scherpunkt system) that harmonised with the identical German system developed a decade earlier (Figure 10.1). Three major ranks of centre are distinguished and related to the two levels of axial links between the centres. He proposes a series of city-regions which, despite the efforts at integration, somewhat paradoxically recognise national boundaries. Thus the 178 cities classed as development centres, which include 47 metropolitan centres, 35 primary centres and 96 secondary centres, are able to formulate plans in relation to this wider supra-national view of urban development.

Another practical example of the supra-national character of urban planning is the case of Geneva which is surrounded by a 103 kilometre boundary with France and only a 4.5 kilometre-wide strip linking it to the rest of Switzerland. It is a city that attracts 25,000 frontier workers into the city from the French *départements* of Ain and Haute Savoie. Here in 1973 a Franco–Swiss Consultative Committee was formed to study topics such as frontier workers, transfrontier planning, transport and noise pollution from the international airport (République et Canton de Genève, 1977). As a result of these initiatives the 1975 Plan Directeur Cantonal does include reference to developments on French territory that are an integral part of the overall structure of the Geneva city region (République et Canton de Genève, 1975).

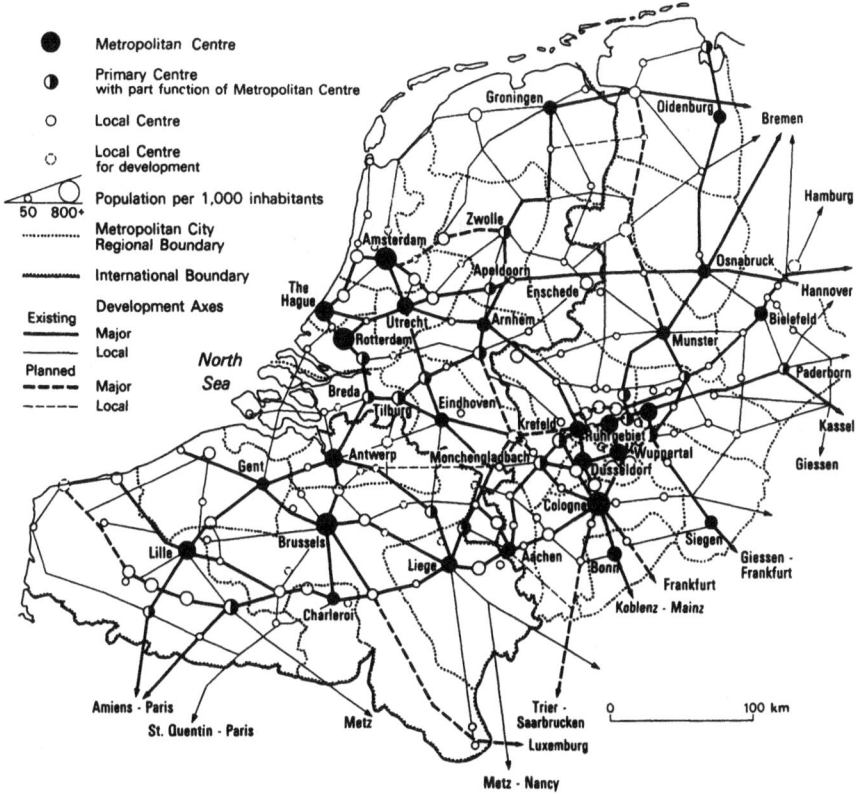

Figure 10.1 The axis–central place system in north-west Europe (after Michels, 1976) (reproduced by permission of Bundesforschungsanstalt für Landeskunde und Raumordnung, Bonn)

A third example of co-operation can be seen at the heart of the European Economic Community where the plans of four countries acknowledge the common problems, needs and aspirations of the peoples living within the area of the Saar–Lor–Lux Triangle (Burtenshaw, 1976a). Here the development programmes for the Saarland, Rhineland Palatinate, the Grand Duchy of Luxembourg and Lorraine all acknowledge the existence of two major axes of development. The first runs south from Luxembourg City, through Esch-sur-Alzette, Thionville and Metz to Nancy, while the second extends from Kaiserlautern through Saarbrücken, Forbach and Saarlouis towards Luxembourg. Between these two axes and beyond them to the west and north lie the recreational zones of the triangle. Given this agreed international view of the directions of growth, the individual cities of the region have a set of unique guidelines for the future so that the plans for the expansion of Saarbrücken, for example, accept the constraints of the broader strategy.

The measures of regional aid dispersed by other international bodies have also had an effect on urban growth and change. The European Coal and Steel

Community (ECSC) and the European Investment Bank (EIB) have both been involved in regional development (Pinder, 1983).

The European Regional Development Fund has not concerned itself directly with urban areas although its funds may be channelled into urban areas. This apparent lack of concern for urban areas is all too evident in the themes adopted by the conferences of European Ministers of Planning; only the Bari conference in 1976 had a specifically urban theme which was pursued at Torremolinos in 1983. However, the accession of Greece, Portugal and Spain has increased the demands for funds for regional development objectives.

Of the more recent integrated regional development programmes the Integrated Development Operations have targeted urban areas, initially Naples, Birmingham and Belfast. It is significant that these operations are funded via the city administrations and not through central government, a fact which displeased the centralist UK government.

Cheshire and Hay's (1989) analysis of European urban problems has pointed graphically to the lack of spatial congruence between the areas qualifying for ERDF assistance and the worst Functional Urban Regions (FURS) (Figure 10.2). According to their analysis. Cheshire and Hay conclude that there should be three types of urban policy established by the ERDF to ensure better targeting of funds to the FURS in greatest need:

● economic aid to problem FURS;
● quality of life policies targeted within FURS by the Environmental Directorate of the EC; and
● social policies directed to poor households within qualifying FURS.

They stress that, at present, funds do not necessarily benefit the urban areas.

National influences on urban planning

Within each country there are overall policy guidelines, some of which are often not intended to have a direct impact on towns and cities, that govern the expansion and development programmes of cities. While it is true to say that central government involvement in city planning has grown throughout West Europe, the actual course of this involvement has varied considerably. In some countries there has been a consistent pursuit of a set of social goals which has resulted in an equally consistent set of policies. Such consistency has been realised in Sweden, for example, through a continuous period of government by one party until recently, whereas as we saw in France in the area of housing policies it has been maintained despite changes in government. On the other hand, policies have oscillated in the United Kingdom with changes in power while policies in Italy have been slow to be created and, more important, slow to be implemented owing to the inability of the administration to make effective decisions.

Figure 10.2   Zones qualifying for European Regional Development Fund aid (after Cheshire and Hay, 1989)

1 Great Britain

Bourne (1975) has drawn attention to the impact of national policies on urban development in both the United Kingdom and Sweden. Early attempts at regional planning in Britain included restriction of growth in the overcrowded south-east, bringing with it a policy of limiting first industrial development and, much later, office development. In line with the recommendations of the Barlow Commission of 1940 the policy of decentralisation implied the restriction of growth of London and the transfer of development to the development regions and, ironically as it transpired since it was so successful, the rest of southern Britain (Barlow Report, 1940). The new towns, partly a reaction to another Barlow recommendation and the result of half a century of campaigning, were another policy decision that has influenced subsequent British urban development. The impact of post-Barlow policies on London has been well documented. For example, industrial employment declined, office rentals rose following restrictions and there was a reduction of over 200,000 commuters a day into the capital. Beyond London the legion of industrial estates in the south-east with factories built to avoid the restrictions of floor space, and later

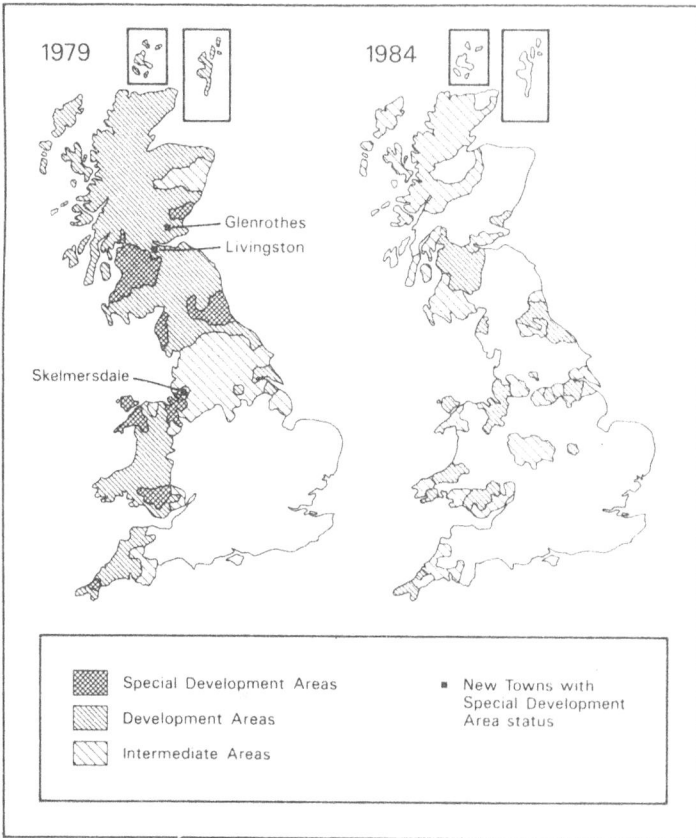

Figure 10.3   Regional Development in Great Britain post-1984 (Pinder, 1990)

the sudden interest of property developers in the outer suburban centres such as Croydon, were tangible effects on urban areas of national policies. The changed central area townscapes of Reading, Ipswich and Swindon also reflect the effects of office policies on medium-sized cities. Further afield the movement of the growing number of civil service posts, for example from London to Newcastle and Durham, was altering the social geography of cities in the development regions.

However, there is a realisation that the inner cities house the most severe economic and social problems. The introduction in 1981 of Enterprise Zones to encourage industrial growth with the minimum of red-tape, which were to be flagships for new industrial developments, were abandoned as a policy in 1987. The Urban Development Corporations in London's Docklands, Merseyside, Cardiff Bay and Manchester have also proved to be less successful than planned because the investments in many of the zones have not matched the high expectations that the developments in London Docklands generated. Other

UDCs have not achieved the same level of private investment that has been generated by London Docklands in Canary Wharf, STOLport, Docklands Light Railway and the extensive areas of new housing and refurbished warehouses.

Governmental influence on urban development in the UK has shifted from that of intervention to enabling, from widespread aid to spatial targeting, and from public to private investment, in keeping with the ideology of the new right. The catalogue of initiatives to alleviate urban decline introduced since 1981 has increasingly involved the participation of private investment. Spatial targeting was a feature of the General Improvement Areas and Housing Action Areas but it was given more focus in the Enterprise Zones and Urban Development Corporations. Task Forces involving the Confederation of British Industry members, compacts between employers and education, Derelict Land Grants and the Priority Estates project have all involved private sector capital. Even new towns are to be developed by private consortia of builders. The role of the state in promoting urban growth and change has been partly rejected. Financial feasibility rather than social desirability provides the new emphasis. The effects on city development throughout Great Britain, and especially the areas of changed status, will take time to emerge, but no doubt, the changed direction of policies will have repercussions on urban development far beyond the immediate policy objectives.

2 Denmark

Denmark provides a further example of the dominance of one particular regional problem in urban development, namely, the overwhelming importance of Copenhagen, which houses approximately one-third of all the Danish population. The original Regional Development Act dates from 1958 but the later Act on National and Regional Planning in 1973 excluded Greater Copenhagen, which was the subject of entirely separate provision in 1974. The reforms of planning initially reinforced the need to cater for the concentration of population in the capital metropolitan problem zone, the three major centres Aarhus, Aalborg and Odense, although more recently four broader regions have become the focus: N. Jutland, including Aalborg, the Thisted area, S. Jutland and the island of Bornholm.

3 Belgium

The liberal philosophy that had created the Belgian state out of the Greater Netherlands did not encourage central planning and Belgium has been slower than its Dutch neighbour to enact regional planning legislation. The Regional Development Acts of 1959 and 1966 and the National Plan of 1962 were, however, a beginning upon which subsequent legislation, such as the 1974 law for open space preservation, has been built.

In Belgium, however, the problem of planning the national capital has dominated the attention of the country's city planners, in this case not only because of its size relative to the other Belgian cities, but also because of the drastic constraints imposed by the country's nationality dispute. Although Brussels lies to the north of the line demarcating Flemings from Walloons, it has been officially designated as a bilingual area since 1932. Legal bilingualism has not, however, prevented immigration, or the aspirations of parents, and the influence of international firms and institutions has encouraged a trend towards the consolidation of the French language at the expense of the Flemish. The intractable planning problem arises with the physical expansion of the city into the surrounding province of Brabant. Such urban spread is unacceptable to the Flemish politicians, who fear the consequences of an extension of French-speaking suburbs into previously Flemish-speaking rural areas. The threat of the spreading Brussels 'oil slick' provokes a determined and sometimes violent response. On the other side, attempts to contain the legally defined city of Brussels within the 'iron collar' of its 50-year old boundaries is seen as unrealistic, and unfair to those French-speaking citizens whose northern suburbs are still legally defined as being within the Flemish language area. Numerous attempts to seek compromises that would allow the existence of some form of 'Greater Brussels' have ended in frustration on both sides, and the fall of more than one Belgian government (Ashworth, 1979).

The nationality dispute has not only hindered the effective planning of the capital but is also always a consideration that haunts the planning of the other Belgian cities, and the designation of Namur as the capital of Wallonia, which has led to considerable growth. The necessity for governments to be seen to be fair in the allocation of resources between the two national groups adds to regional planning a criterion other than need (Riley and Ashworth, 1975). Consequently, the development 'blocs' are neatly packaged into those in Flanders and those in Wallonia. It also enhances the importance of Antwerp and Liège as 'national' capitals, and in one dramatic instance led to the creation of a new town – Louvain-la-Neuve – when the French-speaking faculties of Louvain University were compelled to migrate across the language frontier. National policy has developed a strong urban conservation ethic, partly as a response to the frenetic building boom in Brussels which resulted in protected status for much of the city core (Gay, 1987).

4 Netherlands

In sharp contrast to its Benelux partner, the Netherlands has evolved an urban system in which there is no single primate city but rather a primate urban region composed of three large and many more small and medium-sized towns located close to, but separate from each other. On the national scale, Dutch thinking has focused on two obsessions, the preservation of the unique character of the urbanised western provinces of North and South Holland and Utrecht, and the stimulation of the economically and spatially peripheral regions of the south, east and north.

Detailed discussion of the planning policies for the Metropolitan West Netherlands is found in Chapter 11, but the accommodation within this urban region of around one-third of the Dutch population, and the country's most important industrial concentrations, has given planning policies a national rather than purely local significance. Planning objectives for the west have altered little since 1945, but the chosen instruments for their achievement have varied considerably; they have swung decisively in recent years away from policies of persuasion by providing incentives and financial support for the periphery, towards policies of prevention of growth in the west.

Policies for the stimulation of the peripheral regions have been reformulated many times since the war (Tamsma, 1972), but have consistently focused on the development of urban centres, variously termed, according to the fashion of the time, growth poles (*groeckernen*), growth cities (*groeikernen*) and counter-magnets (*tegenpolen*). There has been considerable uncertainty, however, about the sort of urban hierarchy that will most promote economic growth. Opinion during most of the 1960s favoured the development of a large number of relatively small centres, in accord with the principles of grouped deconcentration initially favoured by the Second Structure Plan (1966) and continued in the 1985–2000 urban policy (Gay, 1987). More recently, opinion has swung in favour of the encouragement of fewer but larger urban centres and the concentration of investment in such urban regions as Groningen and Delfzijl-Eemshaven in the north, Arnhem-Duiven-Westervoort in the east and Breda and Eindhoven-Helmond in the south (Rijksplanologische Dienst., 1976).

The link between regional and urban planning is perhaps more evident in the Netherlands than in the larger West European countries. The size of the country and the spread of efficient public transport make it particularly difficult to distinguish between policies designed for the city-region and those concerned with regional imbalance on the national scale. Similarly the success of Dutch regional planning – and it can be shown to have had some measure of success in reversing the internal migration flow and equalising standards of living and welfare between regions – can be attributed to change at the city-region level. The expansion of the West Metropolitan area southwards into the northern part of the Rhine Delta, north-eastwards into Flevoland made possible by the opening of the rail line in 1988, eastwards towards Arnhem-Nijmegen and south-eastwards into Brabant, has improved the economic situation of the more accessible parts of the peripheral regions. Some growth has also been encouraged in the green heart at Alphen, Gouda and Woerden besides the outer towns such as Almere, Alkmaar, Houten and Hoorn. Only the Westerschelde area in Zeeland, the provinces of Friesland and Groningen in the north, the old textile area in Twente in the east, and the old coalmining area of South Limburg in the south remain outside the metropolitan orbit. Nevertheless the concern for environmental condition in the inner cities and broader environmental issues have focused attention on the quality of life, and urban developments which might result in a deteriorating quality of life stand little chance of success whether they be high rise apartments or obnoxious activities.

5 France

In France the national policy was designed to limit the growth of Paris which, like the capitals of the Scandinavian countries, was dominating the country. The concern was most aptly expressed in the title of Gravier's book in 1947, *Paris et le désert Français*. The regional policy outcome in 1964 was the creation of the eight *métropoles d'équilibre* as counter-growth magnets that would take a share of employment growth, particularly in the tertiary and quaternary sectors, which might otherwise have gone to Paris. The centres themselves were an assorted collection of urban areas ranging from single nodes, such as Marseilles, to 'composite nodes' such as Lyons-St Étienne-Grenoble or Nancy-Metz-Thionville. These provisions within the fifth-plan period were an attempt to distribute services more rationally throughout France.

However, the policy of *métropoles d'équilibre* did not meet with great success, and by 1970 it was being questioned, especially by those who favoured the development of what was termed the *ville moyenne* or town of from 50,000 to 200,000 people. An influential report to the Conseil Économique et Social (1973) suggested that the aims of regional planning, especially the maintenance of the population and infrastructure of the peripheral regions, would be better served by encouraging the development of the smaller towns rather than the less numerous and more widely spaced *métropoles*. It has been strongly argued by the commentators such as Lajugie (1974) that the smaller urban centres are better equipped both to fulfil the goals of French regional planning and provide for a richer quality of life for citizens. For whereas the population of the Paris agglomeration grew by 1.3 per cent per annum in the 1960s, and the population of towns under 80,000 by only 1.1 per cent, the *villes moyennes* grew by an average of 2.1 per cent per annum.

On the other hand, there was a feeling that the role of Paris as an international centre had to be strengthened (DATAR, 1973). Therefore the old strategy was no longer given priority in the Sixth Plan and the emphasis shifted to the medium-sized towns with contracts to develop signed with the state. In all, just over 50 contracts were signed (Chaline, 1978). Even this policy, designed to spread growth, was felt to be confined to too few urban areas and in 1975 a third strategy for assistance to small towns was initiated. By 1978 agreements had been signed with 140 towns. This latest strategy reflects an increasing polarisation of French regional planning, first to supporting Paris as the international centre (the subsidy for Paris transport is still greater than the regional premiums), and second an increasing emphasis on the rural areas where much governmental political support could still be found. So the medium and small towns' strategies cannot be understood without also noting the policy

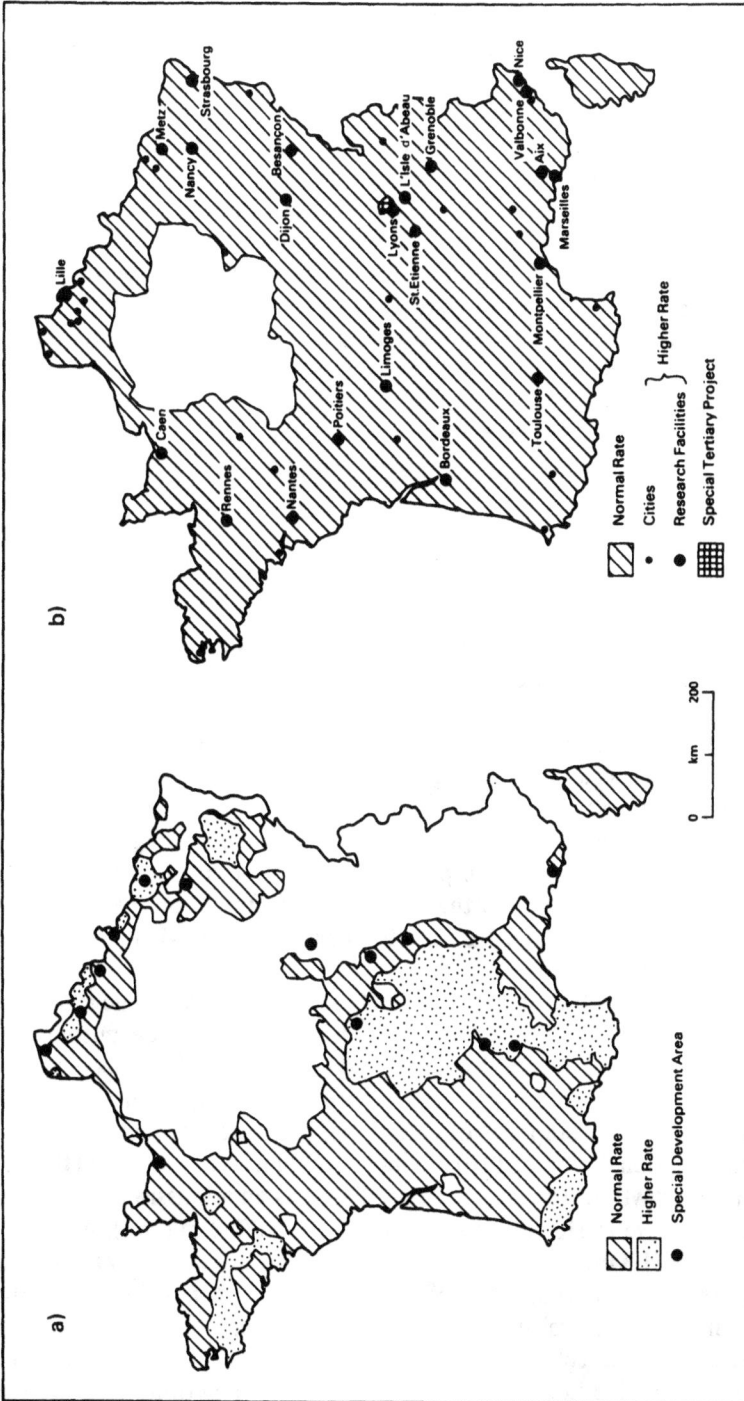

Figure 10.4 France: system of financial aid. (a) Graded assistance for new manufacturing activities. (b) Graded assistance for tertiary development.

for preserving coastal areas introduced in 1976, the mountain area policy of 1977, and the declining rural areas policy.

Since the advent of the Mitterrand administration in 1981 national policy was changed to promote development in the whole of France. Regional development incentives were streamlined in 1982 (Figure 10.4) and spatially distinct for the secondary and tertiary sections. Twenty-three *pôles de conversion* were created in 1984 to target assistance to the major areas of deindustrialisation, e.g. Roubaix and Valenciennes, and to improve derelict areas. Initially, after 1981, control over the service sector aimed to control growth in the Paris region, but the globalisation of many service sector companies does appear to have brought about a relaxation of controls. This is intended to ensure that Paris remains a global centre and, preferably, the key centre within the European time zone, replacing London (Clout, 1987). At the same time decentralisation, devolution and policies such as that towards *technopoles* have boosted the development of the tertiary and quaternary sectors in cities such as Grenoble, Montpellier and Caen.

6 Italy

Italian urban and regional planning has been traditionally weak in comparison with that of France or the Scandinavian countries, primarily owing to deep-seated weaknesses in the administrative structures, despite many attempts by both the ruling and opposition parties in parliament. The result was that planning too often became an emergency response to pressing problems and responsibility is divided among governmental agencies, a process aptly termed 'Balkanization' by Nanetti (1972). In 1972, town planning became the responsibility of the regions. The problem of the south has dominated the regional programme but much of the help has concentrated on the rural areas, despite the emphasis given to urban development in the Bari–Brindisi–Taranto region.

It was not until 1965 that Italy passed its first National Economic Plan, 1970 before regional administrations had been formed to provide autonomous administration of economic development, and much later before this reorganisation was actually implemented. This reform gave Italy a three-tier local government structure, composed of commune, province and region.

The second national plan of 1970 did include urban areas and noted the need to control the growth of the largest cities. This was an admission that policies had been ineffective in slowing the growth of Rome (2.8 million), Milan (2.1 million), Naples (1.3 million) and Turin (1.2 million), and that the various local controls were powerless to structure effectively urban growth, primarily due to a lack of local finance. Local planning was relatively unco-ordinated except for some intercommunal schemes to create outlying business centres (*centri direzionali*), as in Rome, or to build self-contained peripheral large-scale developments (*quartieri organici*), as around Rome, Bologna and Milan.

The overriding dominance of the north–south distinction makes it difficult to generalise about the character of Italian cities or the nature of their planning. The problems and responses of Milan have more in common in many ways with the cities of central Europe such as Munich or Zurich than with Mediterranean Naples or Syracuse. Nevertheless, despite the inability of Italian regional planning either to slow the growth of the large cities where decentralisation has been encouraged or effectively to reduce the Mezzorgiorno problem, there is still a strong tradition in favour of state control. By the 1980s deindustrialisation in cities such as Genoa and Turin set in motion attempts to solve problems on a city by city basis, which broadened urban planning away from its focus upon conserving historic centres that dominated policy in the 1970s. By 1987 an urban areas ministry (Ministero per le Aree Urbane) had been established to co-ordinate policies (Cheshire and Hay, 1989). The tradition of the sovereign state has its origins in the Roman Empire and is part of the traditions of the Catholic church. It embodies the view that the state is above the individual view or that of sectors of society, and that the state operates more completely and satisfies a more genuine public interest. Thus the reforming zeal of the socialist governments of the cities of the Romagna, such as Bologna, can be seen as sectoral, and state policies which severely limit the power of the local authority as policies in the public interest. On the other hand, the growth pole policy in the Mezzogiorno has had a definite impact on urban development of a few cities, such as Taranto and Bari, which has not always been economically or socially beneficial.

7 Greece

Greek policies which impact on city planning have tended to focus upon the problems of urbanisation, unplanned urban growth, industrial development and traffic congestion and pollution. The solution to the problem has been that of north-west Europe over a decade earlier, decentralisation of activities and some decentralisation of decision making. It was not until 1983–5 that structure plans for Athens and Salonika were approved. Prior to that period there were plans to combat decline in the old city of Athens which took the form of a conservation plan. The impact of the national will on the Greek city has been minimal, which does not become a founder nation of early urban planning.

Federal influences on urban planning

1 Switzerland

The relationship between national economic, social and regional policies and urban development is different again within the federal states. For in countries such as Switzerland, Austria and West Germany the relationship between the

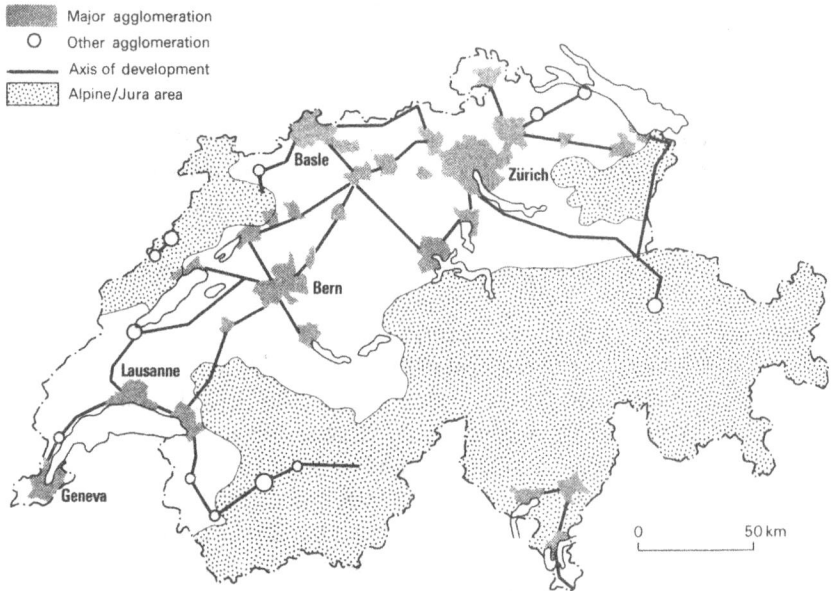

Major agglomeration
○ Other agglomeration
—— Axis of development
Alpine/Jura area

Figure 10.5 The settlement structure of Switzerland

various tiers of government are enshrined in the constitution, with the result that increased central government activity is vigorously opposed at the more local level. Switzerland has often been called a 'referendum democracy' and the 1976 Federal Raumplanungsgesetz (Area Planning Law) was a casualty of such a referendum (Elasser, 1979). The strength in regional planning terms still rests with the 26 cantons and the 3,000 gemeinden, and the federal government confines its regional policies to the mountain region and the Jura watchmaking area (Uhrenregion).

Regional development has concentrated on three issues, the first of which are the problems of the mountain areas. Second come the strong city–town division and the problems of the rural areas in an affluent society. The third problem is that of the cities and agglomerations, five of which – Zurich, Basle, Bern, Geneva and Lausanne – house 30 per cent of the country's 6.3 million people. Here the emphasis is on the rapid growth of the city-regions (750,000 living in the Zurich city-region), the consequent problems of the pressure on the central area, transport and pollution, and increasing segregation of the immigrant workforce. The settlement structure of Switzerland is shown in Figure 10.5. It distinguishes the five major centres, the development bands linking these, the fourteen medium agglomerations and 34 minor agglomerations and urban areas. Also plotted are those areas regarded as underprovided with urban facilities. Swiss planners would like to see a reduction in the concentration of population and employment in the Geneva–Lausanne axis and the Zurich–Basel–Berne triangle, and the development of concentrated-deconcentration in the minor agglomerations and small towns (Chifelle, 1987). But they realise that as long as cantonal powers exceed those of the confederation then a truly national rather than 'Vaudoise', 'Genevan' or 'Baslar' policy will be difficult to achieve.

2 Federal Republic of Germany (pre 1990)

The history of Germany as, firstly, an agglomeration of a large number of important and lesser kingdoms and principalities, and, more recently, a federal state, has meant that West Germany, like Switzerland, developed an urban and regional policy that reflected the imprint of history and present national ideology. In the immediate postwar years there was a dominant mood that saw planning as an evil associated with the Nazi period and as a result any physical planning was purely local and rarely related to national need other than that of re-establishing urban life as rapidly as possible. Once the *Wirtschaftswunder* had passed and some of the realities of late twentieth-century economic and social change began to emerge in the late 1960s coinciding with the rise to power of the SPD (social democratic party), the need for an overall urban and regional programme became more apparent. Besides the definition of regional action programmes that encompassed problem rural areas, the politically sensitive eastern border region and declining industrial centres, the federal programmes also defined the major population concentration and 24 growth areas (*Verdichtungsräume*) which could include development regions (Figure 10.6)

Verdichtungsräume were first designated in 1968 and were the last attempt to define the regions of most rapid urban change and development that go back to Isenberg's 1957 *Ballungsraum* and before that Scott's Agglomeration of 1912. This definition perpetuated a growth oriented policy which defined urban issues in terms of the inadequacies of the transport infrastructure (Konukiewitz and Wolmann, 1984).

Regional policy was developed at a Federal level and has evolved from a reluctant realisation of specific spatial problems through a period of relative widespread assistance to a period of retrenchment in the 1980s when the ability of regional policy to achieve the aims of reducing spatial variations in economic policy was questioned. Thus the current Regionalen Aktions programme (Figure 10.6) represents a realisation that assistance may alleviate the symptoms, but will not cure. Obviously German unification has left the status of the *Zonenrandgebiet*, frontier development zone, with the former German Democratic Republic in doubt and will inevitably lead to a radical redrawing of the map of assisted areas. The effect on the existing areas with a few notable declining industrial regions can be expected to parallel that of the 1984 revisions in the UK, which redrew the map of regional aid.

In the 1970s the growing emphasis on the quality of life resulted in the introduction of further policies for urban renewal and improvement which, like the earlier policies, continued to be administered by the federal *Länder* each

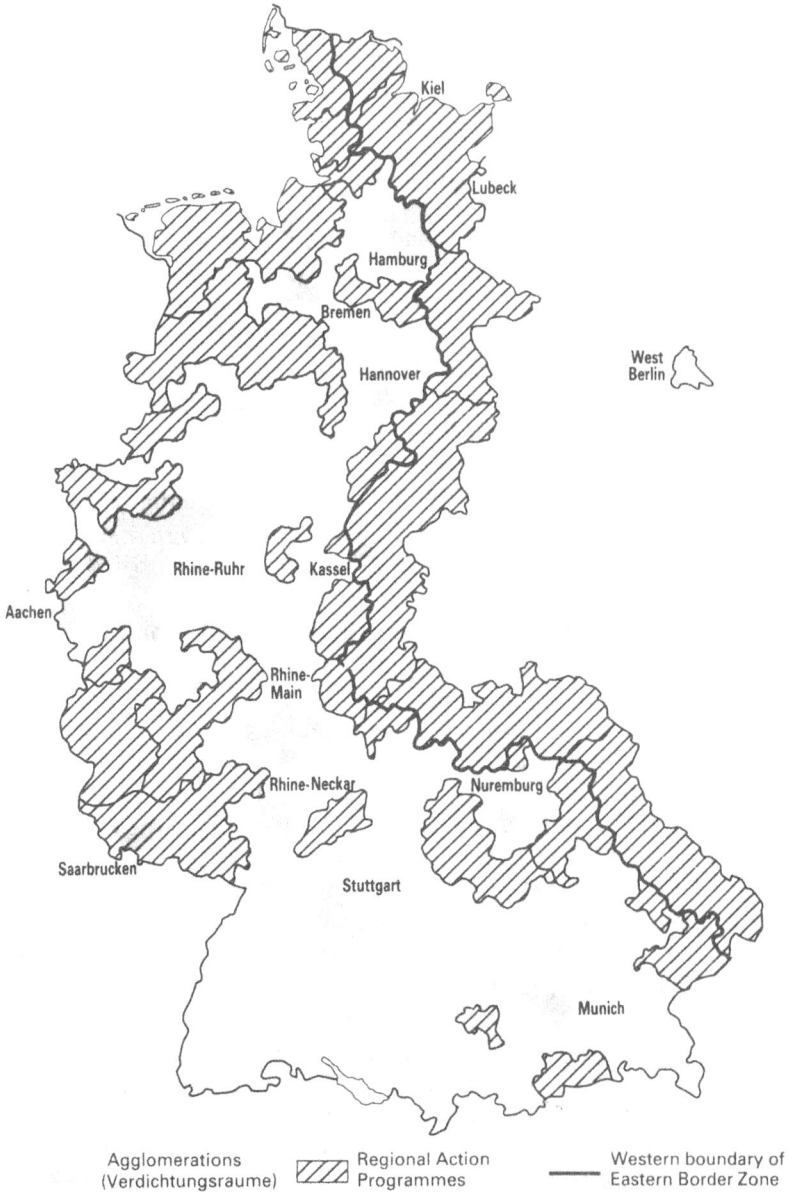

Kiel

Lubeck

Hamburg

Bremen

Hannover

West
Berlin

Rhine-Ruhr   Kassel

Aachen

Rhine-
Main

Rhine-Neckar   Nuremburg

Saarbrucken

Stuttgart

Munich

| Agglomerations (Verdichtungsraume) | Regional Action Programmes | Western boundary of Eastern Border Zone |

Figure 10.6   West German regional development and growth zones, 1989

with their own political complexion. Urban renewal programmes have focused upon the areas of mixed uses in the nineteenth century suburbs, particularly in the Ruhrgebiet. In some cities this has inadvertently gentrified whole quarters, e.g. St Agnes, Cologne. Environmental improvement, greening the city, has become the focus of many schemes which involve traffic calming, increasing tree planting and open space creation and, not least, the provision of a dense network of poop scoop dispensers and bottle, paper and can banks to ensure that all are aware of greening. Re-use of derelict land in cities is given a high priority.

The emphasis on a federal structure has maintained West Germany as a multicentred urban system undominated by any one city as are Britain, France, Denmark or Austria. The largest urban agglomerations, Rhine–Ruhr and Rhine–Main, contain respectively the capital Bonn, only officially recognised in 1971, and the commercial centre Frankfurt. However, the rank–size order was dominated by West Berlin, a city whose unique political and geographical position separated it from the national system. The *Land* capitals have added status and, as a result, a greater diversity of employment, but not all are as high in the rank–size order as Hamburg (2nd) and Munich (3rd). Kiel, for instance, is 20th and Wiesbaden is 23rd. Other centres such as Cologne (4th), Nürnburg (11th) and Mannheim (17th) have remained prominent, with activities of naitonal significance located in them. Thus, in contrast to many of her neighbours, West Germany developed a much more polycentric city system. Whether the future will see a return to a federal state dominated by Berlin only time can tell.

3 Austria

The third federal state, Austria, has also devolved planning matters to the nine *Länder*, although there is an overall co-ordination role for federal regional planning. Urban planning has its origins in Vienna in the nineteenth century, as we saw in Chapter 2, although its modern origins lie in the period after the First World War when Vienna became the capital of a country one-eighth the size of the old Austro-Hungarian empire. Vienna has continued to dominate the urban system, housing 20 per cent of the population, although its dominance has lessened slightly with decentralisation. Only since the signing of the Austrian State Treaty in 1955 has urban and regional planning developed, at first to provide housing for refugees and to equalise growth between the regions and especially the problem Ö Grenz, the eastern frontier zone (Figure 10.7). Since 1989 the status of this region, like that of West Germany's *Zonenrandgebiet*, is now open to question. Thus the growth of *Land* capitals such as Salzburg, Klagenfurt, and Bregenz, which are part of the seven recognised Ballungsräume growth areas, has been encouraged in preference to maintaining people in the rural areas. Decentralisation from the administrative area of Vienna to the surrounding districts is fostered alongside a policy to give the city a new

Economically weak regions

Structurally weak industrial regions

Growth areas (Ballungsräume)

Settlements with renewal problems in
- • < 4 areas
- • 5-25 areas
- • 357 areas (Vienna)

a

0   100 km

Federal/*Land* programmes for industrial job creation

Federal/*Land* programmes to provide employment in tourism

Rural mountain Assistance Areas

——— E.R.P. Programme Area

•••• Agricultural Frontier Zone

b

0   100 km

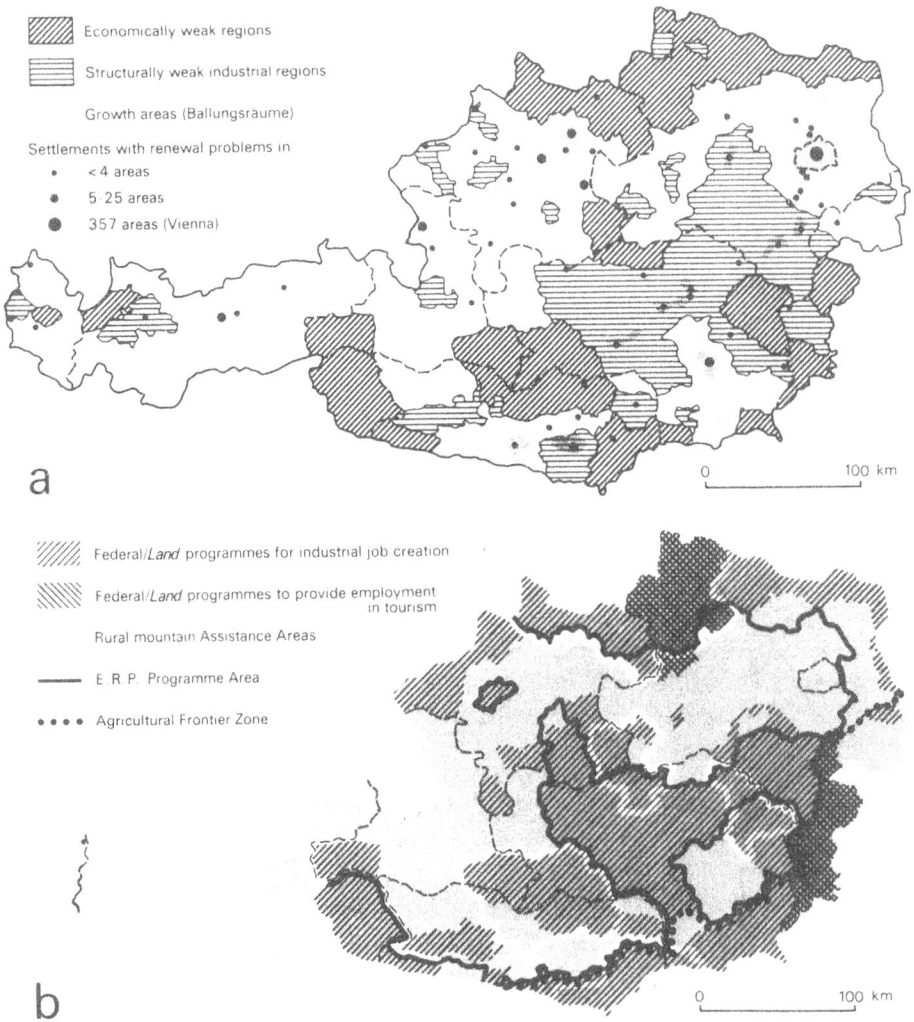

Figure 10.7   Austria: regional differentiation – (a) problem regions; (b) regional programmes.

international role commensurate with the country's neutrality (Burtenshaw, 1987). As in West Germany, history and the search for a new national ideology in the wake of the tragedies of two wars have produced a distinctive national urban and regional strategy.

4 Spain and Portugal

The Spanish experience has been one of a long history of centralisation and authoritarian rule. However the 1978 constitution recognised the right of regions and nationalities to autonomy and, as a consequence, 17 autonomous communities have been established. The autonomous communities have planning powers vested in them and are intended to bring decision making closer to the communities of Catalonia and the Basque country in particular. These new bodies have been able to influence the recent development of their city and region, enabling the construction of technology parks in Madrid, Barcelona, Valencia and Bilbao and promoting Barcelona as an Olympic city (Cheshire and Hay, 1989). Portugal has not moved as fast and had very incoherent policies which were only beginning to gel in time for EC accession in 1986. The aim is to create a regional structure which will decentralise planning power along Spanish lines. In both cases the changes from a centralist state to a more federalist structure are having a profound effect on the future directions of urban development.

In the following chapter the city plans for a selection of West European cities will be discussed. The range of national, regional and urban planning policies outlined here have created a context for urban planning, sometimes encouraging urban growth while at other times and in other locations acting as a brake on urban development. Few urban planning strategies can be viewed in isolation from their national context. The basic distinctions between federal and centralised states, between city-centred and rural support policies will obviously colour the planning initiatives of the West European cities.

References

Ashworth, G. (1979) Language, nationality and the State in Belgium, *Area Stud.*, 1, 28–32.

Barlow (1940) *The Report of the Royal Commission on the Geographical Distribution of the Industrial Population*, Cmd. 6153, HMSO, London.

Bourne, I. (1975) *Urban Systems*, Oxford University Press, London.

Burtenshaw, D. (1976a) *Saar–Lorraine*, Oxford University Press, London.

Burtenshaw, D. (1976b) 'Problems of frontier regions in the EEC', in Lee, R. and Ogden, P. (eds) *Economy and Society in the EEC*, pp. 217–231, Saxon House, Farnborough.

Burtenshaw, D. (1987) 'Austria', in Clout, H. (ed.) *Regional Development in Western Europe* (3rd edn), David Fulton Publishers, London.

Buursink, J. (1971) De Nederlandse hierarche der Regionale Centra, *Tijd. Econ. Soc. Geog.*, **62**, 67–81.

Chaline, C. (1978) *New Departures in Regional Planning - France* (mimeo), Paper presented to Regional Studies Association Conference, 19 May.

Cheshire, P. and Hay, D. (1989) *Urban Problems in Western Europe*, Unwin Hyman, London.

Chifelle, P. (1987) 'Switzerland', in Clout, H. D. (ed.) (1987) *op. cit.*

Clout, H. D. (ed.) (1987) *Regional Development in Europe*, David Fulton Publishers, London.

Conseil Economique et Social (1973) *Rapport 13*, 7 August, Paris.

de Jong, F. (1969) *De Economische Betekenis van de R. V. Groningen voor de Provincie*, R. V. Groningen.

DATAR (1973) *Paris Ville International. Travaux et Recherches de Perspective*, DATAR, Paris.

Elasser, H. (1979) *Siedlungs Struktur und Raumplanung in der Schweiz*, and *Regional Politische Problem der Schweiz* (mimeo), papers presented to Geographische Instituut Rijksuniversiteit, Utrecht.

Gay, F. (1987) 'Benelux', in Clout, H. (ed.) *Regional Development in Western Europe* (3rd edn), David Fulton Publishers, London.

Gravier, J. F. (1947) *Paris et le Désert Français*, Flammarion, Paris.

Hansen, J. (1972) Regional disparities in Norway with reference to marginality, *Trans. Inst. Brit. Geog.*, **57**, 15–30.

Hammond, E. (1968) *London to Durham*, Rowntree Research Unit, University of Durham.

Hollmann, H. (1977) Grenzuberschreitende Landesplanung Bremen/Niedersachsen, *Raumforsch. Raumord.*, **35**, 218–224.

Kormoss, I. (1976) 'Urban extensions in Belgium', in Laconte, P. *et al.* (eds) *The Environment of Human Settlements, Volume 1*, pp. 177–186, Pergamon, Oxford.

Konukiewitz, M. and Wolmann, H. (1984) 'Physical planning in a federal system', in McKay, D. (ed.) *Planning and Politics in Western Europe*, Croom Helm, London.

Lajugie, J. (1974) *Les Villes Moyennes*, Editions Cujas, Paris.

Michels, D. (1976) Das System der Enwicklungsachsen und Verdictungsschwerpunkte in Nordwesteuropäischen Kernraum, *Forsch. zur Raumentwick*, **Nr. 3**.

Ministry of Housing (1976) *Denmark's National Report to Habitat*, Copenhagen.

Nanetti, R. (1972) Urbanization in France and Italy, *Council of Planning Librarians Exchange Bibliography*, **340**.

Pinder, D. (1983) *Regional Economic Development and Policy: Theory and Practice in the European Community*, Allen and Unwin, London.

Pinder, D. (1990) *Western Europe*, Belhaven, London.

République et Canton de Genève (1975) *Plan Directeur Cantonal*, Directeur de l'Aménagement – Division de l'Equipement, Geneva.

République et Canton de Genève (1977) Some Aspects of Town Planning in Geneva (mimeo), Department of Public Works and Town Planning, Geneva.

Rijksplanologische Dienst (1976) *Verstedelijlsnota*, The Hague.

Riley, R. C. and Ashworth, G. J. (1975) *Benelux, An Economic Geography of Belgium, The Netherlands and Luxembourg*, Chatto and Windus, London.

Second Structure Plan (1966) *Second Report on Physical Planning in the Netherlands*, The Hague.

Tamsma, R. (1972) The Northern Netherlands: large problem area in a small country, small problem area in a large economic community, *Tijd. Econ. Soc. Geog.*, **63**, 162–179.

Verbeck, V. (1973) Het Benelux Middengebied, *Stadebouw en Volkhuisvesting*, **10**, 372–380.

CHAPTER 11

City Planning Case Studies

Plans aimed at the control of the growth of major cities are by no means new. In West Europe, however, it is only during the last half century that a number of countries have made concerted efforts to direct the future growth and form of any major city. Such efforts have usually involved the planning of the capital city. In some cases the priority has been to curb its growth in an attempt to redistribute it and revitalise declining areas elsewhere in the country. For instance, the rapid growth of London and south-east England in the interwar period focused attention on the planning of the capital and its region, but at the same time gave rise to plans to inject new growth into ageing industrial areas such as Merseyside, South Wales and north-east England. In other cases, such as Paris, this secondary consideration linked to regional problems came later and plans such as the 1934 Prost Plan sought merely to direct the manner in which the city was growing, particularly because uncontrolled development was rapidly absorbing the surrounding rural area. Copenhagen in the 1940s was presenting just such problems and a plan was introduced to harness the development of the city, which was already overdominant within the urban hierarchy of Denmark. In this chapter urban planning is discussed in terms both of the common characteristics which may be seen to run through many of the individual plans and of the distinctive approaches of a selection of plans.

Peter Hall (1977) has contended that 'Urban decentralisation is now a world-wide force in all highly industrialised countries'. It is difficult to challenge this contention and what is intriguing in a European context is the varying manner in which this force has been met, either through active promotion of decentralisation policies or alternatively by strongly restrictive measures already discussed in the context of tertiary and quaternary activities (Chapter 4).

The planning of major metropolitan areas can be seen to have evolved through a number of distinct phases. Initially came the realisation of the need to plan for future growth as these urban centres became increasingly dominant. Then specific plans were introduced to cater for this rapid urban development. The early plans, such as those for both Paris and London, envisaged the restriction of growth and the imposition of population limits on the metropolitan region under scrutiny. Later plans acknowledged that such approaches were not feasible and concentrated instead on directing growth along rational lines. There was considerable variation, however, in the way the newly acknowledged inevitable growth would be accommodated. In the event, many of these plans have been proved to be somewhat exaggerated in their population forecasts, bringing about a general reappraisal of their important underlying assumptions during the 1970s. This reappraisal marked a stage in the evolution of metropolitan planning, and in some cases has given opportunity for new priorities to be established in the formulation of the revised plans. Priorities concerned with conservation in the broadest sense have replaced those concerned with providing the structure to accommodate vast new populations. In the process, it has enabled the organic tradition of planning to replace more dogmatic and tightly circumscribed forms of planning. The 1980s saw a furtherance of the reappraisal begun in the 1970s reaching, in the extreme cases, the abolition of city-wide planning coordination and, in other cases, the introduction of more permissive development legislation. It should be noted, however, that there are significant differences in the degree to which these new approaches have been adopted within West European states.

Not all cities have gone through these clearly defined phases. There can be little doubt, however, that the need for planning has changed rapidly from the post-war explosion of many of these centres, and the consequent need for stringent control and direction coupled with the desire to rebuild cities fit for the late twentieth century, to the need for a more measured adjustment of the urban system which characterises today's planning of cities. This is not to say that the urgency has been lost, but that it now stems from persisting imbalances within the urban system rather than from explosive growth.

Early planning directions

In the case of many cities in West Europe, the interwar period demonstrates a clear need for measures to be taken to control their growth and logically the economic activities which gave rise to that growth. In Britain, the need to combat economic decline and the stronger economic and demographic growth of the south-east region in general and of the London area in particular, was the context of the Barlow Commission Report 1940 which McCrone (1970) regarded as ahead of its time, because it introduced concepts which were not applied until the 1960s. In particular, it regarded the congestion problem of some cities and the unemployment of the depressed areas as different aspects of the same problem. At the base of the post-war legislation lay the overwhelming dominance of London in the growth sectors of the economy and the need to evolve a more equitable distribution of economic activity.

The physical growth of London during the interwar period had also been dramatic, with the creation of vast areas of medium-density housing tripling the urbanised area between 1921 and 1939. The need for some degree of control was obvious and the possibility of planning the capital in some measure was given added impetus by wartime destruction. The control came in 1943 in the form of

Patrick Abercrombie's County of London Plan which, on the basis of official population estimates, assumed that there would be no appreciable increase in population. It endeavoured, however, to provide for a lower density of urban development and some relief of inner city congestion. Outward growth was restricted by the establishment of the green belt, first defined by a parliamentary act in 1938. This belt was subsequently increased in extent so that it now encompasses a zone some nine kilometres in width. Such control had to be complemented by provision for some growth elsewhere and Abercrombie proposed the establishment of a series of new towns located beyond the green belt. These were to take a significant share of the 'overspill' of London, estimated by the plan to be in excess of a million. The concept of a green belt with satellite new towns is very much in keeping with the utopian tradition of planning, although in practice there were obvious problems.

Abercrombie made incorrect assumptions based on trends of the 1930s on the growth of London. Thus, far from his plan solving London's problems, it could do no more than impose a partially rational structure on the growth of the city as it continued to grow in the post-war period, both from natural increase and, at least in the early years of the period, by in-migration. Added weight was given to the planner to zone development following the 1947 Town and Country Planning Act.

The Parisian experience was not dissimilar to that of London, although its interwar growth was even more anarchic than that of London. London's new housing estates were sometimes elaborate in design, but they were served by improved arterial roads and by public transport routes such as those of the expanding underground and Southern Railway electrified routes. In contrast, uncontrolled development of Parisian suburbs with the purchase and development of individual plots of land (*lotissements*) was commonplace. The uncoordinated development created the chaotic *système pavillonaire*, rarely served by any good public transport service and often not even provided with good made-up road surfaces. As early as 1928 a planning authority, the Comité Supérieur de l'Aménagement et l'Organisation Générale de la Région Parisienne (CARP) was charged with producing a general development plan for the Paris region. The result, the Prost Plan (1934) gained government approval in 1939, its aims including the control of the development of *lotissements* through the designation of building zones. In practice, however, little was achieved before war intervened and eventually created new planning priorities in the whole of post-war France. The Prost Plan was often the sole legal codification of land-use in some communes for almost forty years (Lecoin, 1988).

Note, however, that just as future growth predictions in London had certainly been conservative, so in Paris the relatively modest growth in the period before 1939 had led to expectations of a continuation of the low growth rate. By 1956, this optimism had been proved to be ill-founded and a new master plan was deemed necessary. The new plan, the *Plan d'Aménagement et d'Organisation Générale de la Région Parisienne* (*PADOG*), was approved in

1960 and contained proposals for a growth of a million in the population of the Paris region from 1960–70. It put forward proposals embodying some decentralisation of employment and an improvement in the transport infrastructure, including the Boulevard Périphérique, completed in the early 1970s. In addition, it proposed a massive urban renewal programme linked to the creation of new nodes of development within the existing urban area. However, it was to be supplanted within five years by a new plan for Paris, discussed below. One of the main reasons for the necessity for the new plan lay in the assumption of the *PADOG* that growth could be limited to an increase of population of 100,000 per annum in the decade 1960–70. It was the assumption of low, controlled growth which proved to be false. Indeed, the period 1962–64 saw the Paris region growing by an average of 165,000 per annum, of which 65,000 was by natural increase, 50,000 by internal migration and 50,000 by repatriation from Algeria (Hall, 1977). Plainly, the original aim of containing Paris within the limits of the existing conurbation would never have been achieved.

The post-war period saw the awakening of interest elsewhere. A series of reports in the 1940s and 1950s examined the problems of overconcentration of population within the western provinces of the Netherlands, centred on what came to be called the Randstad. In the case of Stockholm, active planning of the city started somewhat earlier. Municipal action was considerably eased by a policy dating from 1904 in Stockholm of bringing land into municipal ownership, so that by the mid-1960s the city owned approximately 70 per cent of the suburban land within its boundaries. A regional plan was formulated and approved by the city by 1959 and given national approval by 1960.

Stockholm's growth has been relatively slow by West European standards but sufficiently rapid to concern the city authorities: population growth of 1.95 per cent per annum 1965–70 far exceeded that for the country as a whole for the same period (0.77 per cent). A new plan was required within ten years, actually approved in 1970, for the 29 municipalities, which laid rather less stess on predictions and more on policy directions. In the Swedish context it is important to note that planning has been traditionally permissive rather than authoritatively directional. The instruments of planning are restricted to setting constraints on land-use and controlling transport development. Even now the municipalities have a surprising degree of autonomy in the sphere of physical planning.

Finally, Copenhagen illustrates the manner in which the mid-twentieth-century reassessment of the scale of the problem of growth in large cities came about. Until the 1930s the development of Copenhagen was relatively restricted. A parliamentary decision in 1930 led to the electrification of the suburban railways with the first subway trains (S-bane) running on the new system in 1934. The new-found accessibility to the city-centre via the S-bane led to major urban growth, particularly around the new S-bane stations. Private bus lines operating on the periphery of the built-up area further promoted urban growth. Having incorporated many surrounding districts into the municipality at the turn of the century, Copenhagen spread rapidly into the new suburban areas; consequently the population rose rapidly and reached a million by 1940. A clear need for a stricter control and direction of growth was established. Although the city continued to purchase land for future growth in the surrounding areas, the surrounding municipalities themselves wanted to retain their independence. Nevertheless, coordination of urban development was clearly necessary, and in 1939 a metropolitan committee was established with a brief to prepare a constitution to govern intercommunal relations in the Copenhagen metropolitan region. Politically that was no easy task and, in fact, a final constitution did not emerge until 1973. However, the stimulus for action was evidently there and a regional planning committee was formed, with some limited initial results in the area of recreational planning before the war. After the war, in 1947, a regional plan for Greater Copenhagen was produced, the renowned 'Finger Plan', which was to serve as an unofficial blueprint or guideline for the city for the period until 1960, when it was superseded by a new plan. The 'Finger Plan' was a bold attempt to channel growth along particular axes of development, following the S-bane routes. A new plan was eventually necessary, however, as growth exceeded expectations and, as elsewhere, the new mobility brought by private car-ownership created new pressures on the intervening rural land.

The awakening to the problems had taken place and the plans had begun to emerge. Admittedly many were based either on unsound assumptions concerning the ability to restrict development, or on predictions of growth which proved to be inaccurate, or on both, necessitating their replacement by new plans in the late 1950s or the 1960s. The projections on which these plans were based were also questioned, some with remarkable rapidity. A further important lesson learnt in this early period related to the administrative machinery called upon to implement any plan. The multiplicity of authorities, in many instances, rendered effective planning at best inconvenient and at worst impossible. An essential prerequisite for effective metropolitan-scale urban planning was administrative reorganisation and reform.

Local government reorganisation, which took place in almost every major West European country since 1945, resulted in a decrease in the number of small administrative units and the creation of larger administrative units. An extreme example is that of Sweden which reformed its local government structure in 1952 and 1964, progressively reducing the number of communes in the country from more than 2,000 before 1952 to 282 in 1964. In respect of major metropolitan areas, however, the choice has generally been between amalgamation of existing units to form new units or systems involving voluntary cooperation between existing administrative bodies. Very often special attention had been paid to the metropolitan areas quite apart from the rest of the country. Thus, the Greater London Council was brought into being in 1965, nine years before more general reform of local government in England and Wales. In 1971, a new county council for Greater Stockholm was created, although a considerable amount of local autonomy still rested with the local municipalities. Voluntary association may be illustrated by the Regional Planning Council for Copenhagen established in 1967 with representatives from the three metropolitan counties of Copenhagen, Frederiksberg and Roskilde together with the cities of Copenhagen and Frederiksberg. Perhaps the most complex reorganisation was that which led to the new *départements* in the Paris region. Initially, in 1961, the District de la Région Parisienne was established as a coordinating body for existing local authorities, including within it three *départements* of Seine, Seine-et-Oise and Seine-et-Marne. In 1964, two of the *départements* which had major problems on account of their size were sub-divided into more manageable units to produce in the region as a whole a new administrative structure of eight *départements* including the Ville de Paris as a separate *département* in its own right. Later, in 1966, the Paris region was designated as a higher level authority to coordinate regional activity, as part of a national system of planning regions. In 1972, it was renamed La Région de l'Ile de France. Local government reorganisation has been a testimony to the fact that urban planning has had to look further than anachronistic boundaries of urban settlements to encompass the wider areas which need to be planned as a coordinated whole. The creation of Grossraum Hannover and Grossraum Braunschweig in 1963 was further testimony to this fact, as were the metropolitan counties in England created as a result of local government reform in 1974.

Characteristics of the major plans

The second generation of plans may be classified into three broad types, although it should not be thought that these are mutually exclusive categories, since some plans have characteristics typical of all three categories. First, there are those plans which have a series of preferential axes for new growth as their basic structure. An important accompanying advantage is the comparative ease of access to the open space between each major access, as well as the optimum use of transport corridors. A second category of plans may be termed polycentric, usually characterised by attempts to create a new pattern of urban centres through the designation of growth nodes, to counterbalance an overdominant centre. A third, more limited, category includes those where open space retention is the prime aim, governing all other characteristics of the plan, or at least severely restricting the available options. Examples of the first two categories of plans are considered in detail.

(i) Preferential axes of development

In the 1944 Greater London Plan, Abercrombie chose to limit the growth of London by the imposition of a green belt. Later plans for directing growth on a broad regional scale of the whole of south-east England, but dominated by the

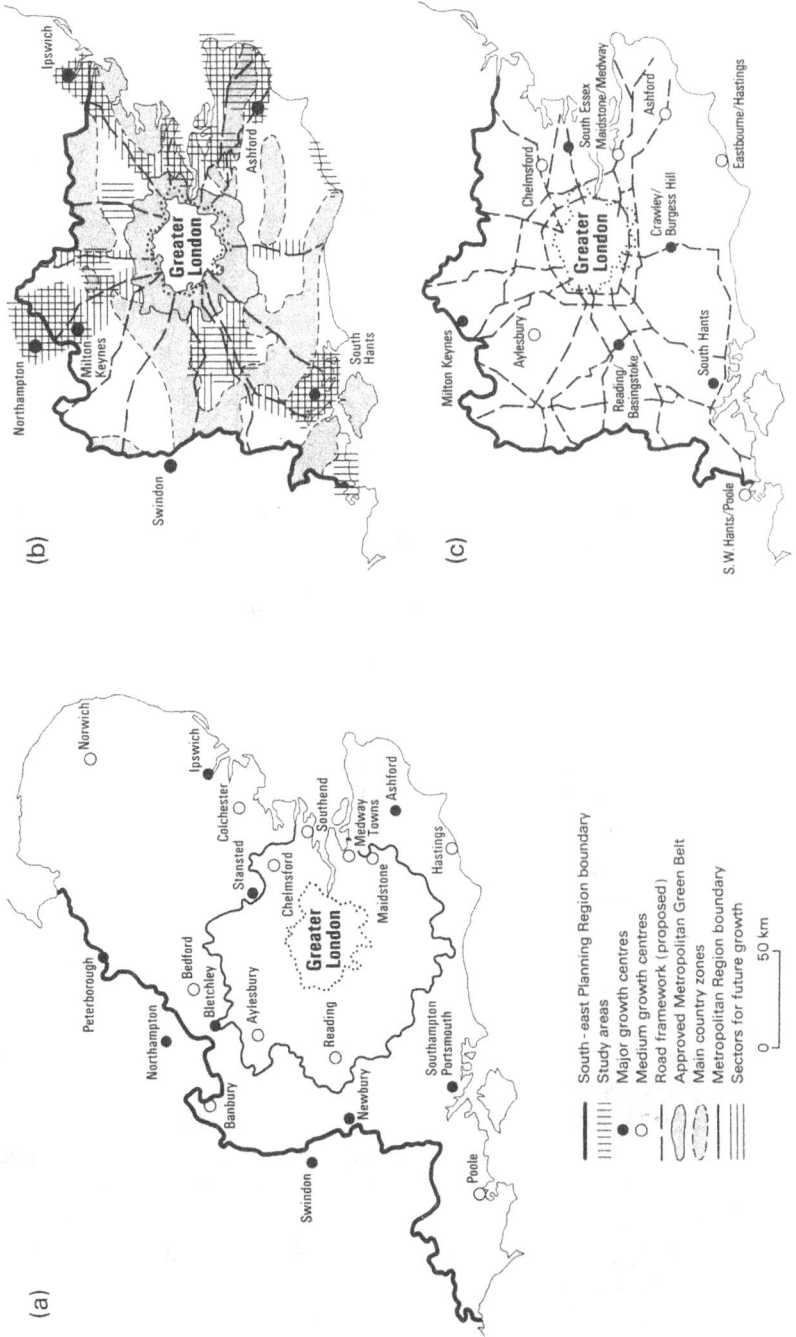

Figure 11.1 Plans for London and South-East England: (a) *South-East Study* (1964; (b) *Strategy for the South-East* (1967); (c) *Strategic Plan for the South-East* (1970) (all reproduced by permission of Greater London Council)

metropolitan area, have generally emphasised the structural device of axial development. The green belt has remained as the most consistent element in the physical plans for the London region. In 1967, the concept of preferential axes of development were contained in the first report of the South-East Economic Planning Council, entitled *A Strategy for the South-East* (HMSO, 1967). It emphasised a series of sectors within the region, not with the intention of their being fully urbanised but to indicate the general direction of growth (Figure 11.1b). The proposals also incorporated the principle of growth centres originally introduced by the *South-East Study* in 1964 (HMSO, 1964) (Figure 11.1a). In the earlier document, new cities had been proposed at Newbury to the west of London, to the north at Bletchley and to the south-west in South Hampshire in the region of Portsmouth and Southampton. In addition, major town expansions were to be implemented at Stansted (Essex), Ashford (Kent), Ipswich, Northampton, Peterborough and Swindon. *A Strategy for the South-East* (1967) took a number of these and incorporated them into an axial structure with each axis terminating at a major node (see Figure 11.1b), some beyond the regional boundary. These plans in the 1960s were designed to house major population growth. For instance, the *South-East Study* was put forward on the assumption of an additional population of 3.5 million in the period 1964–81, two-thirds of which would be through natural increases in the region. Later reappraisals were to revise such forecasts quite markedly.

The *Strategic Plan for the South East* (1970), compiled by the South East Joint Planning Team, put its emphasis on a flexible framework but within the context of continued growth (HMSO, 1970). Individual urban areas were to be given clear identity and a limited number of major growth areas were to develop as more independent city regions. An improved regional transport network would be an integral part of such a plan as indicated in Figure 11.1c. The green belt, together with major tracts of rural land, was to play an important role in maintaining a clearly identifiable structure of separate city-regions. In many respects, therefore, this plan continued to follow the basic principles of the earlier plans.

Rome experienced very rapid growth particularly in the outer suburbs in the period 1951–71. Between 1961 and 1971 the city's population grew by 27 per cent mainly as a result of in-migration from the south and Sardinia. Many settled in illegal settlements in the outer suburbs while most settled initially in the inner suburbs. The plans for future development have included a major axial development. It has been proposed to build a chain of new business centres four kilometres east of the city along a route linking the two ends of the Autostrade del Sole. The 'directive centres' which would upgrade the poorer eastern districts with offices, shops and hotels should relieve the centre, where controls on commercial developments have existed since 1962. The eastern axis plan, despite its ambitious aims, is not meeting with much success, mainly as a result of the illegal alteration of buildings in central Rome.

The original plan for Copenhagen accepted the notion of axial development from its earliest stages. Its 'Finger Plan' dating from 1947 was a clear attempt at confining growth to axes whilst leaving the intervening zones clear of major urban development. Whilst it was partially successful, the axes (or fingers) grew rather longer than originally planned and also broadened, partially destroying some of the advantages of axial development, particularly that of ease of access to open rural land. Thus, a new plan was prepared by 1960 entitled a *Preliminary Outline for the Copenhagen Metropolitan Region*, planning for growth up to 1980. It retained the concept of axes but implemented it in a more restrictive manner, suggesting that the major part of the growth should be located in a main axis running south-westwards along Køge Bay. The growth was to be concentrated into two new towns, each with a population of 250,000. To the north, North Sjaelland was seen as an area to be preserved for recreational purposes. There was some departure from a purely axial structure in this plan. This came, firstly, in the intention to establish two major nodes within the Køge Bay and, secondly, in the proposal to create a new major urban centre of sufficient scale to relieve the city of Copenhagen itself of some of its functions. In the event, this part of the plan proved contentious and was not implemented, and in 1963 a modified provisional plan was agreed upon, termed the 'First Phase Plan'.

This plan stressed the importance of the development of the two axes from Copenhagen towards Roskilde and Køge. New sub-regional centres were also proposed at Høje-Tåstrup to the west and at Lyngby to the north. The Køge Bay development axis to the south-west was retained as an important element in the 'First Phase Plan', but with new growth concentrated not in two nodes but in ten new townships, each with a population of 150,000. Unfortunately, two essential elements in the development of the axis, its rail and road transport links, were subject to considerable delays. The westward motorway, a key element in the entire plan, was not opened until 1973, while the new S-bane line was much delayed in its servicing of the entire length of the axis.

The axial plan, best epitomised by Copenhagen and by Stockholm with its similar pattern of new urban nodes along its T-banen system, have obvious advantages of flexibility. Either axis can be developed in sequence as need dictates, or they may be extended to some degree to accommodate unexpectedly high growth. Recreational provision should be facilitated by ease of access to the interstitial open space. It might be noted, however, that there may be severe problems inherent in maintaining the separation of the axes and also of maintaining a viable farming structure in the face of close proximity of the urban area with associated problems of crop damage and trespass. Obviously, in order to be successful, such a policy requires strong planning measures to limit peripheral growth of the axes and coherent recreational policies to combine both viable agricultural use and recreational use of the intervening land. A further disadvantage lies in the potential for congestion towards the centre, where the axes converge. It could be argued that a strictly axial plan whose axes radiate from the central city is best suited to a medium-sized city such as Copenhagen or Stockholm, and elsewhere, in larger metropolitan centres, other solutions have to be implemented.

Axes of growth were an essential element of Schumacher's 1921 plan for Hamburg which have remained the essential building blocks of Hamburg's plan for 70 years. The basic structure today is of eight regional axes extending to towns in the region which are the endpoints of commuter lines such as Elmshorn and Lüneburg. Nearly all developments in the Hamburg region have been planned to fit into the overall concept of axial development; central place functions, housing densities, transport provision, and new office districts. Even the 1980 modifications maintained the same principles embodied in the 1921 and 1969 plans (Möller, 1985). Similar axial growth plans exist for Bremen, Freiburg, St Etienne and Trier, and for the development of the Metz-Nancy-Thionville *métropole d'équilibre* (Burtenshaw, 1976).

(ii) The establishment of new growth nodes – polycentric plans

Certain plans for major centres have emphasised the need to restructure the distribution of urban functions by employment growth nodes into which a proportion of the new growth may be directed. The best example of this is undoubtedly that of Paris. It could be argued that the plan for Paris more properly fits into the previous section of plans dependent on preferential axes, but there are good reasons for seeing it instead as a prime example of a polycentric plan.

The population growth rates in the early 1960s made it increasingly clear that a policy of restriction on the growth of Paris was impracticable. The *tâche d'huile* (oilslick) of Paris was inexorably spreading and required direction and, internally, a new structure. The new plan aimed at fulfilling these requirements was published in 1965 and superseded the *PADOG* described above. It was the *Schéma Directeur d'Aménagement et d'Urbanisme de la Région de Paris* (Figure 11.2a). There is little doubt that this plan was probably the most ambitious of its kind in the world. Set against the rapid economic growth of France of that period, it reflects a tremendous optimism that order could be brought to the chaotic urban pattern inherited from the interwar period, and that those mistakes would be forgotten as new development was carefully channelled into new and dramatic forms of urban development epitomised by some new town architecture. It assumed a growth of population for the region from nine million in 1965 to 14 million by the end of the century. It predicted particularly rapid rises in the tertiary employment sector and, indeed, the intervening period since 1965 has borne out that prediction to be accurate, since the Paris region now has approximately 40 per cent of French office jobs and about 60 per cent of its workers working in the tertiary sector. It foresaw a doubling of the urbanised area from approximately 1500 square kilometres to 2750 square kilometres. The need for new dwellings was exacerbated by the very high residential and occupancy rates prevailing in Paris, as well as the need for new dwellings for the new growth in the population. Other priorities included an improvement in the transport network, which had already been the subject

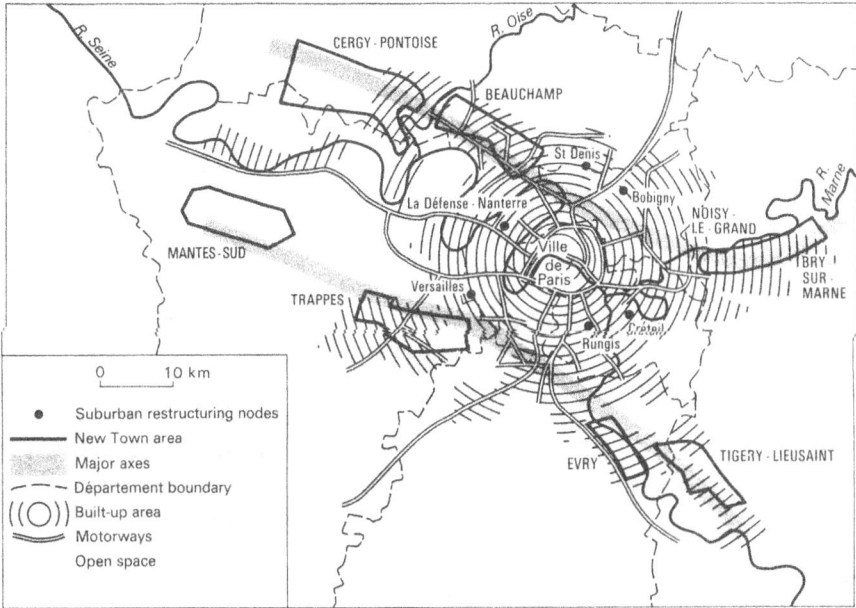

(a)

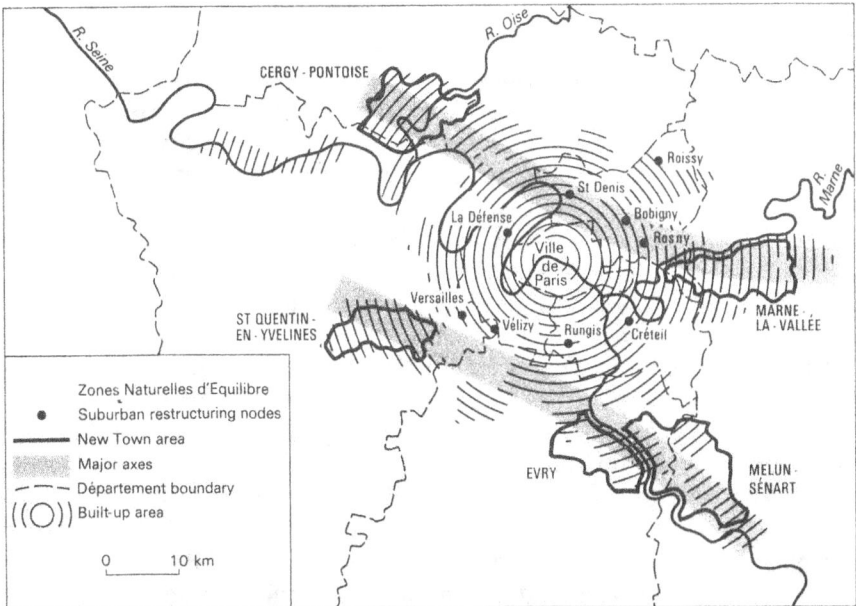

(b)

Figure 11.2 Plans for the Paris region (a) 1965 *Schéma Directeur de la Région*; (b) 1975 Modifications to the *Schéma Directeur*

of close attention under the *PADOG* of 1960. It therefore proposed 900 kilometres of new urban motorways, both circumferential and radial, together with the installation of a new express metro system, the Réseau Express Regional (RER).

The direction for the new development contained two interrelated elements. Firstly, there was a clear attempt to orient development along preferential axes, for the advantages already described above. Secondly, however, the plan retains an element from the 1960 plan in its adoption of new suburban growth nodes. These took the two forms of *pôles réstructurateurs* in the existing suburbs and the more peripherally located new towns. While, as Figure 11.2 shows, the axes were important in giving directional structure to the plan, it can be strongly argued that at least thus far the establishment of the new towns and the suburban growth poles had been dominant in the implementation of the plan. The axes were selected to give a measure of reorientation to the dominant focus of Paris itself. Thus, two axes were designated running tangentially to the existing agglomeration. One runs from the new town of Cergy-Pontoise in an east-south-easterly direction towards the second new town of Marne-la-Vallée. The other runs from the Seine valley near Mantes on the western side of Paris, curving south-eastwards to terminate around the new towns of Evry and Melun-Sénart. Both axes emphasise the broad regional direction of growth and the historic and future role of the Seine in guiding development, and in a sense look westwards towards Rouen, the Lower Seine and eventually Le Havre. Nevertheless, as contended earlier, this is still essentially a plan based on the notion of polycentrism. The suburbs of Paris lacked almost completely the strong hierarchical structuring of service centres so characteristic of London or Cologne. The designation of new nodes has given to the city that essential structure and their creation has been a high priority in the past decade. They included La Défense, Versailles, one of the few obvious suburban nodes predating the plan, and St Denis, an industrial suburb to the north, located close to Le Bourget airport. Perhaps the most dramatic new suburban growth node was Créteil, often confused with a new town because of its striking architecture.

Although the axes can be seen to link these and the new towns together (an exception being, perhaps, La Défense), there can be little doubt that the principal tangible result of the 1965 *Schéma Directeur* was the establishment of the polycentric structure for Paris. In fact, each of the new towns and suburban growth poles have been so effectively linked to central Paris, with new or much improved transport links along the line of the axes, that it is difficult to see them emerging as strong counter-growth zones to the magnet of Paris itself.

The 1965 plan was first modified in 1969, the Chalandon review, when the number of new towns was reduced to five from the original eight. Mantes to the west was dropped from the original plan since Val de Reuil further down the Seine valley to the south-east of Rouen had been designated a new town. Beauchamp to the north was similarly excluded in this first revision, whilst the two new centres at Trappes to the south were modified to form one, renamed St

Figure 11.3 Strategic centres in the *Greater London Development Plan* (reproduced by permission of Greater London Council)

Quentin-en-Yvelines. A final modification was the redefinition of the site of Tigery-Lieusaint, to be renamed Melun-Sénart.

Although plans for south-east England, including London, were based on the definition of preferred axes of growth, a change of scale reveals a polycentric orientation for the detailed plans for London itself. The *Greater London Development Plan* (Greater London Council, 1969) contained from its earlier form the concept of strategic centres into which important new development such as offices could be directed. The distribution of such centres as they were proposed in the modified plan in 1976 is shown in Figure 11.3. They were chosen on the basis of serving a population of over 200,000 and had a 1961 retail turnover in excess of £5 million.

Polycentric growth was built into the 1978 plan for Cologne which based its hierarchy of centres on the application of the principles of central place theory. The result was an integrated structure plan that incorporated the existing city structure into a coordinated view of the future city. However, the proposals were not entirely successful, owing to the problems faced by the major new district of Chorweiler, which had problems in developing the requisite service functions. Nevertheless, the continued adherence to the broad philosophy of defining all types of provision in a nested hierarchical fashion has produced a coherently organised city. Polycentric plans have also been used in the development of Frankfurt am Main, Kiel, and South Hampshire.

The city planning developments in the Lyons and Lille regions, and Glasgow and Newcastle upon Tyne, all relied on a new-town strategy to provide some element of polycentric urban development within a city region. L'Isle d'Abbeau, Lyons, was one element in the structure plan while Le Vaudreuil (renamed Val de Reuil) was to accommodate overspill from Rouen. Similarly, Cramlington and Killingworth outside of Newcastle, and Cumbernauld and Irvine beyond Glasgow, were part of a polycentric growth strategy for the British cities.

Multicentred plans

Polycentric plans are aimed at producing a series of centres, each sharing in growth rather than all growth being directed to a strong central focus. In some areas, however, a large number of centres already exist and require a plan for their future direction.

(i) The West Netherlands

A primate city dominating the economic, political and social life of the nation in the style of London, Paris, Copenhagen, Brussels or Vienna did not develop in the Netherlands. Instead, a tradition of urban independence, and a physical geography that restricted urban development to a number of relatively small sites, produced a group of towns in the West Netherlands that shared the functions of a primate city between them. The slow growth of these towns until well into the modern period preserved this peculiar and untypical pattern of settlement in the western provinces of North and South Holland and Utrecht. This inherited pattern gave the planners a set of both problems and possibilities quite different from those encountered in the great unicentred conurbations.

The idea of the Randstad (Dutch for rim-city) was propounded in detail in the first National Structure Plan of 1960. The concept had four main elements. First, the conurbation was to remain a distinct physical entity and not coalesce with the growing metropolitan areas of neighbouring countries. In the first 15 years after the war there appeared to be little threat to the substantial, if largely undefined, green belt that stretched all around the West Netherlands from Zeeland and Brabant in the south to Overijssel and Friesland in the east and north, that separated it from the towns of the Rhineland and the Schelde. Secondly, the towns of the West Netherlands were envisaged as lying along a narrow but almost continuous horseshoe that stretched from Amsterdam and the IJmond towns in the north, through the Gooi towns to Utrecht, and then down the great rivers through the *Drechtsteden* to Rotterdam and The Hague (Figure 11.4).

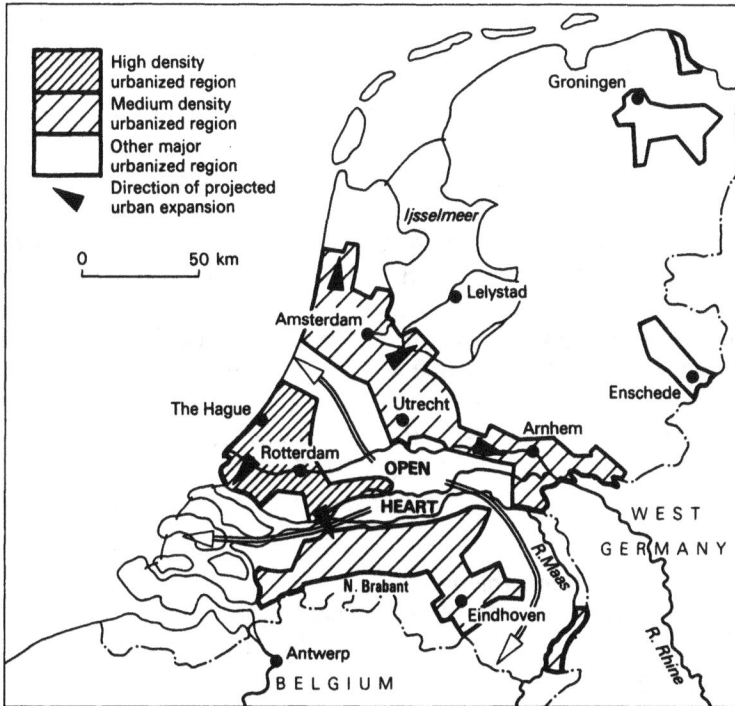

Figure 11.4 West Netherlands (Randstad), 1966: Second structure plan

Thirdly, although the Randstad implied the existence of an urbanised belt, each of the constituent towns was to maintain its separate political, functional and physical identity. Finally, the urbanised *rand* enclosed a less densely peopled area with important agricultural, horticultural and recreational functions, to form what has been called, somewhat misleadingly, the 'greenheart metropolis' (Burke, 1966).

The Randstad may have been created by historical accident but it had to be defended as a political necessity. The threat to it came from the demands for space of a growing and increasingly prosperous Netherlands. The most fundamental pressure derived from the rapid growth in the Dutch population in the 20 years after 1945. West Europe's most densely peopled country also had its fastest rate of natural increase, and substantial net immigration. The West Provinces received the bulk of both the international and domestic migration. The demand for housing was fuelled by the population increase, an even faster growth in the rate of household formation and rising expectations of greater living space. As well as housing nearly half the Dutch population, the West Provinces also accommodated an even higher proportion of the country's industrial capacity, and provided sites for the port-orientated industries that formed the vanguard in the drive for industrialisation after the war. An increasingly affluent population also demanded space for leisure. Provision had to be made for the main centres of the country's foreign tourist industry, the main domestic beach holiday resorts, and day recreation for five million residents. Finally, the most pervasive threat to the integrity of the Randstad idea came from the rapid rise in car ownership which encouraged the dispersal of housing, employment and recreation, thus threatening to 'fill up' the 'open heart', cause a physical coalescence of the individual 'rim' towns, and eventually merge the West Netherlands with the Brussels-Antwerp conurbation to the south and the Rhine-Ruhr upriver to the east. In fact, it could be argued that the survival of the unique Randstad characteristics was only a fortuitous result of the remarkably low car-ownership levels that existed in the Netherlands until the early 1960s, and that this good fortune could not persist.

The second National Structure Plan of 1966 (Rijksplanologische Dienst, 1966) was designed to counter these threats and was, thus, essentially defensive. Physical expansion was to be tolerated in five directions according to a policy of 'grouped deconcentration' (*gebundeld deconcentratie*). These were (Figure 11.4):

(1) northwards into Kennermerland;
(2) north-eastwards across the Flevoland polders to Lelystad;
(3) eastwards from Utrecht up the rivers to the Gelderland towns of Arnhem and Nijmegen;
(4) south-eastwards to link with the fast-growing industrial towns of Noord-Brabant that formed the *Brabantse Stedenrij*; and
(5) southwards from Rotterdam into the Delta islands.

The urbanised horseshoe had now become two much broader east-west urbanised zones, called 'wings' (*vleugels*). The northern wing stretched from the coast at IJmuiden through Amsterdam and Utrecht to Arnhem, while the southern stretched from the coast at The Hague-Scheveningen through Rotterdam to the Brabant towns. The 'open heart' and the buffer zones between the urban nodes remained theoretically inviolable, but the 'open heart' was now more accurately seen as a series of open corridors between the wings.

It appeared, therefore, that, on paper at least, the Randstad had been saved and the West Netherlands was not to succumb to the pressures and become 'merely another vast urban sprawl – a Dutch Los Angeles' (Hall, 1977). In reality, however, doubt could be cast not only on the policies that were intended to conserve the characteristics of the Randstad, but also on whether the Randstad had ever existed except as a concept in the imagination of the planners. The so-called 'open heart' was neither a sort of gigantic central park nor a rural agricultural region. It was in fact a district of small historical towns, like Woerden, Alphen or Gouda, suburban commuter settlements, intensive horticulture and some 'honeypot' recreation. Even its relative emptiness became increasingly questionable as, despite the structures of the planners, many of its municipalities were growing much faster than the national average (Steigenga, 1968). Similarly, it was increasingly difficult to recognise the limits of Randstad, especially in the south and east. The expansion of Rotterdam southwards onto the Delta islands, and Antwerp down the Schelde, make it easy to conceive of the emergence of a single Delta port (Riley and Ashworth, 1975). Similarly, in the east, the urbanised area had reached the German frontier at Arnhem-Nijmegen and threatened an eventual junction with the urbanised Rhine-Ruhr.

There was clearly a discrepancy between the concepts of the planners as expressed in the national planning documents and change as it was actually occurring. Dutch town planning has a longer tradition than most national planning systems in Europe but is stronger on ideas than on enforcement. There are, in fact, fewer restraints on the location of economic activity in The Netherlands than in France or Britain. The existence of three tiers of planning authority – national, provincial and municipal – was a further complication, especially when the interests of a municipality in growth contradicted a national proposal for restraint. (The Swedes face a similar problem in planning for Greater Stockholm.) Despite these misgivings, the town planners of the West Netherlands had by 1970 not only accommodated an increasingly numerous and demanding population, but had created an environment for living that was rightly the envy of their colleagues in Europe's unicentred conurbations.

(ii) The Ruhr

The Ruhr region is the major metropolitan area where the problems of coalescing industrial centres were producing a massive urban industrial region stretching from Hamm in the east westwards to beyond Duisburg. Unlike the West Netherlands, the major problem has not been one of a rapidly increasing population within the area but one of a declining regional economy dependent on coal and metallurgy, a generally low-quality urbanised environment that had sprawled over the past century and a half, and a poor transport network. The Siedlungsverband Ruhrkohlenbezirk (SVR), which was designated in 1920, did attempt to create order within its boundaries in its original strategy published in 1960. But it was forced to recognise that, in planning for the coalfield, it was planning for only a part of the large urbanised region that extends over a roughly triangular-shaped area from Bonn in the south to München Gladbach in the west and Hamm in the east including, therefore, the rapidly diversifying Rhine towns of Düsseldorf and Cologne. The Ruhr plan of 1966 (Figure 11.5) had a basic emphasis on the role of the green zones and belts separating the seven major north-south urban-industrial areas of the core, and a green buffer zone separating the core area from the newer urban industrial centres such as Marl, Wesel, Dorsten and Dinslaken in the northern region (Siedlungsverband Ruhrkohlenbezirk, 1966). Around the periphery are a series of recreation zones. Pollution has been a major concern and the SVR has sponsored legislation to control atmospheric pollution.

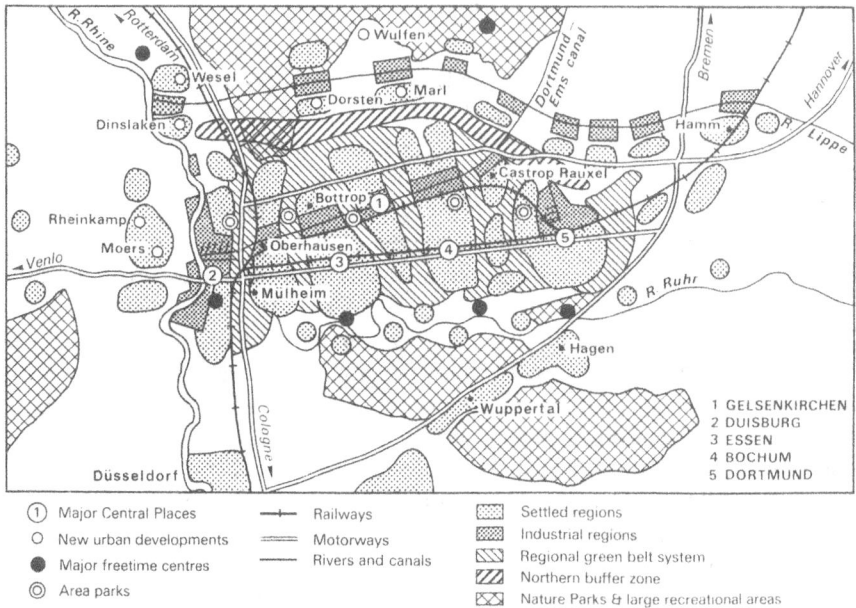

Figure 11.5   The Ruhr region structure plan (1966)

Other aspects of the Ruhr plan contained similar concepts discussed earlier. A hierarchy of central places and proposed growth centres for industry and services was part of the 1969 revision of the plan. These central places were the foci for the integrated transport proposals, including the Stadtbahn, using existing rail-lines and the tramlines which pass underground through the major centres. Other public transport feeds these major routes. In addition, a dense network of urban motorways now laces the region, some utilising the opportunities for access provided by the green zones, and others such as the Ruhrschnellweg and Emscherschnellweg partly driven through the built-up areas as they traverse the region from east to west.

Today, however, the planning powers of the SVR have been dismantled and the current situation is discussed below (page 264).

Reappraisals of the 1970s

The plans for major centres emphasised planning for growth during the 1960s, based on population projections predicting a continuing expansion of the urban population. In many cases, however, a reappraisal of these plans and a review of underlying assumptions was necessary. Predictions based on contemporary car-ownership trends foresaw a vast increase in personal mobility unhindered either by a shortage of petroleum or a slowing-down in economic growth rates. The *Schéma Directeur* (1965) for Paris took as one of its assumptions a fivefold increase in spending power, without being able to foresee the world-wide recession and stabilisation of real spending power which have affected the western industrialised countries to a greater or lesser degree. Visions of urban population growth, coupled with an ever-expanding urban lifestyle necessitating large-scale planning, have had to be reviewed.

Other changes have been evident, both in terms of attitudes and contemporary processes, within the urban system. On the one hand, for instance, a stronger awareness of the environment, coupled with a more responsible attitude towards the exploitation of resources of all types, has been evident during the past decade.

We may certainly add that, despite the urban renovation, European attitudes towards conservation at all levels have still attained a more practical result in terms of their reflection in urban policies. A final reason for reappraisal lay in the unexpectedly rapid decline of the inner city, particularly noticeable in the case of London and the Ruhr towns and cities. A policy over the post-war period of economic decentralisation in Britain, fostered initially by industrial development certificates, and from the mid-1960s by office development permits, had begun to take its toll. Dennis (1978) shows that population declined in Greater London by an estimated 9.4 per cent between 1966-74 while, in the Inner London boroughs, the fall was 17.3 per cent. Even more startling was the decrease in manufacturing employment in Greater London with a 27 per cent decrease entailing the loss of 390,000 jobs to leave only 900,000 by June 1974. These figures may be compared with a national decline of 703,000 or 8.4 per cent over the same period. Obviously, reappraisal was necessary and the impact of these changing circumstances on the plans for London and three other centres, Paris, the West Netherlands and Stockholm, is reviewed.

London – some fundamental problems in the 1970s

The early 1970s saw a degree of conflict in the two major plans published. One was at a regional level – the *Strategic Plan for the South-East*, published in 1970 (HMSO, 1970). The second, dealing with Greater London itself, was the *Greater London Development Plan* (GLDP) (GLC, 1975). While the *Strategic Plan* emphasised dispersal, clearly preserving the basic strategic approaches of *The South East Study* and the *Strategy for the South East* (HMSO, 1964 and 1967), the GLDP was concerned with measures to counter the decline both in employment and population which was becoming increasingly evident in the 1970s. The principal problem was one of balance, since within the South-East it was Greater London which was losing relative to the outer metropolitan region. The prolonged enquiry into the case for the GLDP rejected the case that policies should be implemented to retain population and jobs in London to counteract the drift away from the city.

The scene was set, therefore, for a continuation of the outmigration of jobs to towns located at a medium range of about 60km from London, while the new growth centres envisaged in the *Strategic Plan* would also grow to populations in excess of a million. This process would certainly have produced a multi-centred urban region as the importance of the centre decreased and that of the outer nodes increased. Migration outwards was selective both in terms of people and jobs. The possibility of social polarisation as middle-income earners moved out to the new office complexes in Reading, Basingstoke or Southend was a real one.

The 1976 *Strategy Review* cut the population estimates from 19.8 million to 17.1 million people (HMSO, 1976). It recognised that London would have a smaller share of the region's population than had previously been assumed. Policies for both offices and large-scale regional shopping centres looked towards attracting functions back to the capital and particularly to inner city locations such as Stratford and Clapham. In this sense, GLC policy was closer to the original GLDP than to the *Strategic Plan* of 1970 or the *Strategy Review* of 1976.

A sentence from the GLDP Written Statement of 1976, following its modification illustrates the change of views in the 1970s. 'The plan is not directed at an end-state for London in terms of bricks and concrete, but at imposing standards of life in London, a goal whose attainment will require continuous effort' (Greater London Council, 1976). The attainment of an organic planning framework had been achieved through this approach.

Paris – a decline in growth rates

Both population growth and increases in economic activity began to decline after the 1965 *Schéma Directeur*. Population growth running at 2 per cent per annum between 1954 and 1965 fell away to 1.4 per cent per annum in the period 1968–1975. The core of the city was losing population to the outer suburbs and the major problem lay in the balance of growth between east and west rather than on merely accommodating growth. In the employment sector, manufacturing was giving way to tertiary employment, though once again the task was becoming one of correcting the inequalities between the relatively more prosperous west and the less well-endowed east. By 1969 the original *Schéma* had been modified to designate five rather than eight new towns. The 1975 *Schéma*, however, remained broadly faithful to the 1965 plan. It was proposed to strengthen the transport links between the nodes, so enhancing the chances of the two broad axes becoming a reality. Population shortfall gave to Paris the opportunity to redress imbalances and to make minor adjustments as the city moved towards a polycentric form.

Normally it could be argued that a shortfall in a predicted population should not invalidate a plan since, in time, as population does increase, the original provisions will still be required. This view carries with it the suggestion that the underlying assumptions and objectives of the plan remain unaltered over time, a fact which experience in the 1980s has shown to be true.

Reappraisal in West Netherlands

The 1970s was a period of change in both the realities of the city in the Netherlands and in attitudes towards planning and the city. Most fundamentally, the rapid decline in the birth rate successively lowered the predicted population upon which the previous plans had been based, although this was compensated to some extent by the rate of household formation which continued at a high level. Concern for the conservation of the natural environment, for depleting energy resources, and for the renovation of inner areas of the large cities, were not new in the Netherlands but they received a renewed emphasis. There was also a profound change in the philosophy and machinery of planning not unlike that occurring elsewhere on the continent. The previous plans for the West Netherlands had been detailed, visionary blueprints of a fixed future at a stated time. The plans of the 1970s were more flexible and less detailed. They were indications of directions that could be followed and would need continuous monitoring. In addition, public participation in the planning process was mere consultation, limited to an involvement of citizens in the choice between alternative strategies.

The general pattern of urban and 'open' areas differed little from the 'wings' of the second Structure Plan (Figure 11.4). Quite different, however, were the four more detailed strategies presented to the public for consideration. These offered the consequences in terms of urbanised and open space in the three West Provinces of pursuing alternative policies of residential concentration or dispersal, dependence on public or private transport, renovation or demolition of the older city-centres, and degrees of severity in the containment of urban expansion.

In a country as small as the Netherlands it had become difficult to distinguish between town and regional planning, and the objectives and practice of the two had become inextricably linked. The goal of regional planning since the 1950s had been to stimulate the economic development of the peripheral provinces by encouraging counter-magnets (*tegenpolen*), in Breda, Helmond, Zwolle and Groningen. The physical expansion of the western metropolitan area and the growth in car ownership drew much of the south and east of the country into the orbit of the west. When viewed on the scale of Greater London or Paris almost all of the Netherlands, with the exception of Limburg, Groningen and Twente, can be regarded as a single city region, the regional growth centres as overspill towns on the urban periphery, and the remaining outer areas, such as the Friesian Lakes, Drenthe, the Veluwe, Zuid Limburg and the Brabant heaths, as an extensive recreational green belt.

Stockholm – the new priorities

The 1978 Regional Plan for Stockholm represented a marked departure from the earlier priorities of achieving an optimum location for residential development in relation to employment and transport. Economic and demographic changes necessitated a fundamental change of attitude. Therefore, the 1978 Regional Plan attempted to even out the imbalances within the regional structure in terms of growth potential and resources. In broad terms, this centred around a centre-periphery contrast and a north-south gradient of resource endowment. The plan, which took the form of a careful deconcentration of activity, was not accepted by the Stockholm County Council and two alternatives were produced. The first plan, Alternative A, was the original. It provided for a move of up to 40,000 jobs from Stockholm to four growth nodes over a fifteen-year period. This plan gained the backing of the conservative and liberal members of the ruling majority and some social democrats. The alternative, Alternative C, was advocated by the centre of the ruling coalition and the communist minority party envisaged a larger-scale decentralisation of up to 70,000 jobs over the same time period. It was most distinctive in its move towards achieving an employment-residential balance in each of the municipalities of Greater Stockholm by means of redistributing jobs.

Thus the plan was against large-scale centralisation and favoured small-scale development and self-containment. In this it was somewhat idealistic since, taken to its ultimate, agglomeration economies would be lost; and in any event such a pattern could not be achieved other than by major changes to the existing patterns. This experience of Stockholm illustrates two major points concerning planning for large urban areas. First, previous assumptions concerning objectives and priorities have been constantly questioned and, second, planning at this scale will always be more than a technical exercise, since it will always be a major political issue.

Reappraisal in the Ruhr – why plan?

In 1976 planning in the Ruhrgebiet had become the responsibility of three administrative districts; Arnsberg, Dusseldorf and Munster. The SVR has been replaced by the KVR (Kommunalverband Ruhr), an association of local authorities, whose jurisdiction is mainly in the spheres of environmental protection, recreation, waste disposal and local planning. Since 1976 planning for the whole region has become patchy because the plans are now the sum of the local city plans, and many, such as Hotker (1988), conclude that planning is no longer an instrument of control. The reason for this apparent dismissal of planning is that the population of the Ruhr fell by 400,000 between 1970 and 1984. Population decline has ensured that the development plans of the Ruhr cities, which were based on proposed expansion rather than contraction of facilities, are anachronistic. No city wishes to declare that it is planning for reduced services because they fear that both the Federal and *Land* governments will cut their financial support accordingly. It is also much more difficult to plan for the quality of life and resource conservation in a region whose quality of life manifestly does not match that of most German cities. Over 200,000 emigrants from the Ruhr between 1975 and 1985 can testify to this view. Planning for quality of life is more difficult when those cities depend on resource exploitation. Federal economic assistance from the Regional Action Programme and *Land* assistance for eight technology parks – one *technopole* in the former Krupp steelworks administrative area at Rheinhausen and others at Dortmund, Essen, Hagen and Gelsenkirchen – might diversify the economic base of the cities, but they will not replace all the jobs or radically change the multiple images of deindustrialisation that planning finds it difficult to address.

Why plan? The new right and London

The political tensions evident in Swedish planning in the 1970s emerged in many cities throughout Europe in the 1980s. In the case of London the tensions between a city which elected a socialist government in 1984 and a right wing national government have affected the development of the city and the South East region. By 1984 at the scale of Greater London, the Labour administration had revised the GLDP with the aims of redressing the imbalances that existed between the centre and the developing periphery, and between west and east (GLC, 1984). The alterations to the GLDP proposed to constrain severely the growth of activities in central London, ensure the development of a hierarchy of centres for employment and retailing which would include ten office nodes, and implement policies to enable there to be equal opportunities and access to goods and services. The policy aimed to end the haemorrhage of people and jobs. The plan was accepted by central government, who had other ideas for the government and planning of the capital. The GLDP alterations were also accepted by SERPLAN, the loose grouping of local authority planning representatives for the South East who had proposed their own planning guidelines for the region. They were also concerned by the growth west of London along the M4 corridor, and the relatively poor prospects for the east of the region. They, like the GLC, wished to maintain the green belt. However, the only means that this body had to enforce its principles was that of consensus among the local authorities.

Despite the desire of the GLC to implement its policy and the mandate that it held to do so, central government had been elected in 1979 and re-elected in 1983 on a mandate which proposed rolling back state intervention, pegging back socialism and reducing public funding of development. Already the establishment of the Enterprise Zone on the Isle of Dogs and the founding of the London Docklands Development Corporation (LDDC) in 1981 had enabled the government to introduce planning by non-elected and non-accountable authorities to replace the system of local authority control. It was the 1986 Abolition Act which brought *laissez-faire* ideology into the field of metropolitan planning. The act abolished the GLC and six other metropolitan counties in England on the basis that they were an additional and unnecessary tier of government, often supported by right-wing propaganda about the waste of resources by such authorities on fringe political activity. The GLC was replaced by 33 London boroughs each of which had the authority to develop its own Unitary Development Plan on the basis of strategic guidance from the Secretary of State who, in turn, would seek guidance from the boroughs. If the guidance could be agreed among politically opposed boroughs, there was and is no guarantee that the strategic guidance will follow the borough guidelines. Thus for a period central government sought to relax green belt legislation in order to encourage the development of 'new villages' or private enterprise new towns in the green belt, and it altered the planning guidelines to enable developers to have greater freedom to develop land. However, when the proposals such as Tillingham Hall and Foxley Wood raised the hackles of the government supporters, there was a gentle shift of policy away from the support of rampant self-interest on the part of the developers.

By 1990 the fact that consensus was supposed to affect the development of the South East was a chimera. Local authorities were unable to agree on how to accommodate future housing needs, with Hampshire begging to differ from the consensus. In London itself the boroughs did not necessarily agree with the development in areas such as Docklands where the plans for Canary Wharf were not 'called in' to enable a public enquiry to take place, much to the annoyance of Greenwich. Thus the 8.8 million square feet of office space was given the go-ahead without any discussion in the formal channels of local government.

The stimulus of the deregulation of business activity in London in 1986, a decision which we shall see influenced planners in the Netherlands, stimulated a whole series of major development proposals throughout central London, most of which saw very little in the way of coordinated planning of the transport infrastructure. The failure of private enterprise to raise funds to improve public transport into Docklands has led to a reappraisal of transport provision by central government and a realisation that major schemes cannot be left to the market; the proposed cross-London rail link is one such investment now being state funded. Whether this is a sign that *laissez-faire*, demand led planning with its *ad hoc* decision making is on the wane just as some of the major products of the system are completed, is a matter for speculation. The fact that the management of the capital is split between two ministries, 33 boroughs, a regional transport authority and several other statutory bodies does little to ensure a coordinated vision of the future of London. The London Planning Advisory Committee itself is concerned that the uncoordinated nature of London's development proposals is affecting London's role as an international centre. The concerted planning of Paris and the increased competition from other European capitals highlights the *ad hoc* nature of city and regional planning which has characterised the impact of Conservative policies on London since the abolition of the GLC.

Faith in the system – Paris in the nineties

Paris is still the city which places the greatest faith in the planning system to create and enable the city and its region to progress harmoniously towards a new millenium. Nevertheless, that progress is not without the tensions that afflict all cities and particularly that between the state with its view of the capital and the city with its own aspirations. The tensions that surfaced in the eighties were those between the Presidency, which has always viewed Paris as a city reflecting the global ambitions of the state, and the Mayor of the Ville. They were tensions which have been between a socialist President, Mitterand, and a right wing mayor, Chirac. That is not to say that these tensions did not exist when the President was right wing. In fact a similar tension underlay all the redevelopment of the Forum Les Halles between the wishes of Giscard d'Estaing and Chirac, his conservative rival (Bateman and Burtenshaw, 1983).

Presidents in Fance have long patronised and supported those aspects of urban change in Paris which have best reflected their vision of France and the ideology of the times. As we have seen De Gaulle (La Défense and Roissy airport), Pompidou (Centre Pompidou) and Giscard d'Estaing (Forum Les Halles) have all made their mark on the townscape. Mitterand's impact has exceeded all of these in that he has enabled the construction of several *grands projets*, ranging from the Pei pyramid in the Louvre forecourt to the vast museum complex on the site of the old abattoirs at La Villette (Burtenshaw and Moon, 1985).

Other state investment programmes of the past two decades also reached fruition in the eighties. La Défense was completed with the construction of the Grande Arche, the new towns began to function as towns and to attract new functions, such as the research and development district La Cité Descartes at Marne-la-Vallée, and the TGV express rail routes provided Paris with an ever widening catchment area. All of these investments were commensurate with the goals of the state to make France, especially through the role of Paris, the centre of European business, research and culture. Even the proposals to develop Euro-Disneyland on the eastern fringes of Marne-la-Vallée were part of this almost fanatical quest for dominance of the new Europe.

In addition to these continuing imprints, it is possible to detect a shift in emphasis on the proposals which have and are being made. The first shift has been the increasing recognition of the environmental lobby and the subsequent greening of proposals. In the eighties it was the creation of a new park at Belleville and the extensive park surrounding La Villette which marked the start of this shift. By the time the *Livre Blanc* for the region was published in 1990, the open space of the region was seen as one of the three interrelated systems alongside the urban poles and the transport network. The green web of the region is formed of the open spaces of the city, a series of green wedges separated by the major lines of development and the agricultural and forested areas beyond. Alongside this policy guideline for the years until 2015 is one

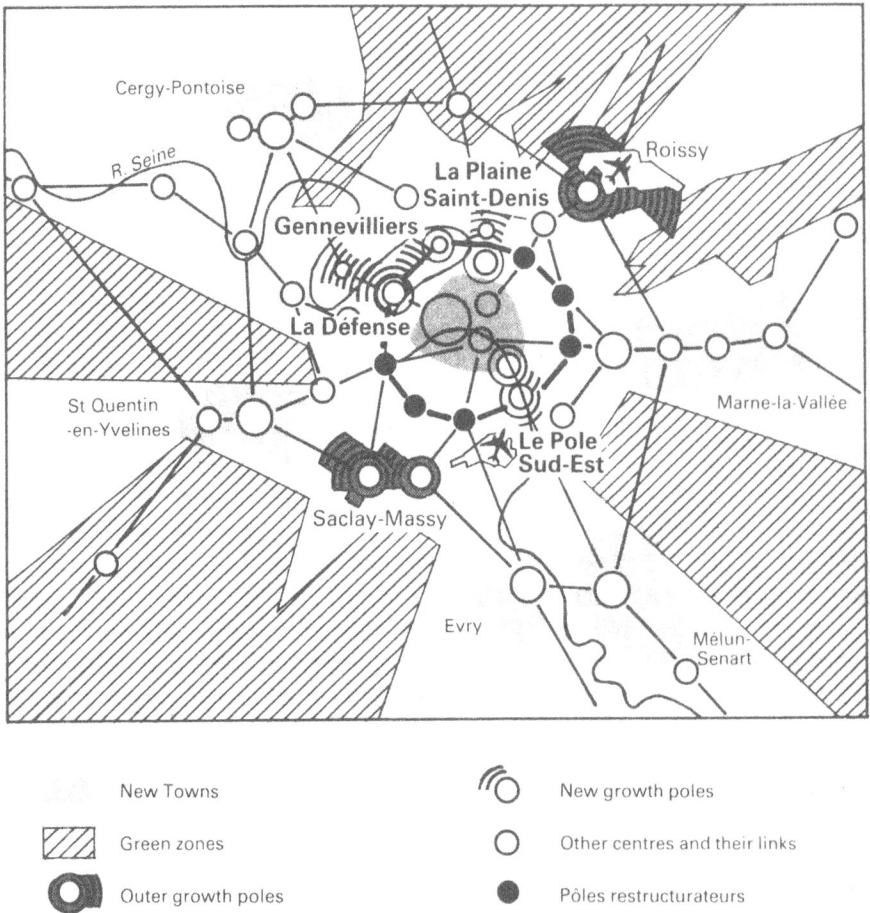

Figure 11.6   1990 proposals for Paris

which proposes to improve the quality of life partly by the improvement of waterside areas and also by the careful control of urban architecture. Notable among the proposals is one to extend the great axis of Paris that runs from the Louvre to La Défense, westwards to a new La Défense, an employment centre for the next millenium (*Livre Blanc*, 1990).

The second shift has been that to a much more technologically based city. Once again the antecedents were there in the previous decade in the investments in the TGV, the RER, the Museum of Science and Technology at La Villette and, at the micro scale, the diaphragm-like, light-sensitive windows of the Arab Institute. The *Livre Blanc* (1990) proposes a new *technopole* in the Saclay–Palaiseau area of the south-west outer suburbs based on higher education establishments and connected to the TGV system. Even buildings of the fifties are being subjected to refurbishment for the age of electronic technology. The old exhibition halls of CNIT at La Défense have been converted into a new centre for high technology which incorporates all the needs of businesses in terms of computer sales offices and consultants, conference rooms, time-share offices for visitors and a trade centre, neatly renamed CNIT (Centre National d'Information et Technologie)!

The proposals of the *Livre Blanc* (Figure 11.6) have yet to be ratified but the history of planning in France would suggest that this is a foregone conclusion. The main thrust of the proposals is to create three strategic urban sectors which continue developments already in progress. The first sector is La Défense, Gennevilliers and Montesson, which might be similar to La Défense or represent a contrasting use for the sector. The south east sector extends from the edge of the *ville* at Bercy–Tolbiac through Ivry, Vitry and Charenton to Maison Altfort, a sector for offices extending out from the new Ministry of Finance and the Omnisports arena at Bercy. The final strategic sector extends from the Gare du Nord and Gare de l'Est through La Villette to the Plain St Denis, and is a zone which will capitalise on the new educational role of La Villette and the industrial heritage of St Denis. Within the broader region the proposals aim to maintain the viability of the towns between 30 and 50 km from the city. Nevertheless, two new urban foci will be developed in the outer suburbs. At Roissy a new international pole will continue to develop on the basis of the airport, TGV and RER routes which converge here. In the south the *technopole* at Saclay–Palaiseau provides a counter-balance based on education and the excellent connections to the south west. 'Une ville n'est jamais achevée' says the *Livre Blanc*, but the proposals that it has made do exhibit a faith in the planning system's ability to influence the shape of Paris in 2015. Such self-belief, a characteristic of planning throughout the continent in previous decades, is the product of a consistent approach to planning which has been abandoned in almost all the major urban areas of the continent.

Faith in the system – West Netherlands, fourth report

The fourth report on physical planning marked a radical departure from the earlier national plans (RPD, 1989), reflecting changes in the Netherlands and beyond. Its most remarkable feature was perhaps that it appeared at all, given widespread disillusionment in the 1980s in the Netherlands, as elsewhere in north-western Europe, with blue-print planning and central government inter-vention. Two sets of circumstances motivated the production of the plan and influenced its content: namely, a belief that a diversion of resources from the west to the periphery was largely an unproductive cost, and that the progressive reduction in the barriers between European countries was leading to intensified competition between metropolitan regions. The plan is broad-brushed in tone and aims to be a stimulant rather than detailed and constraining. Its approach and terminology owe as much to marketing science as to traditional structure planning (Ashworth and Voogd, 1988).

The West Metropolitan Netherlands is viewed as the essential motor of the Dutch economy, competing for economic activities with South East England, Central Belgium and Rhine–Ruhr. The role of public planning authorities is to help shape an environment attractive to new economic activities, and to promote the region to potential investors and to stimulate and coordinate private initiatives in the region. Far from growth being constrained in the West and channelled to the regions, it is to be encouraged. This in turn frees the peripheral regions from their previous economic dependency on governmental subsidies and branch plants. They are to generate a more self-sustaining set of economic activities based on their own resources, centred around groupings of major urban nodes such as Groningen/Assen/Drachten, Arnhem/Nijmegen, and Maastricht/Heerlen.

The questions that will be answered in the 1990s are: can economic growth be sustained in the West Metropolitan area with unacceptably high costs of congestion and competition for space, leading to a drastic decline in the quality of life? And can the regional cities achieve locally developed economies without creating an east–west divide in living standards that would have severe political repercussions?

The lessons learned from the plans

Without the plans based upon the ideals of individuals and the aspirations of nations, European cities would have followed the path of cities in North America in the post-war period. Hindsight does enable us to see that not all plans succeeded and that all failed in some detail, although the basic framework of control was an important element in shaping cities over the subsequent half century.

The postwar period has clearly demonstrated the dangers of relying on long-term forecasts based on current trends. It has also demonstrated that the powers vested in the politician and the technocrat are often the cause of subsequent problems rather than the solution to the needs of the time. Therefore it is not surprising, in the volatile atmosphere of population fluctuations, economic crises, energy crises and major political change, that large scale blueprint planning has been questioned and, in some states, rendered redundant. The blueprints for the millenium are broad policy approaches emphasising the whole environmental system and constant monitoring which can best withstand the short-term and long-term fluctuations in the economy and in political fortunes.

References

Ashworth, G. J. and Voogd, H. (1988) Marketing the city: concepts, processes and Dutch application, *Town Planning Review*, **59**(1).

Ashworth, G. J. (1990) The New North in the New Europe, *Rooilyne*, October.

Bateman, M. and Burtenshaw, D. (1983) Commercial pressures in central Paris, in Davies, R. and Champion, A. (eds.) *The Future of the City Centre*, Academic Press, Cambridge.

Burtenshaw, D. (1976) *Saar-Lorraine*, Oxford University Press, London.

Burtenshaw, D. and Moon, G. (1985) La Villette, *Geography*.

Burtenshaw, D. (1991) CNIT to CNIT, *Geography* (2).

Burke G. (1966) *Greenheart Metropolis*, Macmillan, London.

Dennis, R. (1978) The decline of manufacturing employment in Greater London 1960-74, *Urban Studies*, **15**, 63-73.

Greater London Council (1976) *Greater London Development Plan: Written Statement*, Approved by the Secretary of State for the Environment, 9 July 1976.

Hall, J. (1976) *London: Metropolis and Region*, Oxford University Press, London.

Hall, P. (1973) *The Containment of Urban England*, Allen and Unwin, London.

Hall, P. (1977) *The World Cities*, Weidenfeld and Nicolson, London.

Hall, P. (1989) *London 2001*, Unwin Hyman, London.

Hellen, A. (1974) *North Rhine-Westphalia*, Oxford University Press, London.

Hotker, D. (1988) The Ruhr, in van der Cammen, H. (ed.) (1988) *op. cit.*

HMSO (1964) *The South-East Study*, HMSO, London.

HMSO (1967) *Strategy for the South-East*, HMSO, London.

HMSO (1970) *Strategic plan for the South-East*, HMSO, London.

HMSO (1973) *Greater London Development Plan*, Statement by Rt Hon. Geoffrey Rippon, QC, MP, Secretary of State for the Environment, HMSO, London.

HMSO (1976) *Strategy for the South-East: 1976 Review*, HMSO, London.

Lecoin, J.-P. (1988) Paris and the Ile de France, in van der Cammen, H. (ed.) (1988) *op. cit.*

Livre Blanc (1990) *Le Livre Blanc de l'Isle de France*, DREIF, APUR TAURIF, Paris.

McCrone, G. (1970) *Regional Policy in Britain*, Allen and Unwin, London.

Möller, I. (1985) *Hamburg*, Klett, Stuttgart.

Rijksplanologische Dienst (1966) *Tweeda Nota over de Ruimtelijke ordening in Nederland*, Staatsvitgeverij, The Hague.

Riley, R. C. and Ashworth, G. J. (1975) *Benelux, An Economic Geography of Belgium, The Netherlands and Luxenbourg*, Chatto and Windus, London.

RPD (1989) *Vierde nota over ruimtelijke ordening in Nederland, Staatsvitgeverij*, The Hague.

Steigenga, W. (1968) Recent Planning Problems in the Netherlands, *Regional Studies*, **2**, 105-115.

van der Cammen, H. (ed.) (1988) *Four Metropolises in Western Europe*, Van Gorcum, Assen.

CHAPTER 12

New Towns and Urban Planning in Western Europe

The second half of this century has seen the enthusiastic adoption of a form of urban planning which had its origins deeply rooted in the early growth of the British industrial city. For a period of some thirty years after 1945, new towns truly came of age, and indeed it was an age in which they were to flourish, to act as exemplars of contemporary urban planning. In contrast, by the end of the 1980s following a major general reappraisal, only isolated examples of active new town development remained. The process of growth followed by decline is in itself of crucial importance, since the vicissitudes in the fortunes of the new towns reflect dynamic processes and resultant planning priorities in other parts of the urban system. No matter what their eventual fate, however, the fact remains that new towns have left a major imprint on urban planning in general and, more specifically, on certain urban systems and major metropolitan areas in a number of countries of Western Europe.

In many countries, new towns are a testimony to two distinct facets of urban planning. On the one hand, they bear the imprint of the theorists and the philanthropists of the nineteenth century which culminated in Ebenezer Howard's garden city movement at the turn of the century. The second distinctive strand which interweaves with this inheritance is the fact that they reflect in part the national ideology of the society which created them. In this respect they contain a particular blend of authoritarian, utilitarian, utopian, technocratic utopian, organic or socialist traditions which happened to be prevalent at the time and place of their creation. Thus it could be argued that in early postwar Britain, Crawley bore the imprint of Howard and a utopian tradition of urban planning. In contrast, Evry new town (Paris), focused on its modern Agora of commercial and recreational facilities surrounded by high-density pyramidal residential structures, owes more to a utopian technocratic tradition.

Further differences exist between new towns with respect to the functions that they have been required to fulfil. In some cases new urban centres, though very varied in form, have been part of a more general policy of restructuring major urban regions, such as Paris, London or Western Netherlands urban regions, or on a more modest scale, around Stockholm. In other cases new towns have been designated as new economic growth points and to serve as foci for economic revitalisation of declining regions or even as physical symbols and

272

instruments of desired change. New towns have also varied considerably in their internal morphology. This variation is apparent both among countries of Western Europe and as their development has progressed through time. For instance the early emphasis in postwar British new towns on socially mixed, relatively self-contained neighbourhoods has frequently been replaced by other priorities such as one of maximising internal mobility; or attractiveness to external commercial investment.

It could be argued that new towns assumed an importance in the context of urban planning which was disproportionate to their size. Yet the statistics of achievement in some countries are impressive. In the case of the UK, for instance, in 1987 some 1.1 million of its population lived in new towns and similar urban centres created in the preceding forty years.

New towns and urban planning traditions

In the early phases of new town building, the legacy of the utopian tradition of Howard was very evident. His first garden city at Letchworth in 1903 had a strong commitment to urban open space and to the self-containment of the garden city and other proposed satellite communities which would have existed in his ideal plan. Garden city experiments from the first three decades of the century survive in most Western European countries, especially Germany, Scandinavia and the Low Countries. In Britain, the first generation of state-sponsored new towns manifestly bore the imprint of this thinking following the enabling legislation of the New Towns Act 1946. They were to provide new ways of living, rejecting suburban sprawl in favour of more balanced communities where home, work and recreation were in close physical and functional harmony.

The question of balance within the new urban community, most notably that between residence and employment, has been much debated. Certainly the utopian view would favour a high degree of self-containment in this respect. By the 1960s, in Britain, however, greater realism had emerged, and the principle had been all but abandoned with the acknowledgement that a new town such as either Washington or Milton Keynes would be a part of an extended commuting network incorporating both surrounding settlement and occasionally rather more distant urban centres.

Around the same time, the plans for the new towns within the Paris region also showed a realistic aim in respect to self-containment and moved away somewhat from this particular ideal. In these new towns, there was an initial attempt to achieve a high job ratio (i.e. the number of jobs available as a proportion of the employed resident population), but it was clearly acknowledged that this was to be no more than the net situation after commuting both into and out of the new town had been taken into account.

In the case of Stockholm, the role and form of new urban centres has been discussed over nearly half a century. The initial debate in the 1940s centred on the concept and the precise nature of the satellite town including its degree of self-containment. Thus, in 1945, the first broad planning document for Stockholm advocated satellite towns in the form of relatively independent urban units with a balance of homes and workplaces. They were to be subdivided into distinct urban units, with a suburb of 10,000 population comprising a series of smaller units, each with a population of about a thousand. Nonetheless, it was never the intention to produce self-contained satellite towns. It is important to note that in any event, the power of Swedish central government to shift economic activity or to determine its location has always been rather more limited than in Britain. Faced with this fact there was no concerted effort to create new employment growth points in the new urban centres. Instead the new centres were planned with good public transport links to the central city and its employment base. It could be argued therefore, that the utopian tradition of creating new towns as part of a new urban beginning, complete with a variety of employment, but lacking in polluting industries, has been adopted in only a very limited sense.

The garden city introduced open space into new urban developments. Indeed, in Letchworth the original site was almost 10,000 hectares, later increased to just over 11,000 hectares. Of this, only 3,700 hectares was to be used for the town itself and even this was to be built at a low density. The intention was to create a town where trees would grow and open space would be available both for recreation and for its aesthetic value. This facet of Howard's garden city certainly lived on in many new urban centres. Vållingby, the first of the Stockholm new suburban centres, was spacious in layout, utilising natural vegetation cover where possible, considerably aided in its landscaping by the occurrence of natural rock outcrops. In Britain, new towns outside London also adopted this approach. For instance, two new towns in County Durham in north-east England have each paid close attention to the effect of landscaping. In the early stages of its development the new town of Peterlee, set amongst the mining communities of East Durham, commissioned Victor Pasmore as an artistic advisor on the general design of the town and its open spaces. In Washington, a decade later, landscaping and the introduction of open space in an area scarred by a century of coal mining were understandable priorities.

Certainly the utopian tradition has been a strong one but by no means the only one. It could be argued that the high-density development of central Evry or the business node of Noisy-le-Grand in Marne-la-Vallée in Eastern Paris are reminiscent of the limited resurgence of authoritarian planning in Gaullist France, discussed in Chapter 2.

Elsewhere in southern Europe, new town plans have been formulated which may be categorised as authoritarian both in concept and execution, perhaps best illustrated by the action of the Spanish government in 1970 in passing its Urgent Development Act (ACTURS) which gave government wide-ranging powers of expropriation in order to acquire the land on which new urban communities were to be developed. This legislation was a reaction to the almost total failure of earlier planning legislation aimed at curbing the growth of the major cities, especially that of Madrid and Barcelona (Wynn, 1984). The ACTURS legislation short-circuited the existing planning procedures in order to produce a dramatic solution to an acute problem. The scale of the proposals for the new towns was impressive, with eight new towns designated, to contain a planned total population of some 800,000 people. The legislation was certainly authoritarian and was criticised for the manner in which it cut through other planning legislation. In the event, however, its success was extremely limited and only two of the eight designations, Galecs outside Barcelona and Tres Cantos, serving Madrid, made any progress even in terms of land acquisition. Even then, the legislation provided not for the development of the new town itself but simply for the acquisition of the land and the setting up of mixed private and public companies to actually carry through the development. Only in Tres Cantos was any progress made into the 1980s in terms of new urban development.

Authoritarian approaches to planning expressed through new town plans can also be seen in earlier periods in Italy, where the early years of the Fascist government of Mussolini saw the drawing up of plans for new rural townships (Calabi, 1984). The townships were seen as physical reflections of a political movement to re-emphasise the importance of the rural society of the country. In the case of the scheme for the Pontine Marshes, south of Rome, five new townships were proposed – Littoria, Sabaudia, Pontinia, Aprilia and Pomezia – each of which bore a strong resemblance to the original garden city plan of Howard. Later, in 1942, plans were put forward for four new settlements outside Milan to act as satellite towns for the city, although these plans were to be superseded by new plans for the city as a whole in the immediate period following the defeat of Fascism in the country in 1945 (Calabi, 1984).

Despite these examples of new towns which may be seen as the product of an authoritarian approach to planning, nowhere have new urban centres been planned to match the new towns of Central and Eastern Europe under their former Communist régimes, in terms of an authoritarian approach.

In France, many new urban developments owe as much to the utopian technocratic tradition as to any authoritarian tradition. Certainly it is the new town in the Paris region which of all European new towns has moved furthest away from a neatly ordered garden city style of planning. The original plan to link Cergy-Pontoise to Paris by an aerotrain, although later abandoned, suggested the style of planning at the time of the designation of the new towns for Paris in the late 1960s and early 1970s. The public transport orientation of Evry, and more particularly the manner of its translation on the ground in the form of segregated bus lanes passing through the base of pyramidal residential structures close to the centre, linking these and lower-density development further out with the centre and the main railway station, suggests a strong input of the utopian technocratic tradition. A similar scheme was first introduced in Runcorn in North-West England where public transport was given a segregated route system. It took the shape of a 'figure of eight', centred on the town centre, with residential, development and employment centres closely allied to this

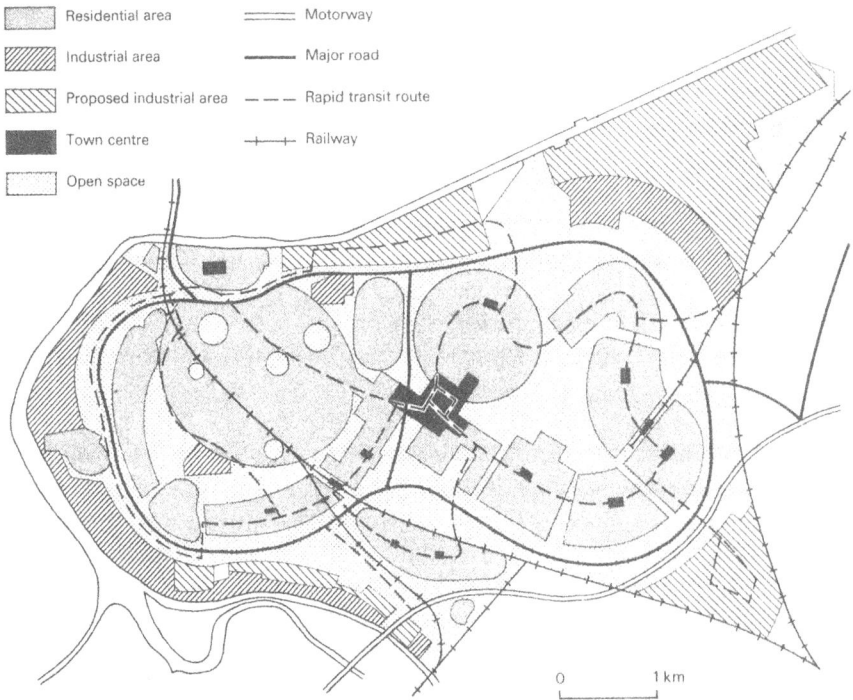

Figure 12.1   Runcorn new town, United Kingdom

segregated system (Figure 12.1); a development that found some imitators, such as Almere in the Netherlands in the late 1980s.

The increasing acceptance of new technology in new town planning has been evident, although it should be noted that despite early expectations no new town has been created with a completely new high-technology rapid transit system. It would appear that the major reason for this is one of economic viability, given the relatively restricted size of new towns.

It is a matter of some debate as to how far a tradition of socialist planning has been evident in Western Europe's new towns. The original ideas of social balance and its implications of social engineering within neighbourhoods in British new towns were soon proved to be excessively idealistic. Indeed today social segregation along socio-economic and occasionally ethnic lines can be seen to be as marked as in any other town. While the early postwar planning of new towns in Britain, under the new Labour government of that period, was frequently criticised as being socialist in concept, in its eventual execution there was no real difference in these terms between a new town and any other urban centre.

A final tradition of urban planning which deserves consideration is that which we previously termed the 'organic tradition'. A continuing problem, but one only recently recognised, has been that of producing a long term plan for a city of perhaps a quarter of a million people. Its eventual completion date may well be a quarter of a century distant, yet attempts have been made to produce a master plan for a greenfield site without any input from the eventual population. Events in the last ten years, however, have frequently created the need for a new approach, involving both monitoring and a flexibility of approach. For instance, the planning of Milton Keynes on a grid-square basis with an expectation of high personal mobility may yet prove to be unfortunate, yet it is difficult to alter the basic plan.

Other new towns, however, have had a greater degree of flexibility built into them. Examples are Val de Reuil, near Rouen in France, and Warrington in North-West England. Certainly the rigid urban blueprint has proved to be an outmoded instrument of urban planning both for new towns and larger existing centres. Increasingly the choice has been taken to keep possible options open for the future in the development of new towns. In this respect, almost by necessity the utopian tradition which was so very strong in the early phases of new town building has been superseded by a more pragmatic organic tradition.

The role and purpose of new towns in West Europe

While many new towns have been developed as satellites, accommodating growth from a major centre, others have had very different functions. These have ranged from stimulants to growth in areas of economic decline to growth points in areas of new economic development. The situation is complicated by the fact that although many of the early new towns had a single clearly defined role, be it a satellite or growth node, later designations frequently had more than one purpose. In still other cases, the original *raison d'être* of the new town has been dissipated since its designation, requiring a change of function for the new town.

New towns as overspill centres or satellites are numerous in West Europe. Both Copenhagen and Stockholm, as a part of plans to channel growth into well-planned new urban sectors, adopted new town satellites. For instance, in Copenhagen in 1947, the so-called Finger Plan served as guidelines for the growth of the city. The 'fingers' of development stretching out from central Copenhagen took the form of linear axes of small townships along the S-bane lines (or subway routes). Later, in 1960, the scale of new towns increased just as it had in the United Kingdom by that time. It was proposed originally that the growth of Copenhagen for two subsequent decades should be housed in two new towns, each with a population of a quarter of a million. Although this proposal was not carried out, it plainly illustrated the strength of the notion of new urban centres as satellite centres.

As suggested previously the new towns, or perhaps more properly new suburban centres, of Stockholm emerged with rather less emphasis on economic structuring than their British counterparts. Nevertheless they were clearly satellite developments, catering for a growing Stockholm. Since the first such developments, however, the new suburban centres changed considerably in form as we shall see in the following section of this chapter.

In the case of London's new towns, they were conceived as overspill satellites for the capital itself. The original new towns had relatively small populations, often less than 40,000, and it was intended that they should develop a strong local economy. Abercrombie in 1944 had proposed satellite towns as a part of his plan for the growth of London. The proposal eventually came to fruition with eight new towns, occupying sites different from those suggested by Abercrombie, but nevertheless designed to channel some of the growth of the capital away from London itself, into economically viable new towns. Their siting was determined by the need to preserve good agricultural land (a high priority in wartime and early postwar Britain), to have good communications links with the capital, and to have a pre-existing centre to act as an initial nucleus for the new town. There was a strong intention to discourage commuting from the new town back to London, characteristic of Howard's garden city. In practice, no new town has ever been economically self-contained, least of all the eight original new towns around London which are now very easily within the commuting zone of the metropolitan area.

In the later regional planning proposals for London and the South-East in the 1960s, further new town designations, albeit on a larger scale, were also intended to perform a satellite function. Milton Keynes, Peterborough and Northampton all fit into this category. Elsewhere in Britain, early satellites included East Kilbride and then Cumbernauld, catering for Glasgow. Telford was designated as a new town to take excessive growth from the West Midlands, but here we see the ambivalent position in which these new towns frequently existed. The local authorities of the West Midlands were initially eager to see the new town relieve some of their growth pressures. By the 1970s, however, when their own economies were being stretched in an economic recession, they were rather less enthusiastic to lose manufacturing employment and their support for the development of Telford weakened.

The situation of Milton Keynes is interesting in that it illustrates the difficulties of planning large scale satellite settlements designed to take overspill economic and demographic growth. Planned for a population of approximately a quarter of a million when it was designated in the mid-1960s, its success was threatened in the late 1970s by a labour shortage since population predictions for the South-East fell short of earlier predictions, although the new city itself saw very rapid growth in the 1970s. By the end of the 1980s, however, it was clear that these problems were largely those of phasing rather than anything more structural. By 1988, it had some 2,800 business enterprises, over half of which had arrived since its designation as a new city. During 1988, some one per cent of the national stock of new housing was built in Milton Keynes, amounting to four per cent of the new housing in the South-East region. Its population had grown from 40,000 in 1967 to 161,000, with a forecast to reach 200,000 by the end of the century. Its early problems had therefore been ironed out such that it was operating very successfully as a significant centre towards which growth pressures elsewhere in the South-East could be directed and absorbed. Indeed Champion *et al.* (1987) showed it to be in fourth place in terms of the booming towns of the UK in 1983.

The new towns of Paris may also be classified as satellite towns, although their role is undoubtedly somewhat more sophisticated than being merely reception areas for excessive growth. The eight original new towns formed one of the essential elements in the emergence of a new bi-axial development for the Paris region envisaged in the 1965 *Schéma Directeur* for the Paris Region. Thus the five new towns that remain of those originally proposed are seen as integral parts of the two new axes. Cergy–Pontoise and Marne-la-Vallée form part of the northern axis and St Quentin-en-Yvelines and Evry, with neighbouring Melun–Senart, are part of the southern axis. Besides accommodating growth, however, they share the responsibility with the new growth poles planted in the inner suburbs (*pôles réstructurateurs*) of bringing to the Parisian suburbs a strong service-centre structure with alternative nodes of employment. It was noticeable that as a part of this policy both Cergy–Pontoise and Evry early in their development had new prefectures for two of the new departments of the Paris region, Val d'Oise and Essonne respectively. In addition, new regional shopping centres were also situated in the new towns, thus emphasising still further their regional role.

Given this level of support, therefore, it could be argued that the Paris new towns were rather less vulnerable than examples elsewhere and that they were well set to achieve their original purpose. To some extent that would be a correct surmise, although the levelling off of population growth meant that by the late 1970s absolute growth was falling behind that originally forecast. The population targets for the new towns of the Paris region were considerably larger than those of the original London new towns, which were often as low as 30,000, although revised upwards by the 1960s and early 1970s to much higher figures, such as the 130,000 target for Basildon by that period. Even later designations such as Milton Keynes, with an original target of 200,000 were planned to be only half as large as the new towns of the Paris region, which had original target populations closer to half a million. By the late 1980s, nearly 400,000 were living in the five Parisian new towns, alhtough it should be noted that the original targets had to be considerably reduced as population growth slackened.

In the Netherlands, the new town of Zoetermeer was a satellite town modelled consciously on the British pattern. Situated two kilometres from The Hague, it was much closer to its 'parent' city than the new towns of London or Paris, each of which were up to 40 kilometres from the main city. With a target population of 100,000, it was never intended that Zoetermeer should be self-contained. Instead it was deliberately planned for commuters to The Hague, with no house more than 500 metres from a train station, and access to the special commuter train the 'Sprinter', linking the new town to its parent city. In both Paris and Amsterdam large-scale developments have been undertaken which are akin to satellite new towns, although not strictly new towns as usually defined. Bijlmermeer with a target population of 80,000 was established to serve Amsterdam, whilst the new suburban growth poles of Paris developed in the framework of the 1965 Regional Plan were similar developments in terms of scale, although much more a part of the existing city.

Serving two functions in a similar fashion to the Paris new towns are the new towns of the Flevoland polders. The early Dutch new towns on the polders, such as Emmeloord on the North-East polder, were purely new regional centres built to service the new settlement of the polders. The newer towns close to Amsterdam, however, have been subject to different pressures. For instance, Lelystad on the Southern IJsselmeer polders (Flevoland) was originally planned as a regional centre, with a population target by the end of the century of 100,000. Almere in South Flevoland was begun in the 1980s and is still under construction, and was never envisaged as anything other than an overspill centre for the north wing of the Randstad, with a target population around 250,000. The approval of the new railway between Amsterdam and Almere and ultimately to Lelystad will underline this role.

While the original concept of a new town was principally as a satellite centre, some new towns have been designated for economic purposes. Wulfen and Marl in the northern Ruhr were to accompany expansion of the mining industry, although with only limited expansion, the *raison d'être* of such new towns was always open to question (Burtenshaw, 1974). A similar problem faced Glenrothes on the Fife coalfield in central Scotland, since a change of policy in the coal mining industry deprived it of its primary role of a new mining town and assigned to it instead a more general role of a new town in a region of economic assistance. In contrast to these new towns, Corby in Northampton-shire was designated in 1950 as a new town in an attempt to widen its economic base, which was heavily dependent on the iron and steel industry. In the event, the town proved to be economically vulnerable in its dependence on one industry. Its iron and steel works closed in 1980, leaving it in the anomalous position of a new town in the southern half of England with very high unemployment rates and requiring considerable economic diversification. Corby was granted development area status, thus qualifying for government economic assistance, in an effort to bring some balance to its employment structure before it was accorded Enterprise Zone status as an even more powerful tool of economic revitalisation. In the case of one of London's five new towns, Basildon, designated in 1949, the opportunity was taken to provide a solution for another planning problem besides that of the growth of London. In this case it was sited in an area of considerable rural dereliction so it was intended that the new town would give a new structure to a poorly planned area.

In Britain many new towns have been established as a part of a regional development policy. Cwmbran (1949) in South Wales, Newton Aycliffe (1947) (later simply Aycliffe), and Peterlee (1948) were amongst the early ones. They were followed by others which had various purposes but still had a major regional policy component in their designation; Skelmersdale (1961), designated in part to provide new housing for Merseyside, and Washington (1964), intended to provide good land for industry, overspill facilities for Tyneside and to set a standard for the urban environment, fall into this category.

New towns have been much more than satellite towns, although it is those which were founded to assist the overspill of major metropolitan centres which have often been the most prominent – partly because of their size and in part because of their success, at least in terms of absolute population growth, as is illustrated below.

The success or failure of the new towns

At this stage it is worth considering the success of new towns in carrying out these varied roles since it can scarcely be claimed that all new towns have been unconditionally successful. In the case of overspill or satellite new towns in Britain, there have been some disturbing conclusions concerning the general ability of new towns to cater for those in greatest need in the 'parent' urban community. For example, it has been suggested that Glasgow's new towns did little to alleviate the problems of multiple deprivation and that by providing houses primarily for those employed in the new town they were selective in taking only one segment of the urban population. Although the argument can become circuitous, in an analysis of urban deprivation the new towns score best since their housing stock, being newer, is automatically of a higher standard, and in addition unemployment is generally lower than in the central city.

A similar view was expressed concerning London's new towns by Deakin and Ungerson (1977). In a study of the impact of the new towns on the inner London borough of North Islington, they offered support for the contention that the inner cities had been socially and economically damaged by the development of new towns. Predominantly it was the young, the skilled and the white population which was attracted to the London new towns. As a result, an analysis of the ethnic composition of the new towns shows that only 1.1 per cent of the population of the eight first generation London new towns were of New Commonwealth origin at the 1971 Census, compared with 5.7 per cent in Greater London. This type of discrepancy is only part of the reason for the more general reappraisal of British new towns which took place in the late 1970s and which is discussed in the final section of this chapter.

If one attempts to measure the economic success of the British new towns then the contrasts are stark. Generally those whose prime function was as an overspill facility for London were more successful than those elsewhere with a similar function or those created for purely economic reasons, although admittedly this is a generalisation which should be treated with some caution.

Crawley, as an example of a London satellite new town, enjoyed considerable economic success. In contrast to many new towns outside the South-East region, it attracted not only manufacturing jobs but also office employment. For instance between 1947 and 1978, 73,986 square metres of office space was completed. Other new towns attracted even more, such as Northampton and Bracknell, with 43 per cent increases in service sector employment between 1971 and 1981, compared to a comparable figure of 5.4 per cent in Aycliffe or 17 per cent in Peterlee (Champion, 1987).

Crawley was particularly fortunate, however, with the location on its northern boundary of Gatwick airport, which emerged as London's second airport. This has generated a considerable amount of service employment in Crawley (an increase of 44 per cent during the 1970s) and, together with vigorous economic growth, gave rise to a major shortage of land for new residential development by the late 1960s. Champion's (1987) analysis of the top local labour market areas in Britain in terms of their economic performance showed three new towns of the South-East in the top twelve – Bracknell (3), Milton Keynes (4) and Crawley (12) – demonstrating the economic success of these new towns within the urban system.

The economic contrasts are all too apparent in a comparative examination of some of the new towns designated in development areas in Britain. Peterlee, in County Durham, for instance, although primarily designated in order to improve the residential stock of the (then) Easington Rural District, also had to attract new industry. In this aim Peterlee met with comparatively little success for a number of reasons. One was its inaccessibility, sited in the east of the county away from the main north–south line of communication. Secondly, it had to withstand the opposition of neighbouring local authorities, such as Hartlepool in the south, which were anxious to improve their own weak economic base. Finally, there were problems of land subsidence arising from its location on a coalfield which was being actively worked (see Thomas (1969) for a fuller discussion of the problems of this and other new towns). The net result is that Peterlee attracted industry providing only 5,067 jobs between 1948 and 1978 compared with the 19,782 new jobs created in Crawley during the same period, and in Champion's analysis referred to above, it occupied a rank of 268 within the UK. The importance of location within a region is illustrated by the fact that Washington, designated in 1966, has always been much more successful than Peterlee in attracting industrial jobs, though it shares the same inability to attract service employment. Washington is on the main north–south axis of the region, alongside the A1(M) motorway, close to Tyneside and Wearside, and in this location attracted 9,200 jobs between 1964 and 1978. By 1984, 226 firms had been assisted to establish in the town (Robinson et al., 1987), and by 1986, it acquired its major employer with the new Nissan car plant, with a workforce of 2,500.

New towns outside South-East England have demonstrated their vulnerability in terms of economic decline at times of more general economic recession. Expanding companies attracted to the new towns in the development areas by a range of financial inducements may well locate only a branch plant in the new town location. At times of economic stress these plants are the most vulnerable to closure. Skelmersdale in Lancashire has had a particularly troubled history in terms of its employment base. Designated in 1961 with the primary aim of solving some of Merseyside's housing problems, and particularly those of inner Liverpool, it developed a vulnerable economic structure, heavily dependent on two major manufacturers, Thorn (making television tubes) and Courtaulds (a major textile company); in 1975 these two companies employed 20 per cent of the town's workforce. A total of 10,000 new jobs had been created between 1965 and 1975, many of these in the small 'nursery' units of less than 800 square metres provided by the development corporation. By mid-1977, however, both major factories had closed and unemployment levels were in excess of 20 per cent. After the closure of one factory *The Times* (1976) summarised the plight of the new town thus:

> It may be that the crisis that Skelmersdale now faces, in terms of both employment and confidence, may lead to fresh thinking and that for some considerable time to come, the effort will have to be directed towards restoring and consolidating the new town within its present limits, rather than looking for big new growth.

Later events confirmed the reappraisal not only of Skelmersdale but of all British new towns. A similar history of branch plant dependence is formed in such towns as Emmen in North-East Netherlands, once again away from the focus of major metropolitan growth.

The success of new towns as instruments of the restructuring of large urban areas varies considerably. The new town or new suburban developments of Stockholm certainly gave a logic to the structure of that city. At the same time they ensured the viability of public transport since they were specifically oriented to the T-bahn (subway) system, in a similar fashion to those of Copenhagen. A more fundamental restructuring such as that attempted in Paris via the new towns has proved to be rather more difficult. The example of Marne-la-Vallée illustrates the difficulties of fulfilling the roles assigned to the new towns.

Situated 10 kilometres from the centre of Paris, Marne-la-Vallée comprises three distinct sectors extending west to east, linked together by the RER line (Réseau Express Regional) and by the A4 autoroute, which runs through the southern sector of the new town (Figure 12.2). The new town was seen as having a major role in correcting the east–west imbalance of the Paris region in terms of economic growth. Thus its western sector, centred on Noisy-le-Grand is essentially a part of the inner suburban ring of communes in eastern Paris. The development of a large commercial complex with a regional shopping centre and a total in 1988 of 882,000 square metres of office space, went some way to providing a better service base for the eastern inner suburbs. The creation after twenty years of over 25,000 new jobs, 16,000 of which were in the tertiary sector, was obviously an impressive contribution to the attempt to create a new equilibrium between the east and west of the region.

Its location, however, within the pre-existing suburban zone, brought with it obvious problems of creating a clear identity for the new town. It was originally intended that the second and third sectors of Marne-la-Vallée, located further eastward, should have progressively lower densities of development and be

Figure 12.2   Marne-la-Vallée new town, France

more akin to 'conventional' new towns with small-scale groupings of new residences and associated facilities separated by recreational land.

The fortunes of Marne-la-Vallée took a dramatic turn in the mid-1980s, with the decision to locate Euro-Disneyland in the fourth sector of the town, with a target completion date of 1993. Such a major economic implant into a town, with a projected employment total of 25,000 people, will obviously make economic comparison with any other new town somewhat superfluous. Aside from the Disneyland project, however, Marne-la-Vallée has succeeded in giving a new structure to the eastern suburbs, though perhaps at the cost of a clear identity for itself.

It is more questionable whether the new towns of the Paris region can effectively help to create the new axes of growth described both in the regional Plan of 1965 and its later modifications in 1975 and subsequently. In the early stages of new town development, it was already apparent that a major initial priority was to forge an effective link with Paris itself rather than with each other or along the lines of the new axes. However, such an assessment may well be premature. The new plan, the outline of which was published in 1990, re-emphasised the axes of growth, and it may be that given time and with the full development of its regional transport system, the Paris region will emerge as a more polycentric region than it has been in the past.

The morphology of the new towns

The internal layout of new towns in West Europe has been the subject of continual reappraisal. Early new towns, following the first British examples, were predominantly oriented towards neighbourhood planning, with a degree of traffic segregation. The neighbourhood size was determined by school catchment areas and each neighbourhood centre had a small range of convenience-goods stores. Increasingly, however, this stereotype of the new town was questioned and new forms of the town emerged. The development and continuous reappraisal of the form of the new suburbs for Stockholm, indicate the degree to which change has been apparent.

The form of the first development was to have small communities of 2,000 population, which in their turn were grouped into suburbs of 10,000 which then went to form larger units of up to 50,000 population. These may be termed successively 'suburban elements', 'suburbs', and 'districts' and the conceptual plan is well illustrated by Vällingby, conceived in the 1940s (Figure 12.3). Developments such as the Vällingby group were termed ABC groups where work (*arbeiter*) combined with residences (*bostander*) and focused on the centre (*centrum*). In practice the work component was rather belatedly incorporated into the plan. Experience with these first-generation new suburbs suggested that some improvements were needed, particularly with regard to the adaptability of these new developments. There were three specific problems. First, it was shown that the age structure differed over time from the original population, requiring

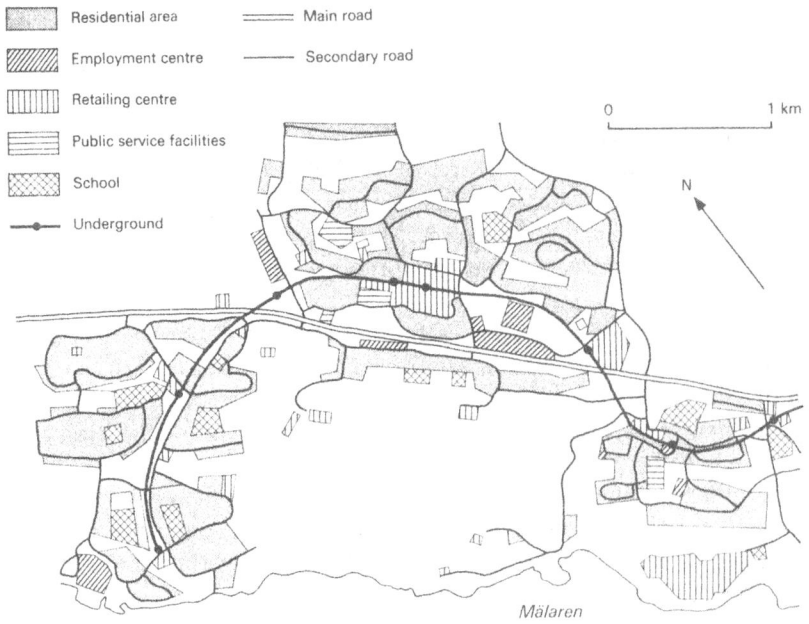

Residential area   Main road

Employment centre   Secondary road

Retailing centre

Public service facilities   0   1 km

School   N

Underground

Figure 12.3   Vållingby suburban group, Stockholm

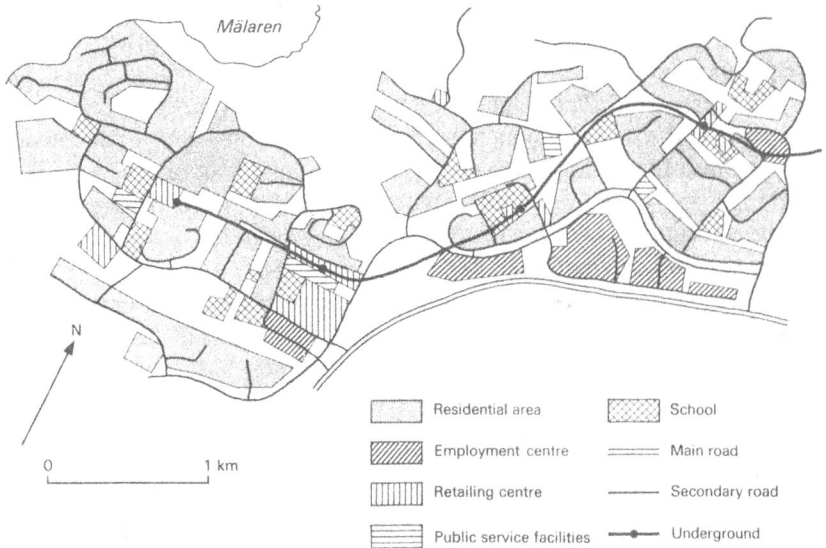

Mälaren

N

0   1 km

Residential area   School

Employment centre   Main road

Retailing centre   Secondary road

Public service facilities   Underground

Figure 12.4   Skårholmen suburban group, Stockholm

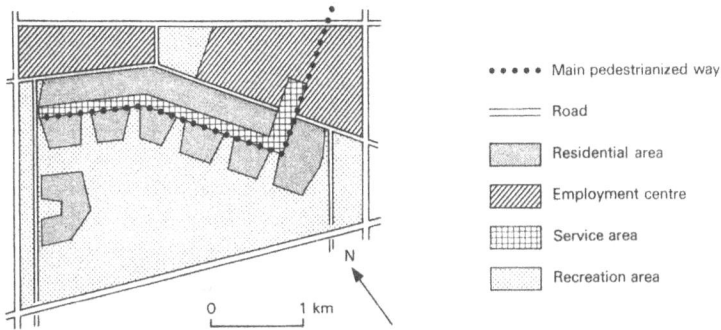

Figure 12.5 Nôrra Järvafältet suburban development, Stockholm

an adaptability in service provision. Second, the Swedish school system was reformed and with it the catchment areas, necessitating a change in the size of the suburban clusters. Finally, the changes in retailing in the 1950s and 1960s, bringing an increase in the scale of operation and the decline of the small shop, favoured the development of larger suburban shopping centres, serving not one but several suburban areas. Thus the 'pearls on a string' development which had characterised early developments in Stockholm were superseded by much larger suburbs. The first-stage subdivisions at the lowest level, between 'suburban elements' and 'suburbs', were omitted and the centre was one of regional importance. Thus the Skårholmen group was developed (Figure 12.4), comprising Bredang, Såtra, Skårholmen and Vårberg, with its retailing centre designed to serve not only these districts but also the neighbouring communes of Huddinge and Borkyrka. In the event, the retailing centres were not as successful as anticipated and further changes in plan design took place.

In 1967, the first inhabitants moved into another type of suburb at Tenstra and Rinkeby, which were the forerunners of a major new development of the 1970s. These two districts were characterised by high-density development, with a belt of service facilities including schools and retailing running through the centre of the suburb along the line of the T-banen route. This axial development adopted in the Nôrra Järvafältet development, contrasts with the small nuclear settlement patterns of the Vällingby group. The Nôrra Järvafältet group (Figure 12.5), comprising Kista, Husby and Akalla, had an axis of service development with high-rise development closest to this axis and progressively lower density development away from it. Although it was intended that this arrangement should ensure that services were equally available throughout the length of the axis, one centre, Kista, had a larger retailing unit than the others, in the form of a covered shopping mall, but with direct access to the T-banen system. The adoption of this style of development following the rejection of the Skårholmen pattern was made partly on the grounds of flexibility but also because of the opposition to the larger-scale planning epitomised by Skårholmen.

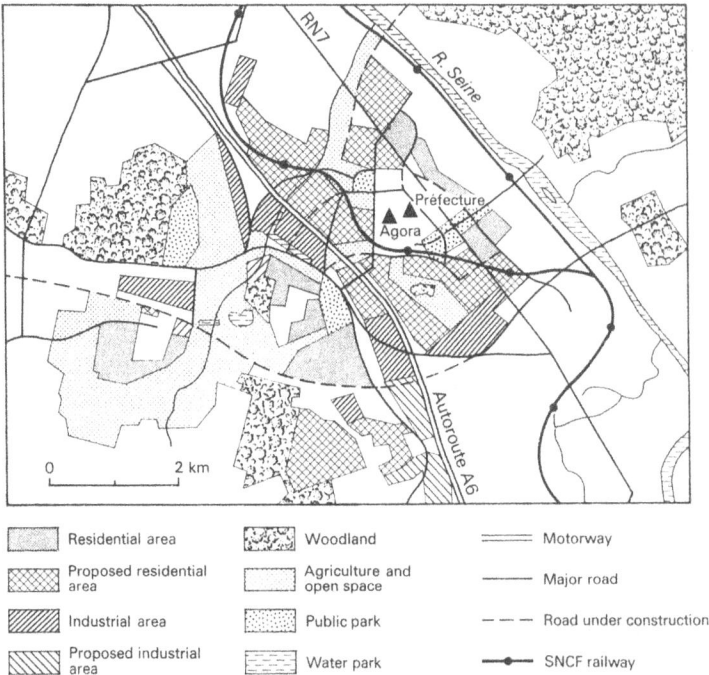

Figure 12.6   Evry new town, France

While the planning of the new suburban centres of Stockholm illustrates changing attitudes in the planning of new urban areas, elsewhere in West Europe new towns have demonstrated a number of approaches, reflecting a variety of priorities which were either socially based or, over time, mobility based. The plan for Runcorn in 1964 was an important milestone in new town planning. No longer was the cellular pattern of neighbourhood units adopted as the basic morphological pattern for new towns. In its place emerged an emphasis on public transport with the adoption of its segregated bus track mentioned earlier. This concept was subsequently extended in the planning of the central zone of Evry, Paris, where the pyramidal apartment blocks frequently straddle a segregated bus route network (Figure 12.6).

A concern for mobility also characterised the planning of two contrasted new towns in Britain, Milton Keynes and Washington. Both date from approximately the same period as Runcorn, and both master plans were prepared by the same design partnership. However, the contrasts in the plans for the two new towns are important, since they illustrate the nature and extent of the reappraisal of the neighbourhood unit which had been so important to the earlier new towns.

A prime design objective of Milton Keynes was to provide freedom of choice for the individual. It was felt that the neighbourhood unit, by providing a

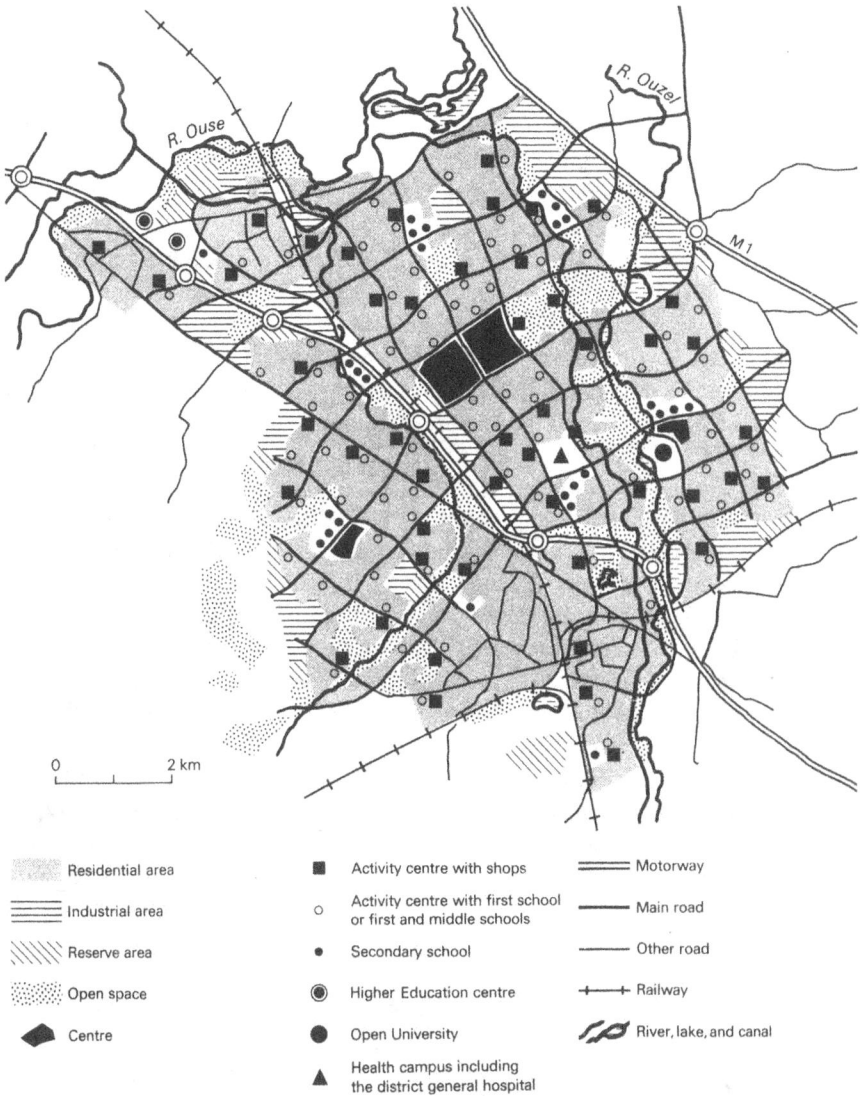

Residential area	■ Activity centre with shops	═══ Motorway
Industrial area	○ Activity centre with first school or first and middle schools	─── Main road
Reserve area	• Secondary school	── Other road
Open space	◉ Higher Education centre	┼──┼ Railway
Centre	● Open University	River, lake, and canal
	▲ Health campus including the district general hospital	

Figure 12.7   Milton Keynes new town, United Kingdom

central focus of services and encouraging social interaction at that level, inhibited such freedom. For this reason, the neighbourhood was rejected. In its place, a grid pattern was adopted, as shown in Figure 12.7, in order to facilitate mobility within the new city. The service facilities were then located on the primary route network rather than being centrally positioned within the residential zones. A high level of mobility for the inhabitants by bus or private car thus ensured their freedom of choice between competing facilities. Such a plan, fostered also by Melvin Webber's notion of 'community without propinquity' (Webber, 1964) is almost a symbol of a high energy-consuming society, particularly unfortunate in terms of its reliance on private transport.

A similar grid pattern of primary distributor roads formed the basic transport network for Washington new town (Figure 12.8). In this case, however, the social planning component was much more traditional. It involved the development of eighteen villages with approximately 4500 people in each. At this level, such a development reflected the local structure of the mining communities of the region in which Washington is located, some of which were included in the designated area of the new town. Each village is subdivided into 'places' of between 200 and 600 families in which it was expected that the inhabitants would best develop a feeling of local identity or 'sense of place'. Finally, the 'places' are subdivided into groups of between 24 and 50 dwellings, often around a common facility such as a shared garden or play space. This closely structured pattern was in sharp contrast to the Milton Keynes design.

In its planning of industrial zones, Washington was also very traditional. Industry was located in a number of small estates, providing opportunity for one village to forge particularly strong links to one industrial estate. Furthermore there was provision for small-scale workshop industry to be incorporated into village centres. Washington, then, was always more a traditional new town than Milton Keynes. It reflected the traditional patterns and residential–employment relationships of its region, although the plan acknowledged the likely role of longer-distance commuting between Washington and the neighbouring centres of Sunderland and Newcastle upon Tyne. Milton Keynes, on the other hand, alongside the M1 motorway, and within a zone with major growth pressures, is perhaps more symbolic of the urban expectations of the last quarter of the twentieth century, although it should be noted that the recurring energy crises of the 1970s called into question some of the basic thinking behind the plan for the new city.

Reappraisal – the new priorities

The 1970s saw a reappraisal of the place of new towns in overall strategies of urban planning. In the Netherlands, a combination of demographic, economic and political changes substituted the 'compact city' for 'dispersed concentration' as the national goal. In Britain the slowing down of population growth raised questions concerning the need for new cities such as Milton Keynes. In France it was acknowledged that the new towns of the Paris region would not reach their original population targets by the end of the century. In both countries there emerged a competition for resources which affected the new towns. For instance in France, it was suggested that the *metropoles d'équilibre* designated in 1966 to act as regional nodes for growth, had been neglected in favour of the new towns and the replanning of the Paris region (Berry, 1973).

Residential area	Parkland, open space and major playing fields	Motorway and primary road
Industrial area	Woodland and areas of landscaping	Secondary road
Commercial area	Village/town centre	Intervillage walkways
Reserve area	○ Primary school	Railway
	● Comprehensive school	Designated Area boundary

Figure 12.8 Washington new town, United Kingdom

In Britain, however, the reappraisal was much more fundamental and dated from government policy statements from the mid-1970s. For instance, in late 1976, the Secretary of State for the Environment (responsible for urban planning), suggested that the emphasis on dispersal to new towns and overspill areas had been directly responsible for 'daunting problems' faced by older urban areas with a declining industrial and economic base.

The major policy change was to come in April 1977 when the government cut back dramatically on the new town programme and particularly on the Central Lancashire new town project. The urban programme with particular responsibility for the inner city was transferred from the Home Office to the Department of the Environment with an increased budget from £30 million per annum to an eventual £125 million by 1979–80. It was felt that the reversal of emphasis away from the new towns to the inner city would assist the revitalisation and economic regeneration of the latter.

During the 1980s, ushered in by the general government dislike for public intervention in urban planning or any other sphere, the new town development corporations in the UK were progressively dismantled. Their assets were either sold or handed over to local government authorities. In practice this meant that there were few outward signs that the new towns were anything other than large scale new urban developments. Elsewhere in Europe, however, the public involvement and intervention characteristic of the new town programme have been maintained, although no new designations of publicly funded new towns have taken place.

In Britain after 1979, the mantle of new town building was returned to the private sector, which was encouraged to propose new settlements often within the Green Belt. On the whole many of these private enterprise new towns have not got beyond the plan stage because of local opposition upheld by the planning process, as was the case at Tillingham Hall and Foxley Wood, two of a ring of thirteen proposals around London. However, where the proposals have been seen in accord with the local structure plan then a new town has begun to be developed, Bradley Stoke outside Bristol being a case in point.

These reappraisals coupled with a considerable slowing down in population growth generally brought to a standstill new town designations throughout Western Europe. Almost half a century of active development of new towns by public authorities in Western Europe has clearly reflected much wider issues of urban planning. The problems associated with large scale urban planning in an era of economic and demographic uncertainty were often profound. Some new towns have reached maturity – such that it is now something of a misnomer to refer to them as 'new towns'. Others, in contrast, are struggling towards a maturity which shows few signs of being attained. Given the history of new towns in general, however, it is hardly surprising that a reappraisal has taken place and that in the last two decades of this century it has been increasingly unlikely that any new towns designations would be made. Economic uncertainty, coupled with a growing distaste for large scale state interventionism, have combined to ensure that new town designations in Western Europe have been confined to the mid-century decades. Their legacy is nonetheless profound and no assessment of urban planning in Europe would be complete without a detailed consideration of their contribution to the urban system, its shape and operation.

References

Berry, B. J. L. (1973) *The Human Consequences of Urbanisation: divergent paths in the experience of the twentieth century*, Macmillan, London.

Burtenshaw, D. (1974) Regional Planning in the Ruhr, *Town and Country Planning*, **42**(5), pp. 267–26.

Calabi, D. (1984) Chapter 3, Italy, in Wynn, M. *Planning and Urban Growth in Southern Europe*, Mansell, NY.

Champion, A. G. *et al.* (1987) *Changing Places*, Arnold, London.

Deakin, N. and Ungerson, C. (1977) *Leaving London, planned mobility and the inner city*, Heinemann Educational, London.

Robinson, F., Wren, C. and Goddard, J. G. (1987) *Economic Development Policies; An evaluative study of the Newcastle Metropolitan Region*, Oxford University Press, Oxford.

Thomas, R. (1969) *Aycliffe to Cumbernauld, a study of seven towns in their regions*, PEP, London.

Webber, M. (1984) *Explorations into Urban Structure*, University of Pennsylvania Press, Philadelphia.

Wynn, M. (1984) *Planning and Urban Growth in Southern Europe*, Mansell, NY.

CHAPTER 13

Conclusion

We began by an examination of the European city as a distinct entity, a product of ideology and history, and we must end by an assessment of how far the cities in Europe have remained distinctly European while retaining within the European urban system a rich internal diversity. How far have the European urban visionaries been forgotten, as the city in West Europe, so rich in culture and fruitful as a place of interchange and exchange, has moved into the second half of the twentieth century, coping with reconstruction, refurbishment and growth? Has the last quarter of a century added to the richness which is European urbanism, or has it systematically procured its dissipation? We conclude with an attempt to identify the trends in Europe which are forming the city of the year 2000. That magical landmark in time is no longer the distant milestone that 1984 was to George Orwell. Nevertheless, prediction is a hazardous process, and we must plead that uncertainty overrides confidence as economic frustration, social change recurring energy crises and changes in political control cloud rather than clarify even the comparatively near future.

The nineteenth and twentieth centuries were periods of rapid growth which transformed many cities. Indeed the scale and pace of change was such that many cities lost their essential character as centres of urban culture and were shaped by the forces of economic rationality alone. The 'property machine' rode roughshod over other aspects of urban culture (Ambrose and Colenutt, 1975). The industrial revolution dehumanised small towns and transformed them into thriving cities, all too frequently characterised by poor housing. Twentieth-century growth frequently gave rise to amorphous, asocial suburbs, the very antithesis of urban centres advocated by the early generation of European planners who had been so anxious not to repeat earlier mistakes. In this sense Europeans cannot claim to have been faithful to the ideas of the urban utopians, many of whom proposed smaller-scale, more intimate urban forms than those created in postwar Europe. Most of them, however, shared the common characteristic of being both expensive and incapable of tackling problems of the scale encountered. In practice, the choice was often between a high quality of provision for the few, or mass-produced standardised facilities for the many. Before condemning out of hand the towerblock and prefabricated estate the critic must show how else the homeless millions of postwar Europe could have been provided with the basic necessities of shelter.

The need to focus attention on the inner city has been taken as an admission of the failures of urban planning in the last few decades, as part of the spatial fabric of the urban system became progressively neglected and then, inevitably, fell into decay. Some towns have found adaptation a difficult process, especially so in the case of uni-functional towns such as the coalmining towns of Europe or other products of the industrial revolution. Even architectural 'gems', such as the eighteenth century spa of Bath developed in a short period of time, have had to face similar problems. These urban centres lack the rich inheritance of many periods of town growth which characterise many West European cities – Hall's 'repositories of the irreplaceable past' (Hall, 1977). This legacy has in many ways equipped them the better for further change. Development has always been the hallmark of cities such as Rotterdam, Lyons, Milan, Munich or Zurich, acquiring new development from successive centuries, including the present. But the twentieth century has brought a more difficult development for Charleroi, Liverpool and Liège and other towns of the industrial revolution. Nowhere is this better illustrated than in the small textile towns of Europe, where the decline of the staple industry has stripped the urban centre of its single dominant function. The towns of Greater Manchester, such as Rochdale and Bolton, or Anny, Enschede and Hengelo in the Twente district of the Netherlands, München Gladbach or Rheydt west of the Rhine in West Germany, and Verviers in Belgium, all demonstrate the specific problems of adaptability in such mono-industrial urban centres. However, for some cities a rebirth has occurred; Glasgow, which we listed a decade ago, has been able to alter its image and habilitate much of its centre on the basis of its new role as a cultural centre.

The scale of the planning response to urban growth has frequently been dramatic. The interwar suburban estates in Britain or, even more marked, the *grands ensembles* of postwar France, were faithful to Corbusier's 'machine for living' rather than to the more human scale of traditional building. Public housing schemes in almost every country have had to pay heed to economic reality and in the process have suffered. All too often they have become second-class places of residence instead of living examples of contemporary urban planning at its best. One of the reasons for this lies in our obsession with access to one resource, that of housing, with scant regard to access to the other facilities which add up to an urban culture. The current movement back into the inner city on a class-selective basis is a clear demonstration that many have realised what the city in its totality offers. Access to all its varied facilities and opportunities is important and was denied to the inhabitants of Sarcelles or Osterholz–Tenever. The reasons for the preoccupation with the provision of housing are not difficult to find. The acute postwar housing shortages, and the continued growth of Europe's largest cities as a result of high natural increase of population and steady immigration, combined to establish the priority of satisfying the massive demand for new residential space. Even in the early new towns the complete package of urban functions and opportunities was all too frequently missing for a full urban lifestyle. In any event, the creation of a new town within one generation cannot hope to provide the cultural variety and the diversity of social and economic opportunity which is the product of many centuries of growth in other European cities. In that sense, although certainly great improvements on a lifeless culturally barren suburbia, they could not hope to complete with established towns. Given the same residential opportunities, there would be few Englishmen who would prefer to live in Stevenage than Norwich, few Frenchmen who would take Evry before Chartres, and few Dutchmen who would not live in Leiden, dominated by its magnificent history, rather than in Zoetermeer, for all its planned precision. So thus both scale of provision and its priorities have been open to criticism.

Until recently it could be argued that materialism in postwar Europe has meant that society as a whole has not been willing to pay for an ill-defined higher 'quality of life', meaning a fruitful and productive urban environment in both social and economic terms. An increase in the scale of building has usually meant a decrease in unit costs and in a period when politics have been increasingly important in the whole sphere of urban planning, costs may be equated with taxes which, when increased, cost votes. Perhaps then, our mistakes are merely a reflection of our communal priorities. We may take this argument somewhat further by pointing to the fact that the mistakes are not necessarily those of the planners, but more usually those of politicians expressing these communal priorities through their vote. A distinctive feature of the city in West Europe has been the degree of democratic control, which brings with it a strong control over the freedom of the planner. The planner prepares a plan which has to be politically acceptable in such a system, and it is then the community which ultimately makes the decisions and must be held responsible for both successes and failures.

To dwell on the problems and failures, however, would be to give a distorted picture, because there is much in contemporary urban planning to ensure the survival of the European city. The realisation of the value of the past, which has given rise to the urban conservation movement throughout Europe, has been not only timely, but vitally important, as demonstrated in Chapter 7. It has meant that despite the increasing omnipresence of Shell, C and A, Marks and Spencer, and now on an intercontinental scale combining to give a 'sameness' to many urban cores, MacDonald's, Col. Sanders', and Woolworth, the underlying diversity will be retained. It may be too late to prevent the emergence of new retailing or office centres which could be equally at home in Chicago, Sydney, Cologne or Paris, but the existing urban cores are being increasingly protected against an encroaching insensitive commercialism, for it is the centre which has been most at risk. Despite postwar developments the towns of West Europe are still more characterised by their variety than their similarity and the urban conservation movement will ensure the continuation of this. There are a number of other reasons for this optimism. One lies in the importance of tourism to so many European cities (Chapter 9). That function can thrive only if accompanied by careful and sympathetic treatment of the urban landscapes which are part of the European cultural heritage that is itself the tourist industries' main resources. The creation of pedestrian precincts in the centres of many cities of major historical interest in the last 20 years has been one technique adopted to enhance their tourist attraction. To be able to walk the medieval streets of York, Chambéry, Bruges or Regensburg, free of vehicles, is an obvious advantage. In these surroundings one can see few signs of the European urban environment being under attack. The main danger is perhaps the opposite of this, that cities, or at least their historic cores, may be preserved as museum exhibits, incapable of undergoing the change and modification which has happened through the ages.

The new town movement in Europe, for all its faults, was a demonstration, to at least the rest of the western world, that modern urban development could be more than unimaginative accretions tacked on to the edge of the growing city. The opportunity was taken in many countries to demonstrate that a high standard of urban environment could be created in the twentieth century side by side with economic growth. The active role of the public sector in the new town movement and the frequent motive of achieving a social balance have often attracted envious glances, particularly from North America. Furthermore the new towns have reflected individual identities in many subtle ways. The carefully toned down appearance of the Scandinavian new town, using much of the natural landscape of evergreens and rock outcrops, is not the same as the colourful vitality, perhaps even the brashness, of Evry or Marne-la-Vallée, which appear to be so much the products of the new technocrats. Similarly the confident planning of Lelystad in the Dutch IJsselmeer polders is very different from Washington or Runcorn implanted into industrial northern England or Wulfen in the Ruhr.

In European cities, the tide of deconcentration generally turned before too many functions flowed from the city to the suburb in the wake of the population. Nevertheless, it was not always the case and the plight of the inner city is a very real one. Even in 1990 the European urban crisis, however, is by no means on the scale of that in North America. There are a number of significant differences which are worth recounting. Firstly, there is little evidence to suggest that any European city has ceased to be governable, with the possible special exception of Belfast, or that any are facing recurrent financial crises which threaten their existence. Financial crises have come to Europe, although these are often compounded by the antipathy between central and local government which fuelled the crises in Liverpool in the early eighties and the confrontation with Lambeth, London in 1990. The general policies of reorientation towards the central city are likely to see such possibilities recede even further. Secondly, the deconcentration of activity from the West European city, eroding its tax base and filtering away its vitality, is still not on the same scale of that in North America. Admittedly within West Europe the strength of policies aimed at preventing the decentralisation of functions such as retailing varies considerably, but the dangers of excessive deconcentration are universally acknowledged. Thirdly, the spatial extent, and therefore the dependence on the automobile, is more restricted in European cities than in those of North America.

The energy crises therefore, although very real, do not have the same inherent dangers.

If these arguments are accepted, are we then correct to talk of an urban crisis, in which the city and especially its inner areas are particularly vulnerable? It would be naive to deny problems that undoubtedly do exist. The poor, frequently neglected housing of so many inner cities, often inhabited by low socio-economic groups or by immigrants, has been a devastating indictment of the ability of our new European society to cope with the problems which manifest themselves within the inner city. However, the policy instruments have often transferred the inhabitants from a run-down inner suburb to a declining peripheral development out of sight and out of mind; from City Nord to Chorweiler or the Gorbals to Easterhouses. In every country in Europe, the gradients of material affluence within its cities can still be remarkably steep. Poverty, violence and racial tension still breed off each other in pockets of deprivation. The West European city is still as much Duisberg's Neumühl or Edinburgh's Craigmillar as Venice's Piazza San Marco or Paris's Place Henri VI.

The crisis exists, then, but it is recognised, and an increasingly large share of public and private resources are being expended to try to reach a solution. Cheshire and Hay have characterised the problems in a statistical analysis of the economic performance of 119 cities between 1971 and 1984. They noted three levels of problem based on their analysis (Table 13.1) which are related simply to growth or decline, although there were occasional confused signals in the results. The problems were conceptualised in terms of 'the impact of forces for change on individual cities in interaction with their adaptive capacity which reflects the regional and national contest; the economic structure, traditions and resources and the social system and the institutional constraints' (Cheshire and Hay, 1989, p. 159). The problems of decline are concentrated in the nineteenth-century and early twentieth-century industrial cities whereas growth problems are characteristic of southern Italy, central and northern Spain, Portugal and Greece. The economic and demographic data which produced the results are the culprit here because the problems are only those revealed by the data.

Rather than focusing upon problems, which is a negative approach, we would submit that some of the analyses which have been undertaken by DATAR have adopted a more catholic approach to the issues which dominate urban development in clusters of cities (Brunet, 1989). Brunet *et al.* note 12 groups of cities in which the broad interplay of international, national and local issues vary both to characterise the city and to place certain broad developmental issues for the city's consideration (Figure 13.1 and Table 13.2). The planning issues are ours and are merely indicative of the broad issues facing cities in the coming decade. Its very existence, however, has demonstrated a new crisis, that of a crisis of confidence in the planning system. In most, although not all countries, has come a realisation that the bold planning approaches of postwar Europe were not solving all the urban problems and may indeed be

Conclusion 299

Table 13.1 Cheshire and Hay's typology of problem cities

	Severe problems		Serious problems		Significant problems
D	Belfast	D/G	Bilbào	G	Barcelona
G	Cagliari	D	Birmingham	D	Cardiff
D	Charleroi	G/D	Dublin	D	Coventry
G	Cordoba	D	Hull	D	Derby
D	Glasgow	D	Lille	D	Genova
D	Liège	D	Messina	D	Le Havre
D	Liverpool	D	Newcastle	D	Leeds
G	Malaga	D	Sheffield	D	Manchester
G	Napoli	D	Le Havre	G	Murcia
G	Sevilla			G	Palermo
D	Sunderland			D	Rotterdam
D	Valenciennes			D	Rouen
				D	St Etienne
				D/G	Teesside
				G	Valladolid
				G	Zaragoza
	G		Athens ——————▶		
	G		Lisboa ——————▶		
◀———— G	———— Porto				
◀———— G	———— Salonica			G	Granada

Note: G = Growth; D = Decline

creating new ones. With this realisation has come a more careful approach involving a continuous monitoring of the impact of the planning system on not only the built environment but also the wider social and economic systems. Some countries, such as France, have had to take a rapid step back from the confident planning of the 1960s, while others such as Sweden and Denmark have realised the need for a more cautious approach. Others, such as the UK, have attempted to dismantle planning bureaucracies in favour of the market, while others are still instituting a stronger element of control.

The decision-making process itself has changed dramatically. The planners of the immediately postwar generation saw themselves as responsible professional experts who were custodians of the 'public interest' – impartial referees who adjudicated between the claims of commercial interests. They operated upon the city for the city's good, but would no more consult the citizen than a surgeon would consult a patient about how he should proceed. However, the word 'participation' took on increasing significance, as a result both of popular demand and subsequent legislation. Official channels of public consultation have been built up in most countries, as a supplement to the normal processes of representative democracy. The cost has been heavy in terms of the time taken to reach decisions, and the opportunities presented by public participation have understandably been exploited most effectively by articulate groups in society who may not be representative of society as a whole. Nevertheless public opinion has become a powerful and immediate influence on planning decisions in a way that was unknown a generation ago.

Table 13.2   Brunet's typology of cities

Type	Title	Examples	Issues (egs)
A	**The Strong Cities**		
1	Strong economic, financial and international centres	London, Frankfurt	Affordable housing, New business centres
2	Strong international R&D, and technology centres	Lyons, Zurich, Utrecht, Bristol, Mannheim, Bologna	Transport, Decentralised services
B	**The Cities with a Definite Role to Play**		
3	International, communications oriented centres with strong economy	Amsterdam, Southampton, Hamburg, Antwerp	Transport, Reuse of areas of former economic base
4	Communication centres with weak economy Some international role	Le Havre, Nantes, Genoa, Bremen, Taranto, Cannes, Palma	Broadening economy, Reuse of old areas
5	Communication but weak in other activities	Strasbourg, Manchester, Kiel, Duisberg	Diversification, Renewal
6	International cultural and research centres	Paris, Rome, Venice, Copenhagen, Athens, Geneva Dublin	Controlling tourism and new enterprise location
7	Cultural, R&D Centres – with weak financial and economic life	Munich, Berlin, Edinburgh, Montpellier, Caen, Münster, Padua, Zaragossa, Eindhoven	Young population, Recreation, Housing
8	No dominant functional indicators	Milan, Madrid, Birmingham, Bordeaux, Ghent, Mainz	Quality of Life, Housing
9	Weak indicators except strong demographic indicators	Naples, Seville, Opporto, Reggio, Messina	Housing, Migration
10	Weak indicators with some economic strength	Hannover, Bilbào, Palermo, Liège, Essen, Liverpool, Cagliari, St Etienne, Thessaloniki	Diversification Declining population
11	Weak indicators with some cultural strength	Tours, Granada, Aarhus, Freiburg	Tourism, Housing, Centre development
12	Weak on all indicators	Lens, Valenciennes, Dortmund, Newcastle	New image, New economic developments, Regeneration

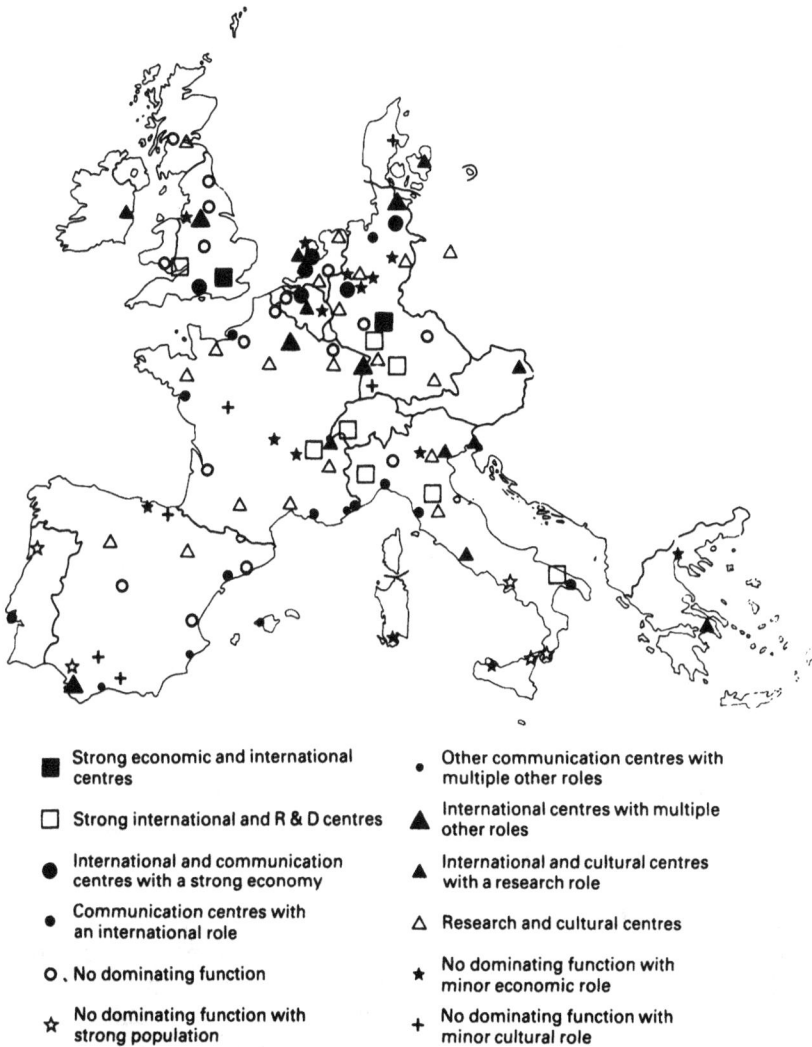

Figure 13.1 The European Urban Profile (after Brunet, 1989)

The need for this reappraisal stems also from our inability in postwar West Europe to make accurate predictions in the vitally important spheres of demographic change and economic growth. Quite apart from that, the social revolutions in even the last two decades in much of industrialised West Europe have emphasised the dangers of formulating long-term plans on the basis of contemporary trends and opinions. Who, in 1945, would have predicted the vast increase of women in the employment market, or the movement of foreign workers into a relatively few German cities? The planners who swept the trams from the streets of so many European cities could hardly have foreseen the cry for their reinstatement as a form of 'light rail rapid transit' 20 years later.

This leads us to the opinion that the planning of cities in Europe may have to follow a more flexible path, sensitive to priorities which may change over time (Burtenshaw, 1985). Long-term goals and objectives may still be desirable, but rigid blueprints may have to give way to written policy statements on the general direction of change. This has already occurred, of course, with the change to structure planning. Indeed in many spheres this approach is being adopted. Nevertheless, it is not without its dangers. Without a major plan, there may well be a reluctance on the part of politicians to cross major thresholds of expenditure. The expansion of a drainage system or the installation of a rapid transit system may well be jeopardised in an atmosphere of step-by-step incremental planning. At times, however, urban government is called upon to take a significant step and to cross such a threshold. In many cases, such expenditure may already have been delayed, thereby depressing the quality of life in an urban environment. Therefore the spending can only make good past deficiencies. In other cases, it may be necessary in order to provide a viable framework for future urban development without the quality of the contemporary city being put at risk. Given the new atmosphere of more careful monitoring the timing of such delicate decision should be made less difficult than it was in the past.

What of the future spatial form of the city of Europe? West European cities have been subjected to a degree of urban sprawl, with minor urban centres being incorporated, with the addition of successively high-order centres, until a multicentral urban system emerges. In some cases, such as Paris, positive planning policies have the polycentric city as an ultimate aim. It seems unlikely, however, that any European city will decant sufficient funds from its core to fundamentally reorient the focus of the city. Nevertheless the agglomerations of Rhine–Ruhr and the West Netherlands stand as good examples of multicentre urban regions. Indeed, on a wider scale a case could be made out for recognising one vast urban region comprising many centres within West Europe. It is possible to construct a map of urbanised Europe whose constituent parts face differing problems (Figure 13.2). Brunet's (1989) study of the cities with over 20,000 people identifies an urban backbone stretching from the English Midlands to the Gulf of Genoa, with its key metropolises in London, Frankfurt and Milan, and dynamic growth in Benelux and the Stuttgart–Munich–Zurich triangle. Paris, which holds a similar key node status, is an exclave. At the second level there are two subsystems:

(1) the 'North of the South', fringing the Mediterranean from Madrid and Barcelona to Lyons and Geneva and to Bologna and Venice. This system is spreading south into Spain and towards Rome and Naples and maybe into Yugoslavia.

(2) The second unco-ordinated subsystem includes the cities of the north and east, Vienna, Berlin, Hamburg, Copenhagen and Edinburgh/Glasgow. These may link eastwards as the barriers fall and contact with the east increases.

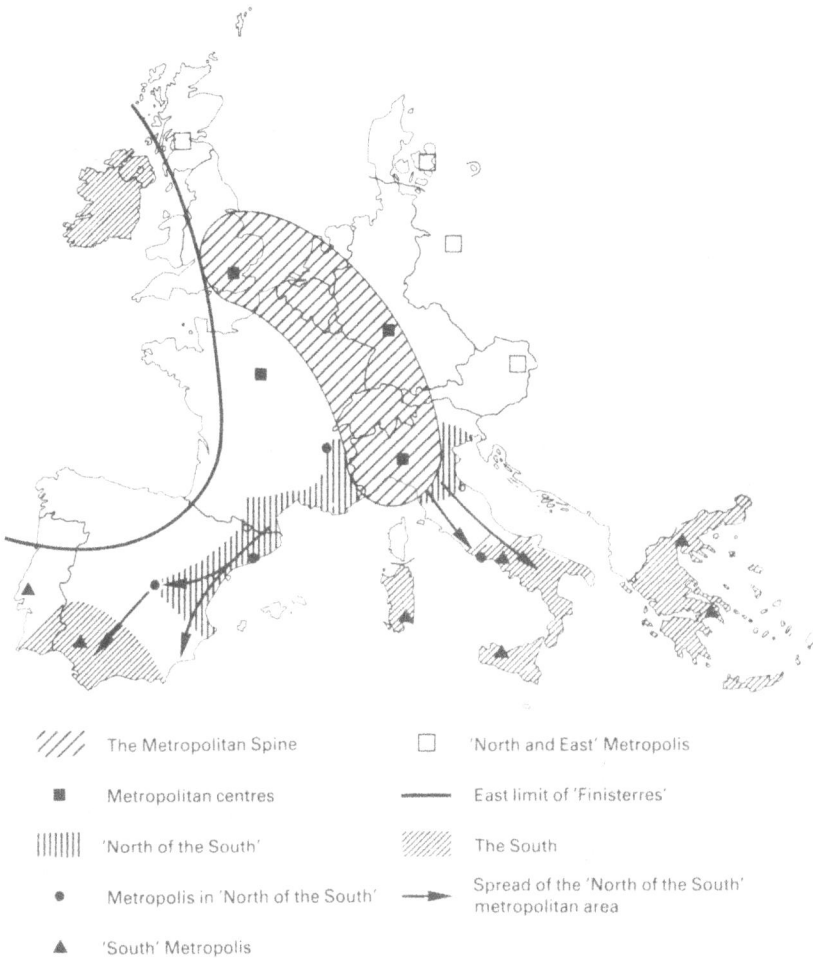

Figure 13.2 Urbanised Europe 1990

The other notable elements in the structure are the peripheral urbanised regions; (i) the *Finisterres* – the centres of the western areas – Bilbão, Brest, Cardiff and Liverpool, and (ii) the *Suds* – the outermost Mediterranean fringe – Athens, Thessaloniki, Palermo, Cagliari, Lisbon, Malaga, together, somewhat surprisingly, with Dublin.

The European tradition of urbanism has been a strong one. The emergence of an urban system spanning national boundaries has become a reality. There would seem to be clear indications that the city in Europe is being controlled and not being overwhelmed by poverty, as are the cities in the Developing World, nor seeing its heritage destroyed by new wealth, as is happening in the cities of the Newly Industrialising Countries. Its traditions of planning are well established and have given a guiding framework to the developing urban system. Despite the past mistakes, the European cities will retain their diversity. The pressures of the twentieth century may have been the catalyst for a reassessment of the strengths of the European urban tradition.

We must end with an affirmation of faith. Despite the concentration of serious social problems in the city and the mountain of literature that describes them, and despite the recent crisis of confidence that has turned planners from olympian idealists into timid fad-blown conformists, the European city remains today, and will remain for our children, as much the democratic fount of civilised values as it was in Socrates' Athens, Dante's Florence, or Erasmus's Rotterdam. Above all it will remain the most congenial living environment ever designed by man.

Our future task is to ensure that our urban culture is not accessible merely to the rich or to the aristocratic as it has been all too often in West European history. Instead the rich resources of the European urban culture which are being enhanced year by year must be made available to all Europeans.

References

Ambrose, P. and Colenutt, B. (1975) *The Property Machine*, Penguin, Harmondsworth.
Brunet, R. (ed.) (1989) *Les Villes "Européenes"*, La Documentation Française, Paris.
Burtenshaw, D. (1985) The Future of the European City: A Research Agenda, *The Geographical Magazine*, Vol. 151(3) 365–370.
Cheshire, P. and Hay, D. (1989) *Urban Problems in West Europe*, Unwin Hyman, London.
Hall, P. (1987) *Europe 20001*, Unwin Hyman, London.

Place Name Index

General Index

For Product Safety Concerns and Information please contact our EU
representative GPSR@taylorandfrancis.com
Taylor & Francis Verlag GmbH, Kaufingerstraße 24, 80331 München, Germany

www.ingramcontent.com/pod-product-compliance
Lightning Source LLC
Chambersburg PA
CBHW070554270326
41926CB00013B/2316